SECOND EDITION

ANIMAL GENETICS

Frederick B. Hutt, Professor Emeritus
of Animal Genetics
*Department of Poultry and Avian Sciences
Cornell University*

Benjamin A. Rasmusen, Professor of Animal Genetics
*Department of Animal Science
University of Illinois*

175 YEARS OF
1807 WJ 1982
PUBLISHING

John Wiley & Sons
New York · Chichester · Brisbane · Toronto · Singapore

Library of Congress Cataloging in Publication Data:

Hutt, Frederick Bruce, 1897–
    Animal genetics.

    Includes bibliographical references and
indexes.
    1. Animal genetics.   I. Rasmusen, Benjamin.
II. Title.

QH432.H87   1982   591.15   81-19671
ISBN 0-471-08497-2

Printed in the United States of America

10 9 8 7 6 5 4 3 2 1

To Randall Knight Cole,
geneticist, skilled poultry breeder,
and, since 1935,
esteemed colleague of the senior author

PREFACE

This edition, like the first one, is written primarily for students in agricultural and veterinary colleges. During the 17 years that have elapsed since the first edition, there have been many advances in our knowledge of heredity in domestic animals. It is not surprising that changes have been necessary in every chapter.

These advances have resulted less from the discovery of new principles of genetics than from the application of the old ones, with new techniques, to the diligent investigation of genetic variation in domestic animals.

Much of this newer research has dealt with variations in physiology. Accordingly, this edition devotes a whole new chapter to blood groups and protein polymorphisms. It was written by Dr. Rasmusen, who has also contributed revisions and additions throughout the book. The former chapter on biochemical genetics has been rewritten and enlarged to cover various kinds of genetically aberrant physiology.

Advances in cytological techniques have extended our knowledge of the chromosomes, which can now be individually identified (Chapter 6). When they go astray, it is possible to identify the ones that have done so and the effects of the mishap (Chapters 18 and 19).

As in the first edition, a special effort has been made to illustrate principles with examples drawn from domestic animals. These are not all cows and horses. Let us not forget that the mouse is a semi-domestic animal, albeit one not usually welcome. That "wee, sleekit, cowerin', timorous beastie" (as Burns called it) is famous for having the most extensive chromosome map of any mammal, including man (Chapter 10). No horse is likely ever to attain similar distinction.

Illustrative examples from plants or from man have been mostly replaced by suitable material from domestic animals. As in the first edition, a special effort has been made to dip only lightly into statistics, and to administer the irreducible minimum of them in small doses. These are scattered through several chapters wherever they best fit in.

We should like to thank collectively the people who have generously provided photographs for the 30 new illustrations or permission to use material published elsewhere. For these last, an effort has been made in the legends to give credit not only to the original author but also to the original publication. When that credit reads *from*, the original illustration is unaltered; when it reads *after*, slight alterations were made (at our request) by the publishers' artists. We are especially indebted to Prof. Ingemar Gustavsson of the Royal Veterinary College, Uppsala, Sweden, who provided several photographs, and to the *Journal of Heredity*, which did likewise. For other acknowledgments, see those for the first edition.

We have not strained any friendships by asking colleagues to read special chapters, but we are greatly indebted to one anonymous reviewer, who, taking his task more conscientiously than do most reviewers, submitted no fewer than 67 suggestions. We accepted 70 percent of them!

Frederick B. Hutt
Benjamin A. Rasmusen

ACKNOWLEDGMENTS
IN THE FIRST EDITION

I must thank, first of all, Professor F. A. E. Crew, for his book, *Animal Genetics*, published in 1925. As it was one of the magnets which drew me to his laboratory in Edinburgh, it played no small role in the chain of events which led eventually to the writing of this second book bearing that same title. Since the two are separated by 39 years, it is inevitable that my *Animal Genetics* should differ greatly from its predecessor, but I hope, nevertheless, that my professor will approve his pupil's handiwork.

I am particularly grateful to my colleagues, Professors C. H. Uhl, A. M. Srb, and R. K. Cole, who read individual chapters of the manuscript, and made many helpful suggestions; and to Professor Clyde Stormont, who read the whole manuscript, suggested several improvements which have been made, and undoubtedly saved me from more than one *lapsus calami* (which, translated freely, means: glaring error), some of which would have sent the reviewers into transports of unrestrained delight.

My thanks also go to all those who aided the cause either by providing photographs or by giving permission to use illustrations already published elsewhere. Rather than list their names here, I have made acknowledgments in each case in the legend of the illustration concerned. With respect to illustrations that have appeared elsewhere, an effort has been made to give credit both to the original author and to the original publication. Thirty-two of them appeared originally in the *Journal of Heredity,* and for many of these which were no longer available from the authors, the editor, Barbara Kuhn, kindly provided the original prints.

Special thanks are due to Miss Donna Hower who drew all the diagrams in Chapters 2, 4, 5, and 7 which show segregation in F_2 generations, and also redrew Figures 5 and 6 in Chapter 12.

Dr. H. P. A. de Boom of the Veterinary Laboratory at Ondestepoort, in South Africa, not only contributed the three pictures of his remarkable "baboon dogs," but also generously allowed me to cite his unpublished findings about them. Dr. H. C. Rowsell of the Ontario Veterinary College kindly contributed the latest extensions in his pedigree of hemophilia B in Cairn Terriers.

I am indebted also to the late Sir Ronald Fisher, and to Messrs. Oliver and Boyd, Ltd., of Edinburgh, for permission to reprint Table 4-4 from their book, *Statistical Methods for Research Workers*.

Finally, I must thank my secretary, Mrs. James L. Morrison, not only for her remarkable ability to turn none-too-legible manuscript into beautiful typescript, but also for keeping in good order the almost endless details about illustrations, tables, references, proofs, indices, and correspondence that lie behind every book. Only other authors and secretaries who have fought with them the battle of a book can fully appreciate how much such help can contribute to the ultimate victory in that battle.

F. B. H.

CONTENTS

CHAPTER

1

Genetics

"Genetics," said Bateson, when he proposed that name in 1906 for the youngest child in the family of biological sciences, "is the science dealing with heredity and variation, seeking to discover laws governing similarities and differences in individuals related by descent."

These terms, *heredity* and *variation*, should be defined according to genetic usage before we go any further. The geneticist's concept of inheritance is a bit more specialized than those of some others who use the term. There is no quarrel with the man who says that he "inherited" his father's farm. That usage, long familiar not only to happy legatees, but also to disappointed ones, and to lawyers who thrive on disputes between the two kinds, has nothing to do with biological inheritance

A dictionary tells us that heredity in the biological sense is the "hereditary transmission of the physical and psychical characters of parents to their offspring." That is a fairly good definition, but hardly a perfect one, for it does not rule out the transmission from parent to offspring of something *not* hereditary in the geneticist's special concept of heredity. In the *genes*, or units of inheritance that a chick receives through the nucleus of its mother's ovum and from the nucleus of its father's spermatozoön that fertilized the egg, are contained all the chick's hereditary potentialities. Some of these, like the color of its downy feathers and the shape of its comb, are evident at hatching. Others, like the color of its adult plumage, its ultimate size, or its capacity to lay eggs, become evident at maturity.

The hen sometimes *transmits* something to the chick that is not inherited through the genes. The egg can carry a bacterium which (in some cases) multiplies and gives the chick pullorum disease during the first week after hatching. Similarly, a mouse can receive in its mother's milk a carcinogenic agent that later causes the mouse to develop mammary cancer. These diseases are thus transmitted from parent to offspring, but are not truly inherited as are traits caused by the genes. Of course, as we shall see later, among those genes are some that determine whether or not the young animal can withstand the transmitted infection or will succumb to it. Conditions passed in a germ cell from parent to offspring, but outside the nucleus and not through genes, are sometimes said to be transmitted by *cytoplasmic inheritance* to distinguish them from heredity through the genes.

Let us not overlook the word *variation* in Bateson's definition. Some of that variation is caused by heredity, but much of it is caused by the environment. The innumerable debates of yesteryear on the relative importance of heredity and environment are less common nowadays, but as geneticists we can be sympathetic to both sides. Most of us have devoted considerable effort to determining how much of the variation in some plant or animal is caused by genes and how much by environment. Sometimes we try to exclude all environmental variation from experiments in order to show up differences that are caused by heredity. At other times, various environments, and even conditions of stress, are deliberately imposed to reveal genetic differences in ability to tolerate them.

The familiar saying, "Like begets like," has long been the guiding principle for animal breeders who select their ideal animals and reproduce them in the hope of getting further generations of similarly ideal descendants. Unfortunately, as those who have tried it know, like does not always beget like. It is true that cows beget calves and sows beget piglets. Similarly, Holstein-Friesians usually beget black and white Holstein-Friesians, but occasionally their calf is red and white. It is even more discouraging when the breeder finds that 300-egg hens do not regularly have 300-egg daughters, and, similarly, that the calf from the cow yielding 30,000 pounds of milk cannot do the same. The geneticist's business is to find out why such variations occur, to determine in what proportions the variant types are to be expected, and to learn how to make more likely the probability that ideal animals will beget ideal offspring.

Scope of Genetics

Genetics is not just a matter of 3:1 ratios. Some of the most important traits with which we are concerned in agriculture do not segregate in any recognizable ratios whatever. Differences in ability to produce eggs, milk, wool, meat, and other characteristics are certainly inherited, but, as we shall see later, the genetic bases for such variations are much more complicated than the basis for the difference between black calves and red ones. Sometimes the environmental influences far outweigh the genetic ones. Part of the geneticist's task is to discover the optimum environment for the maximum expression of the inherited potentialities.

Geneticists are concerned with genes, with the chromosomes that carry them, and with the cells in which those chromosomes are found. Animal geneticists have not yet delved as deeply into cytology as have plant geneticists, but we are equally concerned about the normal behavior of chromosomes and the mechanisms by which chromosomes are transmitted from one generation to the next. We would like to know much more about the extent to which they may interfere with fertility and viability when chromosome behavior is aberrant.

Developmental genetics deals with the question: How does the invisible gene produce the visible character? For example: How do genes for albinism cause some animals to lack melanic pigment? How do genes for inherited dwarfism in various species cause the retardation of growth that the breeder of bantam fowls likes and the breeder of Hereford cattle does not? How do genes for determination of sex make some animals males and others females? These are all good genetical problems, even though none deals with simple 3 : 1 ratios.

Another important field of genetics is concerned with the study of mutations, or changes in the germ plasm. There is ample evidence that these occur in nature. Many of them have been preserved by breeders to differentiate breeds of domesticated animals and horticultural varieties of plants. Most seem to be undesirable, but others have proven useful in agriculture, and some geneticists have studied the feasibility of inducing further mutations that might raise productivity beyond the current high levels to which it has been brought by selection.

Finally, genetics is inseparably bound to the big question of how evolution of species and subspecies has been brought about. This entails a study of the frequencies of various genes in wild animals and the relation of those genes to the "fitness" of different individual animals to survive. Statistical geneticists study the frequency of genes of various kinds in breeds, flocks, and herds of domestic animals, and devise elaborate schemes to show how those frequencies might (theoretically) be altered by different methods of selection.

Importance of Genetics in Agriculture

There are a number of good reasons why people who breed domestic animals, or keep them for profit, or minister to their afflictions, should know something about genetics. Since we are concerned primarily with increasing the yields of milk, fat, eggs, wool, meat, or some other desirable product, we should know how to manipulate to best advantage the genetic variations affecting these characteristics. The risks and results of inbreeding are better understood with some basic knowledge of how the genes behave. Conversely, high levels of productivity in improved breeds and strains can often be raised still higher by the mysterious power of the hybrid vigor that results when suitable strains are crossed. Geneticists would like to know how to make that hybrid vigor more amenable to their control.

Breeders of domestic animals are concerned about inherited defects. These range all the way from inconsequential downy feathers between a chick's toes, through assorted congenital weaknesses, to lethal abnormalities of form or function. Not all such defects are inherited. A little knowledge of genetics helps to distinguish the ones that are from those that are not and facilitates the elimination of undesirable genes. Differences among animals in ability to tolerate disease are clearly inherited, and a start has been made on producing disease-resistant strains of fowls, heat-resistant breeds of cattle, and other animals particularly adapted to environments where some similar stress is inevitable. Genetic ability to withstand stress can also cause problems in agriculture, as is shown by the houseflies and other insects which, by natural selection and without any help from geneticists, have bred their strains so highly resistant to DDT. The codling moths and mites that have become resistant to various toxins illustrate nicely the fact that animals differ in genetic ability to withstand stress of one kind or another, and that such differences can be accumulated by selection.

It is to be hoped that the animal geneticist will also develop a point of view conducive to the genetical improvement of the species with which he works. The veterinarians who guard the health of our domestic animals and the doctors who do the same for man have much in common. Both try to heal the sick. The doctors ministering to man, however, try to keep all of their parents alive to the last possible gasp. This humanitarian practice is usually appreciated by the patient, but it is not always the best thing for the human race. Some of those patients have genetic defects that should not be inflicted on later generations. In such cases it would be ideal if the patient could be kept alive, and in the best health possible, but with some assurance that his bad genes would go no further. However, most twentieth-century doctors seem no more concerned about later generations than was their revered Hippocrates, who lived in an age that antedated the Roman chariot. Their duty, as they see it, is to the individual in this generation, not to those in later ones.

The veterinarian can do somewhat better with his patients. Many of them also have genetic weaknesses and defects that should never be passed on to later generations. Here the veterinarian with a knowledge of genetics can be of great help, not only to his own clients but also to those of his successors, if he will advise against the reproduction of defective animals and do whatever he can to prevent it. There is no reason why genetic defects should not be repaired, or otherwise overcome, provided that this is done to permit raising the animal to a marketable age; but it is to be hoped that such repairs will not be knowingly made by veterinarians in animals that might be used for reproduction. At least one veterinarian has repeatedly refused requests to sew up umbilical hernia in bull calves. That is sound eugenics. On the other hand, a specific form of impotency in bulls was frequently repaired by veterinarians in a European country until they learned that the defect was caused entirely by heredity. Thereafter the curative operations ceased. In short, the veterinarian can think of the individual, but also of the race.

Finally, even a little knowledge of genetics is helpful to those who might otherwise be overawed by some of the sensational tales in the daily press. Newspapers delight in providing the public with heart-rending and hair-raising reports of the routine operations of heredity in the human species. These range from the now-familiar plight of the erythroblastotic baby, through the victims of hemophilia, of muscle dystrophy, or of other inherited disasters, to the perennial enthusiasm over the possibility of controlling sex, and the titillating speculation about the probable sex of children not yet born. On top of all this, there has been added in recent years many a controversial column about the risks of future generations that might be induced in the present one by environmental agents of various kinds. The frustrating sense of fatalism often engendered by some of these harrowing tales is less likely to bother the reader whose leavening knowledge of genetics permits him to read between the lines.

These few applications of genetics just considered do not exhaust the list, and current studies of genetic variations in the biochemical and physiological processes of domestic animals suggest that important new applications recently found are only forerunners of others to follow in the future. Advances in knowledge of the blood antigens of domestic animals have made immunogenetics a useful tool for identifying paternity and verifying pedigrees.

Beginnings

Genetics has flourished as a science only since the turn of the century. Progress in that time has been so rapid that it is difficult to believe how slow was the advancement in knowledge of reproduction and heredity up to 1900. Not till 1672, when de Graaf discovered that mammalian ovaries produce eggs equivalent to the eggs of birds, was it recognized that the female's contribution to the next generation comes from such eggs. Five years later Leeuwenhoek and his student, Hamm, discovered in mammalian semen the innumerable "animalcula" which we now know as spermatozoa.

For about a century thereafter a great controversy raged between Ovists, who held that the new life developed entirely from the ovum (with the seminal fluid acting merely as a stimulant), and the Animalculists, who believed that the individual developed entirely from the spermatozoön. One of the Animalculists, Hartsoeker, looking down his microscope at human sperm, believed that he saw in the head thereof a tiny "homunculus" shaped like a man (Fig. 1-1) and all ready to develop when planted in the proper environment! We must remember that in those early days the microscopes were not as good as they are now. In any case, Hartsoeker should be enshrined in the annals of

FIG. 1-1 The "homunculus," as Hartsoeker pictured it in his *Essai de Dioptrique*, Paris, 1694.

biology as the patron saint of thousands of hopeful microscopists who in later years have looked down better microscopes and seen wonderful things that were not there.

The Ovists and Animalculists had in common the fact that they were all preformationists: they believed that the new individual was already preformed in a tiny cell and that only some stimulus was necessary for development. Even Spallanzani (1729–1799), the first biologist to demonstrate the feasibility of artificial insemination, although convinced that something from the semen was necessary for the development of the egg, did not realize that both the ovum and the spermatozoön are necessary to start development of a new individual. Not until the latter half of the nineteenth century did biologists come to recognize that the all-important thing is fertilization, the fusion of nuclei from both the egg and the sperm.

So far as heredity was concerned, with almost the single exception of Mendel's work, prevailing thought up to the end of the nineteenth century went sadly astray. Most of that period was dominated by the teaching of Lamarck (1744–1829), the celebrated French biologist, that acquired characters are inherited. Even Darwin, who saw so clearly the fact of evolution, believed that environmental influences could be accumulated and transmitted, and thus play a part in the evolution of species. Biologists speculated about possible ways in which characteristics of the individual could be accumulated in the germ cells and passed on to the next generation. In the present century, study has been concentrated on exactly the opposite problem, *i.e.*, how the inheritance in the fertilized egg affects the development of the individual.

One ingenious attempt to account for the transmission to an individual of characteristics present in its parents was Darwin's theory of pangenesis. It proposed that all cells in the body contribute tiny elements called gemmules, which are supposedly carried by the blood to the testes and ovaries, and there utilized in the formation of the germ cells. In fairness to Darwin, it should be noted that he himself never considered pangenesis as anything more than a theory, one for which there was no evidence whatever. However, it did fit well with the prevailing ideas that acquired characters could be inherited. Obviously, if the body were modified in some way by some environmental influence, the gemmules from that part of the body would carry the modification, and it could be passed on.

Fortunately, these beliefs were eventually counteracted somewhat by the teachings of Weismann (1834–1914) that modifications of the body are not transmitted, and that, so far as heredity is concerned, the body, or soma, is unimportant. The all-important things are the variations in the germ cells. Those cells, together with the glands that produce them, he designated as the *germ plasm*, and it was his view that the bodies arising from the germ plasm in successive generations are, in effect, merely devices to ensure the continuity of that germ plasm.

Mendel

At the time of his famous experiments, Gregor Johann Mendel (Fig. 1-2) was both a monk in the local monastery and teacher of science in the high school at Brünn, in Austria. (That city is now known as Brno and is in Czechoslovakia.) Like many another good investigator who is promoted to an administrative position, he was appointed abbot of the monastery and was thus lost to science. It is interesting to speculate whether or not his important discoveries would have lain buried so long if Mendel had not forsaken science to become an administrator.

There are several reasons why his investigations were much more fruitful for the understanding of heredity than any that had gone

FIG. 1-2 Gregor Johann Mendel, 1822–1884. A picture taken at the time of his research. (Courtesy of J. Krizenecky, Mendel Museum, Brno, Czechoslovakia.)

before. For one thing, when he crossed his different varieties of garden peas, Mendel did not consider inheritance as a whole. He studied separately no less than seven different characteristics in which those peas differed, and in each trait he observed contrasting types. For example, the seeds were smooth or wrinkled, the cotyledons were yellow or green, the plants were tall or short, etc. To find how these contrasting characteristics are influenced by heredity, Mendel took as original parents varieties of peas that differed in the traits to be studied. He was fortunate in choosing for study paired characteristics in which one member of each pair proved to be *dominant* and appeared in the first-generation hybrids to the complete exclusion of the other, *recessive* type. Mendel was also lucky because (unknown to him) the genes for each of the seven paired characters that he studied were on different chromosomes. Furthermore, the garden pea has only seven chromosomes.

A second factor responsible for Mendel's success was his careful recording of the exact numbers of each form or color that appeared in the first, second, and later generations. In the second generation, in all seven traits the dominant and recessive partners appeared in ratios close to that of 3 : 1. It was clear that one of the contrasting types was overcome by its partner in the hybrid, but reappeared in about 25 per cent of the progeny from such hybrids.

The results of these interesting experiments were reported to the Natural History Society of Brünn in 1865, and were published in the proceedings of that society the next year. Thereafter they lay buried for 34 years, until rediscovered independently by de Vries, working in Holland, von Tschermak in Austria, and Correns in Germany. It is sometimes said that the importance of Mendel's work was not realized earlier because biologists of the period were more concerned about the controversy over evolution stirred up by the publication in 1859 of Darwin's "Origin of Species." Another factor probably was that the *Proceedings of the Natural History Society of Brünn* were not widely distributed to biological laboratories and libraries elsewhere. Scientists have always had the problem of keeping up with the literature in their fields, and in Mendel's time there were no convenient abstracting journals such as modern biologists find indispensable.

After Mendel

It must have been some satisfaction to Mendel as he struggled with the administrative problems of his monastery to reflect that he had carried out a good experiment and had obtained consistent results. When the world paid no attention to his work, he is reported to have comforted himself by saying, "My time will come." Since the rediscovery of Mendel's work, countless experiments have proven that what he found with garden peas applies equally well to flies, fish, and fowl, to mice, maize, and man, and even to bacteria and fungi.

The first evidence that Mendel's laws apply to animals as well as to garden peas was provided by William Bateson, who, in December, 1901, reported to the Royal Society of London the results of his experiments in crossing fowls with different types of combs (Fig. 1-3). He had found that pea comb and rose comb were both dominant to single comb in first-generation hybrids, and yielded 3:1 ratios in the second generation. This report did not appear in print until 1902, but Bateson's experiments were started in 1898, two years before the rediscovery of Mendel's work. Clearly, like Mendel, he was on the right track. It is sometimes said that, if it were not for Mendel, Bateson would be sleeping in Westminster Abbey. Be that as it may, further confirmation was forthcoming in 1902 when the French zoologist, Cuenot, reported that crosses of his grey mice with albinos had yielded

FIG. 1-3 William Bateson, 1861–1926, whose studies with the fowl provided the first evidence that Mendel's laws apply to animals. (Courtesy of R. C. Punnett.)

only greys in the first generation and a ratio of 198 grey : 72 albinotic in the second.

In that same year, William Sutton, working in the laboratory of Prof. E. B. Wilson at Columbia University, pointed out that the behavior of the chromosomes during the formation of germ cells is such that it could account for Mendelian inheritance if the units of that inheritance are carried on the chromosome.

During the next two decades biologists demonstrated Mendelian inheritance in many organisms. Exceptions to simple 3 : 1 ratios soon appeared, and the explanation of these necessitated special experiments and further studies, by which knowledge of heredity was greatly expanded. Many of the exceptions to simple Mendelian inheritance are treated in later chapters of this book. For the present it will suffice to say that geneticists now know that inheritance is carried by units in the chromosomes, known as *genes*. These are arranged in linear order and have the capacity to reproduce themselves. They can change, and in many cases, if not in all, the gene is divisible into smaller units within itself. In fact, some geneticists say that the gene is made up of subgenes, which might better be called mutational sites. Finally, we know that the gene consists largely of a complex substance—deoxyribonucleic acid, or DNA. We have come a long way from Hartsoeker's homunculus.

Selected References

BATESON, W. 1902. Experiments with poultry. *Repts. Evol. Comm. Roy. Soc.* **1**: 87–124. (Significant as the first evidence that Mendel's laws apply to animals.)

BATESON, W. 1902. *Mendel's Principles of Heredity.* Cambridge. Cambridge Univ. Press. (Contains the Roy. Hort. Soc. translation of Mendel's paper and a vigorous defence of it against early critics; interesting for comparisons between this first text on genetics and later ones.)

ILTIS, H. 1932. *Life of Mendel.* New York. W. W. Norton & Co., Inc. (An excellent biography, originally published in German, translated by E. and C. Paul.)

MENDEL, G. 1866. Versucheüber Pflanzenhybriden. *Verhandlundgen des naturforschenden Verein in Brünn.* Abh. IV. (This paper was reprinted in facsimile in *J. Heredity* **42**: 1–47 (1951). A translation into English made originally by the Royal Horticultural Society of London is sold by the Harvard University Press. It can also be found in the fourth and fifth editions of *Principles of Genetics*, by Sinnott, Dunn, and Dobzhanksy, published by the McGraw-Hill Book Co., Inc.)

PUNNETT, R. C. 1928. Ovists and the Animalculists. *Amer. Naturalist* **62**: 481–507. (An interesting review of the controversy between them.)

THOMSON, J. A. 1932. *The Great Biologists.* London. Methuen & Co., Ltd. (Their stories so well told that few readers will stop at Darwin, Mendel, or Weismann.)

For more information about Mendel, his work, and that of his contemporary breeders of plants and animals, readers should consult *Folia Mendeliana*, a journal edited by V. Orel and published at irregular intervals since 1965 by the Moravian Museum, Brno, Czechoslovakia. Recent issues are published as part of the journal *Acta Musei Moraviae*. Articles are in English or German.

CHAPTER

2

Simple Mendelian Inheritance

Like other sciences, genetics has a special vocabulary all its own, and prospective geneticists should learn right at the beginning some of its indispensable terms. They can best be understood by following through an example of simple inheritance of two contrasting characters.

Red Holstein-Friesian Calves

A case suitable for analysis was provided by the following item in a Canadian newspaper:

> The Holstein bull, Maple Leaf Reflection Governor, one of the greats of the breed, is to be retired because of a recessive genetic factor causing him to sire red and white instead of black and white progeny.
>
> The Waterloo Cattle Breeding Association's annual meeting yesterday voted to eliminate the bull from its stud.

Red and white calves are occasionally produced by purebred black and white Holstein-Friesian cattle with impeccable pedigrees. There was a time when such calves were considered just as good as the others, but now they cannot be registered and hence are undesirable. More than one breeder of purebred Friesians has had to make adjustments because a bull that he had sold at a good price later sired red calves.

However, if one wished to do so, it would be a fairly simple matter to establish a herd of red and white cattle. They could hardly be called Holstein-Friesians, even though descended entirely from that breed, but at least one such herd has been established. Similar red "sports" from Aberdeen Angus cattle have been used to build up a herd of Red Angus.

Simple Terms

When a red cow of this sort is mated to a Holstein-Friesian bull (of the kind that cannot sire red calves), the resultant calf is black. In genetic parlance, black is thus shown to be *dominant* to red, and red is *recessive*.

The black and red parents are called the *parental generation*, and their black calf is said to belong to the *first filial generation*. To abbreviate, the contrasting parents are called the P_1 generation and their progeny the F_1.

The colors that we see are referred to as genetic *characters*; in this case they are also *unit characters* because they are dependent on a single pair of the units of heredity that we call genes. That pair, of course, acts within the combined action of all the other genes.

One could continue to refer to these particular paired units as "the gene causing black coat color in cattle" and "the gene causing red coat color in cattle," but a convenient shorthand device commonly employed by geneticists makes such lengthy designations unnecessary. We assign *symbols* instead. These apply both to the characters that we see, and to the invisible genes that induce them. The symbols consist of one to three letters—the fewer the better—and usually give some clue to the character.

Thus, B = black; b = red. Together these genes constitute a pair of *alleles*. The dominant member of the pair, in this case the gene for black, is always given a capital letter, and the recessive allele is indicated by a small letter or letters. Alleles may be defined as alternate (contrasting) forms of a gene, or of the characters induced by a pair of allelic genes.

It has been recommended that symbols for "wild-type" genes, those found in the wild ancestor of the domesticated species, or in its "standard" form, be given also a plus sign, as a superscript. Thus the symbols Cp and cp^+ applied to the fowl would designate, respectively, the dominant creeper gene (which shortens the limbs) and its recessive allele (which does not do so), and the $^+$ tells us that the White Leghorn's ancestral jungle-fowl was not a creeper. In formulae, where several genes are considered, the letters can even be omitted and the wild type designated as $+$. This is all very well for the fruit fly, whose wild relatives can be caught in any garbage can when comparisons are desired, but it is hardly necessary for domestic animals. For most of them, as for man, there is no certainty about the appearance of the wild ancestor, no record of its genes, and no agreement on what should be considered a "standard" form. To simplify matters, the plus signs are omitted from this book, but the reader should remember what they mean when he encounters them elsewhere.

The genes are located in the chromosomes, and normally each gene occupies its own special place in the particular chromosome to which it belongs. That position is referred to as its *locus*, or position, in the chromosome, from the Latin word for "place." By special techniques which we shall examine later, it is possible in many cases to determine not only to which chromosome a given gene belongs, but also the relation of its locus to the *loci* (plural) of other genes in that same chromosome. We do not yet have that information about B and b in cattle.

Nearly all domestic animals carry in each cell (except the later stages of germ cells) two sets of chromosomes, one set from the sire, another from the dam. (We shall come eventually to one domestic animal that is a non-conformist.) Each chromosome from one parent is matched by a *homologous* chromosome from the other parent. (Certain important chromosomes not always conforming to this rule are discussed in Chapter 6.) Homologous chromosomes are alike in size and shape when seen under the microscope. Each of them carries the same gene loci as the other, but not necessarily the same allele or form of the gene.

It follows from all this that a Friesian getting from each parent the gene *B* cannot carry *b* at all, and hence will "breed true," producing only black and white calves. Its gene formula, or *genotype*, is *BB*. Similarly the red calf has received from each parent only the recessive allele *b*, and its genotype is therefore *bb*.

Segregation in F₂ Generation

We are now almost ready to return to our cross of black × red Friesians, but before doing so, there are still three terms to be defined.

Reproductive cells, including oöcytes, or eggs, and spermatocytes, or spermatozoa, are called *germ cells* to distinguish them from the *somatic* cells that make up the body. Mature germ cells that unite in fertilization are known as *gametes*, and the cell formed by the union of two gametes is a fertilized egg, or a *zygote*.

The terms *gamete* and *zygote* are usually considered by geneticists in terms of the genes that they carry. Thus, while a hen's egg is at one stage a gamete, and later (if fertilized) a zygote, the fowl geneticist doing the family shopping will ask for a dozen infertile eggs, rather than for a dozen female gametes of *Gallus gallus*, no zygotes to be included.

Returning now to our cross of black × red Friesians, the transmission of genes for these colors is shown diagrammatically in Fig. 2-1. For that illustration, it is assumed that several animals are produced in the F_1 generation, and that these are mated *inter se* (among themselves) to produce the second filial (F_2) generation.

The symbols ♂ and ♀ , used by biologists to designate male and female, respectively, were borrowed from the astronomers, who long ago adopted the spear as their symbol for the planet Mars, and the looking glass for Venus.

From the explanations given above, it should be easy to follow through the diagram and to see why the F_2 generation contains black and red calves in the ratio of 3 : 1. One need only remember that the P_1 black parent can produce only gametes carrying the gene *B*; his mate's gametes must all carry *b*. The resultant F_1 animals must therefore be *Bb*.

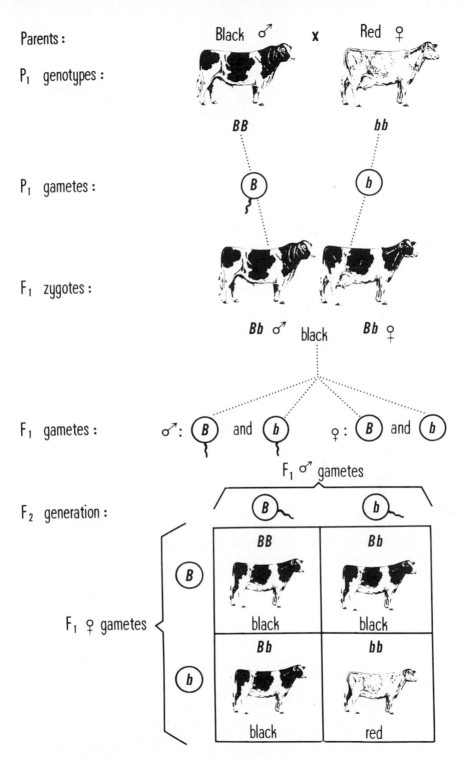

FIG. 2-1 Results in the F_1 and F_2 generations from the cross of pure black (and white) × pure red (and white) cattle.

Now, when these black F_1 cattle are mated together, each animal produces gametes of two kinds and these occur in equal numbers. Half carry B; the other half carry the recessive allele b. To determine all the kinds of calves that can be produced from the F_1 parents, we must assume that each type of gamete produced by the cow (with respect to B and b) is fertilized by each kind produced by the bull. That is the purpose of the square at the lower part of Fig. 2-1. Such squares are called Punnett squares, after their originator, R. C. Punnett. This one shows that only four combinations are possible and that the resultant F_2 generation consists of animals in the proportion of:

$$
\begin{array}{lll}
1 & BB & \text{black} \\
2 & Bb & \text{black}
\end{array} \Bigg\} \ldots 3
$$
$$
1 \quad bb \quad \text{red} \ldots\ldots 1
$$

As the two kinds of black animals are indistinguishable, the visible ratio in F_2 is 3 black : 1 red. It should be clearly understood that if the F_2 generation consists of four animals only, these will not invariably be 3 black and 1 red. All four might be black, or all might be red, although the latter is less likely. With larger numbers, the expectation of about 3 black : 1 red is more nearly realized.

Mendel's Law of Segregation

In the F_2 generation, the red cattle are just as red as their grandparent. They do not show any sign that both their F_1 parents were black, and they are indistinguishable from cattle produced by two red parents. In cases like this, when the two contrasting parental characters are recovered in a later generation, with each showing no indication of its causative genes having been influenced or contaminated by association with their alleles, the geneticist says that the two characters *segregate* cleanly, or sharply. Actually, it is the alleles which segregate, but it is common usage to speak of the resultant segregation of characters in a 3 : 1 ratio, or whatever the ratio may be. For convenience we follow that custom in this book.

Mendel, who studied the inheritance of different characters in garden peas, was fortunate in selecting for analyses contrasting characters which segregated sharply (round or wrinkled seeds, yellow or green endosperm, tall or short plants, etc.). The uniform reappearance of the two forms in 3 : 1 ratios in F_2 generations led him to formulate a principle which is sometimes designated as Mendel's law of segregation. In modified form it may be expressed thus:

The F₁ generation from parents differing by two contrasting characters shows one of them or the other, but in the F₂ generation the two characters reappear in definite numerical proportions.

Genotype and Phenotype

In the F_2 generation from our cross of black × red cattle, the ratio of visibly different colors is 3 black : 1 red. However, as Fig. 2-1 shows, one of the blacks is *BB*, and two are *Bb*. These two kinds have the same *phenotype*, or appearance, but they differ in *genotype*, or genes. It is impossible to determine from inspection which of the blacks are *BB* and which *Bb*, but they can be differentiated by breeding tests.

Such a test is not necessary for the red animals. Since *b* is recessive to *B*, a red calf must be *bb*. As a student once put it: "If the red calf carried *B*, it would be black."

The genotype is the genetic constitution, usually considered with reference to a few genes only, the remainder being unknown. It is often not evident in the phenotype, but can be revealed by suitable breeding tests. Progeny tests for evaluating a sire's or dam's ability to transmit complex characters are attempts to determine the genotype—the extent to which the tested animal carries the genes desired.

The phenotype is the appearance of the organism with respect to the characters under consideration. It may manifest the genotype (as in the red, *bb* calves), or it may not (as in the *BB* and *Bb* black calves). The phenotype often depends upon the environment to which the genotype is exposed. For example, fowls genetically susceptible to leukosis cannot demonstrate that susceptibility unless exposed to the causative virus. Cows with genotypes good for more than 30,000 pounds of milk will yield far less than that amount unless properly fed.

Before going into details of breeding tests, two more terms should be mastered. Animals that carry the same allele in both members of a homologous pair of chromosomes are said to be *homozygous* with respect to the pair of genes involved. Thus, of our four F_2 animals under consideration, the red calf is obviously (from its phenotype) homozygous for *b*. Similarly, one of the three black F_2 animals should be homozygous for *B*, but two received *B* from one parent and *b* from the other, and hence are *heterozygous* with respect to the *B-b alleles*.

Thus, a *homozygote* may be homozygous for the dominant allele of a pair or for the recessive one, but the *heterozygote* must have both. Individuals that segregate in the F_2 and show the phenotype of a recessive character (as do the red calves) are sometimes called *extracted recessives.* They are *homozygous recessives*, but this term is not peculiar to those that segregate from parents of a different type. Red calves from two red parents are also homozygous recessives.

Backcrosses and Test-Cross

In the foregoing example, F_1 animals were mated *inter se* to produce the F_2 generation. When such F_1 heterozygotes are mated back to animals of either parental type (in this case, black or red), such a mating is called a *backcross*.

When the F_1 *Bb* is backcrossed to the homozygous dominant parent (*BB*), the latter's gametes all carry *B*, hence all progeny of this backcross must be black. About half of them will carry the gene *b*, which they get from their F_1 parent, but none can show it (Fig. 2-2). Their genotypic ratio is 1 *BB* : 1 *Bb*, but in phenotype they are all the same.

In contrast, when F_1 animals are backcrossed to the homozygous recessive parent, or to other red cattle, the gametes of these last can carry only *b*. Among the gametes from the F_1 heterozygotes, about half will carry *B*, and the remainder *b*, so the resulting zygotes should be *Bb* and *bb*, with approximately equal numbers in each class (Fig. 2-2). The resultant phenotypic ratio is 1 black : 1 red.

The terms *backcross* and *backcross to the* P_1 mean merely a backcross to individuals of the same genotype as the parents, with

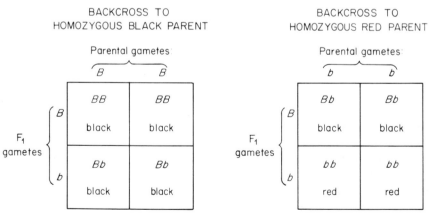

FIG. 2-2 Diagrams showing all possible combinations of gametes in backcrosses of F_1 (Bb) animals to the two paternal types. The one on the left yields a phenotypic ratio of 1 black; 0 red; for the other the ratio is 1 black : 1 red.

respect to the genes under consideration. It is not implied that the F_1 offspring must be mated to their own parents. As with the 3 : 1 ratio, the backcross ratio of 1 : 1 is realized only approximately, and it does not follow that if the first calf is black, the next must be red.

Since Bb blacks × bb reds yield a ratio of 1 black : 1 red, while BB blacks × bb reds yield only blacks, a backcross type of mating in which a black of unknown genotype is mated with reds will serve as a *test-cross* to reveal whether the black animal is BB or Bb.

Such a test-cross is the standard procedure for detecting heterozygotes, and hence for eliminating simple recessive characters that are not wanted. These tests are most easily made in species that mature rapidly and reproduce in large numbers (such as fowls, mice, rabbits, or dogs) but are somewhat more difficult in the larger domestic animals. A test of rose-combed Wyandottes to find those that carry a gene for the unwanted single comb is easy, but to determine which black cattle carry red is a more costly procedure. If the suspect is a heifer, the test may take several years, since, even if she produces three or four black calves, there is still a chance that she might be Bb and that the next calf might be a red one. A bull can be tested more quickly by mating him with eight or ten red cows.

Simple Probabilities

Breeders sometimes ask how many offspring must be produced from a simple test-cross before one can be sure that the animal tested does not carry the undesirable recessive gene for which the test is made.

The rule providing the answer to that question is the same one that tells the probability that a coin will turn up heads in 2, 3, 4, 5, or more successive tosses. When the probability of throwing a head on the first trial is 1 in 2, or ½, and the probability of its happening in a second independent trial is also ½, then the chance that two successive trials will both turn up heads is the product of the probabilities for each trial, *i.e.*, ½ × ½ = ¼, or 1 in 4.

When the coin is tossed a third time, the chance of getting a head is again ½, but the probability of getting in three consecutive throws becomes $(½)^3 = ⅛$. For four tosses it is $(½)^4 = 1/16$. For n tosses it is $(½)^n$. Eleven coffee-drinkers of Salina, Kansas, who met regularly and matched coins for the bill, were amazed when on one occasion all eleven coins turned up tails, but that result was to be expected in about $(½)^{11}$ trials, or at one kaffeeklatsch in 2,048.

Similarly, if we assume that in dogs the ratio of males to females at birth is 1 : 1, the probability that any one pup will be a male is ½. The chance of having a whole litter of male puppies is $(½)^n$, when n is the number in the litter. Among many litters of five, we should expect about 1 litter in 32 to be all males. Since the probability is the same for females, about 1 in 32 litters of five should be all females. The chance of having a litter of five all the same sex is thus $^2/_{32}$, or 1 in 16.

Actually there is an excess of male puppies at birth; hence the chance of any one pup's being a male is slightly more than ½ and being a female slightly less than ½, but, for litters of five, the probabilities given are close enough, unless we wish to deal in fractions of puppies.

The relation of pennies and pups to test-crosses for simple recessive genes is that all three deal with a 50:50 chance, a probability of ½. The coin should fall head or tail with equal frequency; half the pups should be of each sex; and the heterozygous black bull (as an example) should produce gametes half of which carry B, the other half b.

Test-Crosses in Practice

Returning now to the question, "How many progeny are necessary from simple test-crosses to reveal whether or not the animal being tested carries the recessive gene for which the test is made?" we see that the answer depends on how great or how little are the odds with which we are satisfied. If the black bull being tested by matings with red cows produces six calves none of which is red, we recall that, among many Bb bulls thus tested, the chance that one of them would sire six black calves in a row is $(½)^6$, or 1 in 64. We might, therefore, be fairly certain that the particular bull being tested is BB, but to make assurance doubly sure, we might reserve decision until four more calves have been added to the test. If no reds have appeared among the ten calves, the probability that the bull is Bb is reduced to $^1/_{1024}$, and most of us would say that odds of 1,023:1 against his carrying b are enough.

Animals suspected of carrying an undesirable gene sometimes have to be tested in matings other than the backcross to the recessive trait that we have here called a test-cross. When the homozygous recessive genotype is lethal, as sometimes happens, such simple tests are impossible, and the genotype of the suspect can then only be proven by matings with known heterozygotes. From such matings, if the suspect is heterozygous, only about 25 per cent of the progeny will show the recessive character, whereas from backcross test-crosses about half should be of that type.

This situation can be nicely illustrated by a mutation in the dog that the breeder aptly called "bird-tongue." The puppies affected could not suckle or swallow and made no attempt to do so, so all died soon after birth. A distinctive visible abnormality was that the outer edges of the tongue were folded inward, thus making it narrower than usual. Hence the name "bird-tongue" (Fig. 2-3).

In the kennel in which this mutation occurred, there were 12 litters over 7 years in which one or more puppies had this lethal defect. The ratio of normal to abnormal animals in those litters showed that bird-tongue is a simple recessive defect. Every parent of such a pup was thus identified as a *carrier*, *i.e.*, a heterozygote. Accordingly, whenever the owner wanted to use some outstanding animal for breeding, it was first tested by matings with known carriers, and not used extensively until such test-crosses had yielded enough normal pups to ensure that the suspect was not a carrier. By this procedure the defect was eventually eliminated from the kennel.

Among domestic animals, undesirable simple recessive characters which might be eliminated more quickly by the use of test-crosses include the black fleece that crops out in white sheep, the dwarf calves in beef cattle, and the horns that persist in breeds and herds of polled cattle. There now seems to be general agreement that under most conditions cattle fare better without horns, and the development of Aberdeen Angus, Polled Herefords, Redpolls, and other hornless breeds is an indication of that fact. Instead of the caustic potash and cautery now used on calves to prevent horns from forming, and the saws, shears, and gory operations used to remove them from older

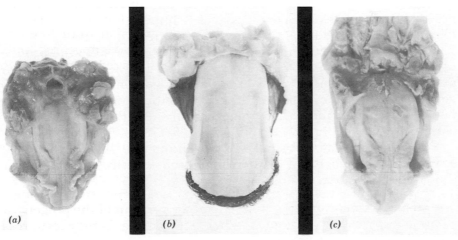

(a) (b) (c)

FIG. 2-3 "Bird-tongues" (a and c) and a normal one, from puppies of comparable ages. In the affected tongues, inward folding of both sides of the distal halves is conspicuous. (From Hutt and de LaHunta in *J. Heredity* as cited. Copyright © 1960 by the Amer. Genetic Association.)

animals, the cattleman might better substitute the dominant gene *P*, which, acting in fetal stages, eliminates horns without any need of tranquilizers for either the animal or the veterinarian. Heterozygous males sometimes show scurs (vestigial horns), but the desirable *PP* homozygotes can be identified by breeding tests.

Yellow fat in sheep, undesirable because consumers of mutton are accustomed only to mutton with white fat, is a simple recessive mutation that has been reported (thus far) only from Iceland. It can occur equally well elsewhere. Test-crosses to detect carriers and so to eliminate the causative recessive gene would undoubtedly be simpler than any attempt to convince the public that yellow fat is just as good as white.

Similarly, most rabbits have white fat, but a recessive gene causing yellow fat is not uncommon. Whether it affects the value of the meat in this species is not clear, but, if it were desired to eliminate the heterozygotes, there are few mammals in which test-crosses could be more easily made.

In the useful tables of Warwick (1932) we see at a glance that, with the simple backcross, five normal progeny (without any recessives) reduce to 0.03 (*i.e.*, 3 chances in 100) the probability that their sire could be heterozygous. To get the same degree of assurance when the suspect is mated to heterozygotes, one would need twelve normal offspring, with no recessives. Obviously, it is easier to test a Holstein-Friesian bull for recessive red by mating him with a few red cows than it is to test a Red Danish bull for the lethal paralysis not infrequently found in calves of the latter breed. As no paralyzed calves survive to breeding age, a backcross to recessives is impossible and cows known to be carriers would have to be used for the test.

A point worth remembering is that the test-cross is the *progeny test* in its simplest form. Later on we shall deal with more difficult progeny tests, but in them, as in the test-crosses for simple recessive genes, the objective is to determine from the appearance or performance of the progeny what genes are being transmitted by their parents.

Fortunately, as we shall see in Chapter 3, heterozygotes can sometimes be recognized by their phenotypes, or by laboratory tests, thus making test-crosses unnecessary.

In Defence of Recessivity

Many hereditary characters have a more complex genetical basis than the ones mentioned thus far, and, as we shall see in Chapter 3, some are neither dominant nor completely recessive, but in every species there are many characters which appear in two contrasting forms, one of which is dominant over the other in heterozygotes.

In most of the examples cited earlier, the recessive allele was the unwanted member of each pair. It is necessary therefore to correct any possible impression that all recessive genes and characters are undesirable. In many cases the recessive is the type preferred by breeders or (even more important) by markets. Although breeders of Wyandottes try to eliminate single combs, a single comb is indispensable to any chicken aspiring to rank as a respectable Plymouth Rock.

Similarly, there are two alleles—W and w—that cause white or yellow skin, respectively, in the fowl. English markets like white skin (WW or Ww), but Americans prefer the recessive yellow (ww). As these genotypes suggest, the American poultry breeder finds it easier than the English one to satisfy his customers. Yellow-skinned birds mated *inter se* can produce only yellow-skinned progeny, barring the rare event of a *mutation*, or sudden change in the germ plasm. They cannot carry W. In contrast, some of the English breeder's birds may be heterozygous (Ww), and these, when they happen to mate with others of the same genotype, can throw yellow-skinned progeny.

Far from being undesirable, recessive genes are currently the stock-in-trade of many a mink breeder, whose pastels, platinums, and sapphires have until recently brought higher prices than the wild-type black minks from which these *mutant* forms were derived. The White Leghorn, perhaps the most efficient egg-producer of the day, is homozygous for a whole string of recessive genes, the dominant alleles of which (in varying numbers) are indispensable breed characteristics in Houdans, Wyandottes, Frizzles, Black Sumatra Games, and other breeds.

Tests for Recessivity

The routine breeding tests to determine whether or not some new variation or "sport" is a simple recessive trait, and the proportions of the new types expected if it is, are as follows:

Mating	Mutants Expected (per cent)
1. Mutant × normal	0
2. $F_1 \times F_1$	25
3. F_1 backcrossed to mutant	50
4. Mutant × mutant	100

When all four matings can be made, the results should answer the original question beyond doubt. However, when the variant form cannot survive, or cannot reproduce, Matings 1, 3, and 4 are impossible. In such cases, a clear 3:1 segregation in the F_2 generation is a good indication. It can be confirmed by further breeding tests with the normal animals in that generation. When a single pair of genes is involved, two-thirds of the F_2 normals should prove to be heterozygous (Fig. 2-1).

Dominant Mutations

When a newborn calf, chick, or child shows some trait not evident in its parents or in any known ancestors, we cannot assume that the abnormality is a simple recessive character. It might be non-genetic, even though congenital (*i.e.*, present at birth), or it might even be the result of a new dominant mutation, or of interactions between genes.

Some congenital abnormalities are caused by mishaps during development of the fetus. Such conditions, like amputations and other injuries after birth, are not transmitted to succeeding generations. They are non-genetic acquired characters. The tailless calves that are occasionally born seem to belong in that class.

New mutations are rare events, and dominant mutations are less frequent than recessive ones. It has been calculated that some mutations in man occur only once in about 50,000 germ cells, and others even more rarely. If we are familiar with genetic and non-genetic variations in animals, we may be able to identify some unusual condition in a young animal as one that is not hereditary, or perhaps one already well known to be a simple recessive or dominant mutation. Thus, albinism is recessive in all cases studied genetically. When an albinotic child is born, we know at once that both parents are heterozygous for the causative gene, and that it could have been handed down for several generations on both sides of the family without any albinos appearing.

When horned cattle produce a polled bull calf,[1] we are fairly safe in assuming that it resulted from a new dominant mutation in the germ cells of either the sire or the dam, but, because of the rarity of such mutations, not in both. The calf is most likely to be heterozygous for the *P-p* alleles. To be certain about it (because there might possibly be a recessive type of hornlessness even though none has yet been reported), we might wait until breeding tests can be made. Tests for

[1]Yes, of course, but this one stayed polled.

identifying recessive traits were outlined in the previous section. It is simpler to prove that a new mutation is dominant.

If the polled bull is really *Pp,* as we suspect, then, when mated to horned cows (*pp*), about half his progeny will get *P* from their sire and be polled; the other half will get *p* from both parents and be horned. The result should be the familiar 1:1 ratio of phenotypes that comes from the backcross of a heterozygote to homozygous recessives. Since the polled sire had horned parents, but transmits the dominant type of hornlessness to half his progeny, there can be little doubt that he carries a new mutation of *p* to *P.*

When chickens with normal tails produce a rumpless chick, we cannot be certain that this variation is dominant, or recessive, or non-genetic. The last of these possibilities is most likely, as accidental rumplessness occurs in one bird in about 1,500 (Fig. 2-4). However, this type cannot be distinguished in living birds from the dominant rumplessness that is characteristic of certain breeds. The recessive type is easier to identify, but only breeding tests would tell for certain whether or not the rumpless fowl could transmit its distinction, and, if so, how. As we have just seen, such tests are simple, especially in animals that reproduce rapidly.

FIG. 2-4   A congenitally rumpless Leghorn from parents with normal tails; only a breeding test could reveal whether or not the condition in this bird is hereditary. (From F. B. Hutt in *Genetics of the Fowl,* McGraw-Hill Book Co., Inc.)

Pedigrees

When some variant appears for which the genetic basis is unknown, a convenient way to visualize the probable manner of inheritance is to construct a pedigree chart showing (in as many generations as possible) (a) the animals manifesting the new form, (b) the "normal" ones that do not, and (c) the relationships of all concerned. This method is not necessary when experimental breeding tests can be made, but in some species such experiments, even if they could be arranged, might be eyed askance (man), and in others they are difficult and costly (horses, elephants, etc.) In such cases, a glance at a pedigree chart will sometimes tell whether the variant form is dominant or recessive and whether or not a single pair of genes is responsible for two forms.

Pedigrees Showing Dominance.   Calves showing tightly curled hair in all parts of the body except the face and lower parts of the legs (Fig. 2-5) have been found among Ayrshires in Kansas. A pedigree showing the relationships of eight such "Karakul" calves in four generations (Fig. 2-6) tells us that this variation is dominant to the normal condition of straight hair; we know this because crosses of curly × normal yielded curly calves. Except for II-2, whose sire and dam were of unknown type, every Karakul calf had one parent of the same kind,

FIG. 2-5   An Ayrshire calf showing Karakul-type curly hair caused by a dominant mutation. (From F. E. Eldridge *et al.* in *J. Heredity.*)

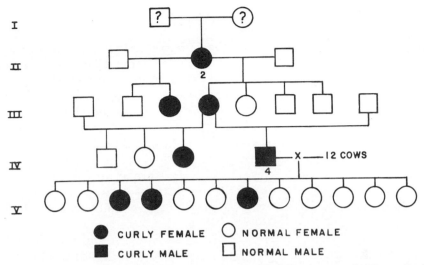

FIG. 2-6 Pedigree of curly coat in Ayrshires, illustrating standard methods of designating generations and individuals and of showing sexes, normal and affected animals, and their relationships. (Redrawn from F. E. Eldridge *et al.* in *J. Heredity.*)

and none segregated from parents both normal, as can happen with recessive characters. Since none of the curly parents produced only curly offspring, it is improbable that any of them were homozygous for the causative dominant gene. From matings of heterozygous curlies × normal, the expectation is 1 : 1 of each type just as in the test-cross of *Bb* black × *bb* red.

Most of the curlies had few offspring, but, among twelve female calves sired by the bull IV-4, there were 9 normal : 3 curly. If the Karakul type of coat were caused by a single dominant gene for which that sire was heterozygous, we should expect a ratio of 6 : 6. By tests to be considered later, one finds that the observed numbers do not differ significantly from those expected, so, pending further information, we conclude that the Karakul coat is probably caused by a dominant gene.

With this information from the pedigree, the breeder knows that if he wishes to retain the variant form he can get more curlies by crossing curly × normal. On the other hand, if the curly coat is considered an undesirable defect, it is necessary to eliminate only the curlies and none of their normal *siblings* (brothers and sisters).

Pedigrees Showing Recessivity. A different genetic basis is evident for two almost completely hairless calves born one year in a small herd of Guernseys (Fig. 2-7). Although their pedigree is comparatively

FIG. 2-7 Incomplete hairlessness in a Guernsey calf, showing hair on tail. (From F. B. Hutt and L. Z. Saunders in *J. Heredity.*)

short, there is enough of it to show how the causative gene was transmitted (Fig. 2-8). The sire (III-3) and the two mothers of the hairless calves (III-2 and IV-2) all had normal hair; the defect is therefore clearly recessive. The normal condition is dominant.

As the difference between hairless calves and normal ones was extreme, without any intermediate forms, it seemed most likely that the condition was caused by a recessive gene in the homozygous state. A calf would be homozygous for hairlessness only if it had received the causative gene from both parents, hence the two affected calves proved that their two mothers and their sire were all heterozygotes, or *carriers* of the gene for hairlessness. As such a situation is most likely to occur if the carriers are descended from some common ancestor, a search of the pedigree is the next step in the investigation. It shows that the sire's dam (I-3) was also the grand-dam of III-2 and that she, in turn, was the mother of IV-2.

Thus, although there were only two hairless calves, the information in their pedigree is in complete accord with the assumption that the defect is a simple recessive character. It shows which animals have been proven to carry the gene, and these can be indicated by a

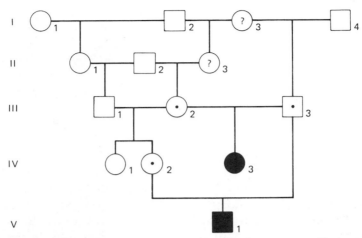

FIG. 2-8 Pedigree showing inheritance of hairlessness in Guernseys as a recessive character. Symbols for affected animals and sexes are as in Fig. 2-6. Dots for III-2 and III-3 and for IV-2 show proven carriers; 1–3 and II–3 are indicated as possible carriers. (From F. B. Hutt and L. Z. Saunders in *J. Heredity.*)

distinctive mark, such as the central dot in the square or circle of a phenotypically normal bull or cow. It also shows two animals that probably carry the gene for hairlessness, although neither produced any hairless calves. One of these is the common ancestor on both the sire's and dams' sides of the pedigree. The other (II-3) is suspect because she is in the direct line of descent from that common ancestor to the mothers of the hairless calves. Such suspects can be conveniently designated by question marks.

The fact that one of the hairless calves was a male and the other a female gives us additional information about the inheritance of the defect, as we shall see in Chapter 7.

The 7:1 Ratio

Thus far we have considered the following Mendelian ratios of dominant to recessive phenotypes:

3:1—With segregation of a simple recessive character in an F_2 generation, or from any mating of heterozygotes *inter se*.

1:1—From the mating of a heterozygote to the homozygous recessive.

1:0—Whenever one parent is homozygous for the dominant allele.

Another ratio which, though seldom mentioned in textbooks, is likely to be encountered in studies of hereditary defects in cattle, sheep, and horses is that of 7:1. It results from the combination of the first and third of the ratios listed above, with four offspring allowed for each.

It sometimes happens that a herd sire carrying some undesirable recessive gene is replaced (when his daughters reach breeding age) by another sire that is a carrier of the same undesirable gene. To illustrate, let us assume that a Holstein-Friesian bull of the genotype *Bb* is followed by another of the same kind. No red calves will appear among the progeny of the first sire if all of the cows in the herd are *BB*, but about half of the calves will get *b* from their sire and the remainder *B*. From their dams they get only *B*; hence, among the whole crop of heifers from the first bull, approximately half will be *BB*, the rest *Bb*. When these are bred to the new heterozygous sire, the *BB* heifers will have only black calves, but the heterozygotes will yield a ratio of 3 black:1 red. The minimum number theoretically necessary to manifest this latter ratio is four calves. Allowing the same number from the *BB* heifers, the two types of matings combined yield a ratio among calves from the second sire of (4:0) + (3:1), or 7:1.

An example of this 7:1 ratio was provided by a herd of Ayrshires in which there appeared in one year four calves showing raw areas lacking normal skin on the knees and above the hooves on all four legs (Fig. 2-9). There were similar areas devoid of skin on the muzzle, and sometimes inside the ears. Investigation revealed that the breeder, having originally a grade herd, had introduced in succession two registered bulls to improve his stock. Daughters of the first were bred to the second, to avoid inbreeding. Both bulls happened to carry a recessive gene (*ep*) causing (in homozygotes) the epithelial defects described above. Among the 30 calves sired by the second heterozygous bull, the ratio was 26 normal:4 defective. That is as close as one can get—in whole calves—to the theoretical expectation of 7:1, or 26.25:3.75.

This same ratio of 7:1 is also to be expected when daughters of a sire that is heterozygous for some recessive trait are mated back to him.

The 7:1 ratio will be distorted if a sire carrying some recessive gene (for example, *ep*) is introduced to a herd in which some of the females are already heterozygous for that same gene. It is quite possible that no calves with epithelial defects would appear in the first

FIG. 2-9 Epithelial defects in an Ayrshire calf. Raw areas devoid of skin show above all four hooves and on the knees. (From F. B. Hutt and J. N. Frost in *J. Heredity.*)

generation. The chance of getting one from an *Ep ep* × *Ep ep* mating is only ¼, and even as many as three or four cows carrying *ep* might by chance produce only normal calves. Among these normal calves, however, the proportion of *Ep ep* genotypes would be ⅔. (The phenotypic ratio in F_2 is 3 : 1, but, as is shown in Fig. 2-1, two of the three showing the dominant character are heterozygous.) It follows that, in our hypothetical case, among all daughters of the first *Ep ep* sire there would be—*not* approximately equal numbers of homozygotes and heterozygotes—but an excess of the *Ep ep* type. If a second sire carrying *ep* were mated to these heifers, his offspring would show a corresponding excess of calves with epithelial defects above the frequency of one in eight.

In the Ayrshires just considered there was no such excess, and the perfect fit of the observed numbers (26:4) to the ratio of 7:1 indicates that none of the original grade cows in the herd was likely to have carried *ep*.

In many cases where 7:1 ratios are expected, an excess of the recessives is reported. Since that type is likely to be abnormal, or at least to be so considered, it is remembered and reported; normal animals are not. It is much easier for a cattle breeder to remember the four calves with skin defects than the exact number of normal ones. This difficulty applies in most studies that draw information from several herds, or from one herd over a period of several years.

The 7:1 ratio is most likely to be encountered in studies of cattle or other species that usually produce only one offspring at a birth. With swine, dogs, and other multiparous species, segregation of recessives in 3:1 ratios will be reported more frequently because the breeders are likely to consider only the litters containing the variant form, and these come from parents both of which are heterozygous.

Problems

The problems in this book are designed to make you think—not to test your memory. As you ponder over them, think in terms of any new words encountered in each chapter, such as genotype, phenotypic ratio, homozygous, gametes, carrier, etc., in this one.

2-1 At the experiment station in Charlottetown, Prince Edward Island, R. C. Parent found that his polled Ayrshire bull, Clover Crest New Design, sired 17 polled and 21 horned offspring, all from horned cows that had no polled relatives. Which of these contrasting characters is recessive?

2-2 What was the genotype of the bull with respect to the P-p alleles affecting horns? Of the cows?

2-3 What was the expected phenotypic ratio?

2-4 In genetic parlance, how would you describe or classify these matings?

2-5 If C.C.N.D.'s polled son, Charlottetown New Burton, were bred to his polled half-sisters (from the matings considered above), would they "breed true to type"? If not, how many horned offspring would you expect among 28? What proportion of the polled calves should be PP?

2-6 Some Ayrshire breeders prefer horns. Would it be safe for one of them to buy horned progeny of C.C.N.D. for breeding, or could that result in unwanted polled animals in later generations?

2-7 Mr. Smith, who has never had any red calves in his herd of Holstein-Friesians, on mating the heifers from Bellowing Dan to his new herd sire, Glamorous Lad, gets in the resultant crop of 26 calves five that are red and white. When he demands of Mr. Brown a return of the money paid for Glamorous Lad, Brown accepts partial responsibility, but claims that Glamorous Lad is not solely to blame. For corroboration of his story, Brown refers Smith to his veterinarian. What would you tell Smith if you were that veterinarian?

2-8 What evidence could you cite to absolve Glamorous Lad from full responsibility?

2-9 If Brown's explanation is correct, what numbers of black and of red calves were to be expected among the 26?

2-10 Mr. Smith now asks the veterinarian, "What is the chance that any one of the black heifers from Glamorous Lad will carry the gene *b* that causes red?" What would you answer?

2-11 Theoretically, how many of the 21 black and white calves are likely to be *Bb*? How many *BB*?

2-12 Supposing that Smith and Brown can still do business together, and recognizing that both now have a common problem, can you suggest any way in which the red calves, three of which were heifers, might be useful?

2-13 Would the cows that produced the red calves be equally useful? If so, why? If not, why not?

2-14 If a Holstein-Friesian bull suspected of carrying *b* were tested by matings with *bb* females, how many black offspring would be necessary to yield odds of 63 : 1 against his being *Bb*?

2-15 A bull, X, carrying red (*Bb*) introduced to a herd of Holstein-Friesians all pure black (*BB*) leaves 40 daughters. These, when mated to the next herd sire, Y, of the genotype *BB*, produce in two years 32 heifers that are eventually mated to a third sire, Z. He happens to carry red, as X did. How many red calves, if any, would you expect among his 32 offspring, one from each of the heifers?

2-16 At the Royal Veterinary College in Stockholm, Björk and others studied in Fox Terriers a nervous disorder that began at four to six months of age and later prevented normal movement of the dogs but was not fatal. Among 91 pups in 23 litters, 25 dogs showed the defect, which was designated cerebellar ataxia. Ataxic pups in different litters were traced to common ancestors. Both sexes were affected. What genetic basis for the ataxia is indicated?

2-17 In the pedigree in Fig. 2-10, which shows how ataxia could appear in litters from normal but related parents, it is assumed that II-1 and II-6 do not carry the gene *a* for ataxia.

List by number all the dogs for which the genotype is known with certainty and give the genotype for each.

2-18 If III-1 be mated to III-6, what is the chance that the firstborn of their litter will become ataxic?

2-19 What is the chance of II-4 being a carrier?

2-20 When she is seven months old, what is the probability that IV-3 is not a carrier?

2-21 Some ataxic dogs become sexually mature. If IV-4 be mated with III-7, what are the chances (a) of their first pup becoming ataxic, and (b) of their producing no ataxic puppies, even in several litters?

22-22 What is the chance that the first pup from III-7 × IV-6 will become ataxic?

2-23 If the owner wishes to test III-7, hoping to find him free of the gene causing ataxia, list three females in Generations III and IV that might be used for such tests. Put them in order of preference, with reasons for that preference.

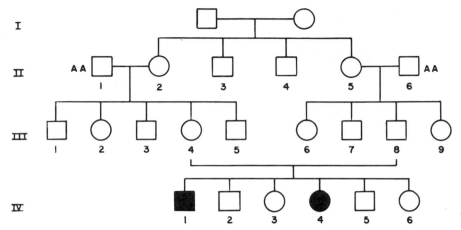

FIG. 2-10 Hypothetical pedigree for two ataxic dogs in one litter from normal parents.

2-24 In Sweden, Eriksson found that among 143 foals sired by the Belgian stallion, Godvän, from 124 mares, there were 65 afflicted with aniridia (complete lack of the iris). These developed cataracts about two months after birth. Godvän had aniridia, but his sire and dam were both normal.

How would you account for the appearance of this rare defect in the 65 foals? In Godvän?

2-25 As this condition prevented the use of affected horses for work, most of them were slaughtered, but Godvän's normal progeny were kept. Would it have been dysgenic to use them for breeding?

2-26 A congenitally tailless rat in the colony maintained by a nutrition laboratory at the University of Minnesota was crossed with normal rats and yielded 49 F_1 progeny, all with normal tails. In the F_2 generation (157 rats) and the backcross of F_1 × tailless (23), no tailless rats appeared. How would you interpret such results?

2-27 At Michigan State University, 16 mares, sired by the stallion Treviso, were bred to Sir Laet. All were purebred Percherons. Among their 42 foals were five with beautiful curly coats of a kind not seen in any of their relatives. (Fig. 2-11.) The sire and dam of Treviso were the maternal grandsire and paternal granddam of Sir Laet. How would you account for the curly coats?

2-28 On that basis, how many straight-haired foals were to be expected among the 42?

2-29 What proportion of these straight-haired foals should carry the gene for curling?

2-30 Some Percheron breeders might like Sir Laet's progeny, but prefer one not carrying curly. For any one smooth-coated foal, what is its chance of being free of that gene?

FIG. 2-11 Curly coat in a Percheron foal. (From L. H. Blakeslee *et al.* in *J. Heredity.*)

2-31 In the 12 litters that contained bird-tongued puppies, the ratio of normal:bird tongue was 54:22. About how many of the normal puppies would be expected to carry the gene that causes the lethal defect, and how many would not be expected to do so?

Selected References

BJÖRK, G., S. DYRENDAHL, and S.-E. OLSSON. 1957. Hereditary ataxia in smooth-haired Fox Terriers. *Vet. Rec.* **69:**871–76. (Account of the disorder considered in Problems 2-16 to 2-23.)

BLAKESLEE, L. H., R. S. HUDSON, and H. R. HUNT. 1943. Curly coat of horses. *J. Heredity* **34:** 115–18. (Reports genetic study of the condition shown in Fig. 2-11.)

COLE, L. J., and S. V. H. JONES. 1920. The occurrence of red calves in black breeds of cattle. *Wisconsin Agric. Exper. Sta. Bull.* 313. (A review of the problem.)

ELDRIDGE, F. E., F. W. ATKESON, and H. L. IBSEN. 1949. Inheritance of a Karakul-type curliness in the hair of Ayrshire cattle. *J. Heredity* **40:** 204-14. (Source of Figs. 2-5 and 2-6.)

ERIKSSON, K. 1955. Hereditary aniridia with secondary cataract in horses. *Nord. Vet.-Med.* **7:** 773–93. (Details of the defect considered in Problems 2-24 and 2-25.)

HUTT, F. B., and A. DE LA HUNTA. 1971. A lethal glossopharyngeal defect in the dog. *J. Heredity* **62:** 291–93. (Source of Fig. 2-3 and Problem 2-31.)

HUTT, F. B., and J. N. FROST. 1948. Hereditary epithelial defects in Ayrshire cattle. *J. Heredity* **39:** 131–37. (Condition shown in Fig. 2-9 and used to illustrate the 7:1 ratio.)

HUTT, F. B., and L. Z. SAUNDERS. 1953. Viable genetic hypotrichosis in Guernsey cattle. *J. Heredity* **44:** 97–103. (Source of Figs. 2-7 and 2-8.)

WARWICK, B. L. 1932. Probability tables for Mendelian ratios with small numbers. *Texas Agric. Exper. Sta. Bull.* 463. (Chances of getting various combinations of dominant and recessive phenotypes in any of nine different ratios with numbers from 1 to 50.)

Incomplete Dominance

In the examples considered in the previous chapter, dominance of one allele over the other was complete. No effect of the recessive gene was evident in heterozygotes carrying it, and such animals could be distinguished from homozygotes only by breeding tests.

In many cases dominance is only partial, and when the dominant and recessive homozygotes are crossed, the resultant F_1 animals are intermediate between two parental types. Their heterozygous genotype is evident from their phenotype; hence, breeding tests are unnecessary.

Sometimes the heterozygote has a more attractive appearance (or one more bizarre) than either homozygote, and, as a result, its phenotype is set up by fanciers as the standard for a distinct breed. Since such animals cannot breed true to type when mated *inter se*, the constant segregation of the unwanted homozygotes during the formative periods of such breeds has provided for many a breeder both frustration and inspiration for further efforts. Eventually, genetic analysis forces recognition of the fact that the old dictum, "Like begets like," does not apply to heterozygotes.

A few examples will illustrate some effects of incomplete dominance that are important to breeders of domestic animals.

Roan Shorthorns

The attractive roan color commonly found in the coat of Shorthorn cattle, and others, results from a mixture of red and white hairs (Fig. 3-1). It seems probable that all Shorthorns are basically red; even those appearing to be pure white usually show some red hairs in the ears and eyelashes.

Crosses of red × white usually yield only roans, and when roans are mated *inter se*, the resultant F_2 generation shows approximately 1 red : 2 roan : 1 white. Most investigators agree that roan is caused by an incompletely dominant gene N which, in the heterozygous state, converts red to roan by removing the red pigment from some of the hairs. In homozygotes all pigment is eliminated except in a few hairs in the ears and eyelashes.

On this basis the expectations in various types of matings, assuming 100 calves from each, are as shown on p. 44.

Among 856 progeny from such matings, Jones (1947) found only eleven not conforming to expectation. Of these, nine were found to be genotypically roan (Nn) but so dark that they had been classified as red. Information about the other two was unavailable. At the other end of the range in the variation of roan, some apparently white bulls have produced red calves, thus indicating that they were Nn rather than

FIG. 3-1 Shorthorn bulls, showing (top to bottom) red, roan, and white coat colors, all with exceptionally good conformation. (Courtesy of American Shorthorn Breeders' Association.)

Parents		Progeny		
Phenotypes	*Genotypes*	*Red*	*Roan*	*White*
Red × red	*nn* × *nn*	100	—	—
White × red	*NN* × *nn*	—	100	—
White × white	*NN* × *NN*	—	—	100
Roan × red	*Nn* × *nn*	50	50	—
Roan × roan	*Nn* × *Nn*	25	50	25
Roan × white	*Nn* × *NN*	—	50	50

NN. Undoubtedly, other genes can influence the variation in roaning between almost all red and almost completely white. Some of these may actually obscure the roan, but, as Jones's data indicate, such exceptions are not common. The white spotting that is characteristic of many breeds sometimes lightens the coat in roan Shorthorns but is usually recognizable.

In cattle that would otherwise be black, the gene *N* produces attractive "blue-grey" roans, and crosses of White Shorthorns with Aberdeen Angus and Galloways are commonly made for that purpose. Blue Albions and other breeds of blue-grey cattle are also probably *Nn.*

Breed standards for Shorthorns accept both homozygotes and the heterozygote, and all three colors can be registered. A peculiar type of sterility that is found more often in white heifers than in the other two types has led to some preference for reds and roans. (See index.)

The Frizzle Fowl

The bizarre appearance of the Frizzle suggests that it has been pulled backward through a knothole (Fig. 3-2). Fanciers like birds on which the feathers stand up stiffly, their shafts curling outward, so that the tips of the contour feathers seem to point forward to the head. The wing and tail feathers are less curved, but the barbs, which in normal fowls form a solid web, are often twisted, and as they eventually break off, these feathers frequently show no web at all, but only shafts. After a moult and before any of the feathers get broken, those in the neck region seem to form a sort of ruff in which the head is framed.

This peculiar mutation was brought to the attention of geneticists when breeders asked why their Frizzles did not breed true, but always threw some chickens with normal plumage. Investigations showed that the kind of Frizzle preferred by fanciers is heterozygous for a dominant gene *F.* Since all exhibition-type Frizzles are *Ff*, matings of

FIG. 3-2 The heterozygous Frizzle; an almost perfect specimen from the viewpoint of the fancier. (From F. B. Hutt in *J. Genetics.*)

such birds *inter se* yield 3 genotypes in the ratio 1 *FF*:2 *Ff*:1 *ff*. The recessive homozygotes have ordinary plumage, but *FF* birds show feathers so curled that the plumage resembles a woolly fleece (Fig. 3-3). No feather has a flat web and the flight feathers are completely useless. Since these extremely frizzled birds cannot fly to the roosts, they crowd together on the floor at night and their feathers are thus broken off. A few months after a moult these birds are almost bare (Fig. 3-3).

In this case, the homozygotes are not considered as respectable members of the Frizzle breed, and are usually discarded as soon as they can be recognized. The breeder of showroom Frizzles has now become reconciled to the fact that when breeding from prize-winners, half their progeny will be wasted. Paradoxically, by mating together the two outcasts, *FF* × *ff*, he can get chickens that all show the respectable phenotype of the heterozygote.

The gene *F* affects primarily the structure of feathers, but interesting secondary effects follow because the insulation normally provided by the plumage is lost in the homozygotes and they become bare. To maintain body temperature, production of heat is increased, the heart is hypertrophied and beats more rapidly than in normal

FIG. 3-3 Homozygous Frizzles. Above: in best "woolly" plumage, as newly acquired after a moult; below: typical lack of plumage several months after the moult, when many feathers have broken off. (From F. B. Hutt in *J. Genetics.*)

fowls, and feed consumption is increased. Homozygous males are slow to reach sexual maturity, and *FF* females do not lay as early or as well as their sisters with normal feathers. A single dose of gene *F* does not destroy the insulating value of the plumage and apparently causes only slight secondary effects, or none at all, on the physiology of heterozygotes.

The Frizzles thus provide an interesting example of incomplete dominance, of preference by the breeder for a genotype that cannot breed true, and of a gene that affects primarily structure but also prevents normal function.

Blue Andalusians and Other Fowls

Breeds and varieties of fowls that are commonly designated as blue are really blacks in which the melanin is in round granules that are restricted to little clumps. Blues can be considered as diluted blacks. The blue feathers usually have a black lacing, and in blue males the lacing is so wide that feathers in the hackle, wing-bow, and saddle appear more black than blue.

The Blue Andalusians have long provided for texts in genetics the classical case of incomplete dominance. When mated *inter se*, blues produce blue-splashed whites, blues, and blacks in the ratio of 1 : 2 : 1. As with Frizzles, the fanciers have set up the heterozygous blues as the standard for the breed. Homozygous blues have so much melanin removed that they appear to be dirty white with a few blue feathers (Fig. 3-4).

FIG. 3-4 Blue Andalusians. Left: the heterozygote, approved as true to her breed; right: the unwanted blue-splashed white homozygote.

The incompletely dominant gene *Bl* is widespread in domestic fowls and has been used to establish (insofar as heterozygotes can ever be established) blue varieties in several breeds. Mating black × blue-splashed whites would yield progeny all blue, but most fanciers dislike to produce blues from parents that cannot qualify as respectable members of the breed. This psychological barrier need not bother the fly-tiers, who, preferring hackle feathers of blue males for their most enticing trout flies, sometimes subsidize youthful poultrymen to raise Blue Andalusians for that purpose. The easiest way to get a large stock of blues is to mate blue-splashed cocks with any kind of solid black females. The latter need not be black segregates from blue parents.

Other breeds of fowls in which the distinguishing characteristic that makes the breed is caused by an incompletely dominant gene in the heterozygous state are the short-legged Creeper, which is considered in Chapter 9, and the beautiful Erminette (p. 57).

The Palomino Horse

The type desired in Palominos is a golden-yellow body color, with mane and tail almost white (Fig. 3-5). Unfortunately, this attractive combination is the expression of a heterozygous genotype, and Palominos mated *inter se* produce offspring among which only 50 per cent show the same color as their parents. Of the remainder, about half are light chestnuts, or sorrels, and half are so light in color that they are commonly called albinos.

The Palomino is basically a light chestnut, or sorrel, to which has been added a gene, c^{cr}, that dilutes the normal reddish, golden brown of light chestnuts to a yellow or cream color. In double dose, c^{cr} causes extreme dilution. Accordingly, when the C-c^{cr} alleles are superimposed on the combination of genes that would otherwise cause the light chestnut color, three different genotypes and phenotypes are possible:

$$c^{cr}c^{cr} \quad \text{albino, "blanco," or cremello}$$
$$Cc^{cr} \quad \text{Palomino, yellow, or cream}$$
$$CC \quad \text{light chestnut, or sorrel}$$

The $c^{cr}c^{cr}$ homozygote, though generally called an albino, is really a pale cream, not white, and the eyes show some pale blue (from melanin in the iris), which is not found in the pink eyes of animals that are true albinos.

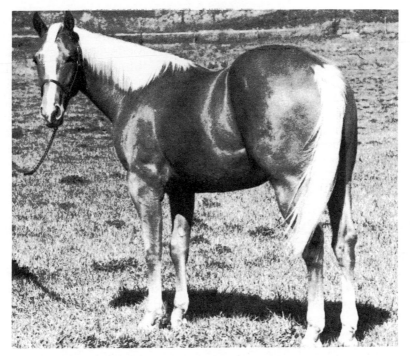

FIG. 3-5 A Palomino mare, Temptest Storm, owned by J. Janowitz, Littleton, Colorado. (Courtesy of Palomino Horse Breeders of America.)

Records of two Palomino breeders summarized by Castle and King (1952) yielded the results shown in Table 3-1. Closer fits of observed to expected ratios will seldom be found, except in the last of the four matings, but the deviation there from the expectation of 7 : 7 is not statistically significant. Adalsteinsson (1974) had similar evidence from Icelandic horses.

Studies by Tuff of the inheritance of this diluting gene in the Vestland horses of Norway showed not only that it converts chestnuts to yellows, as outlined above, but also that horses which would otherwise be brown or bay become "buckskin dun" when it is present in the heterozygous state. (The symbols, Dd, used by Castle, have some advantages, but Tuff's cream Vestlands were Ff, and, according to other investigators, Palominos are Cc^{cr}. Students of genetics might as well learn early in the game that such confusion over different symbols for one and the same gene is a normal situation for many species until some committee, with one eye on propriety and the other on priority, attempts to bring order out of chaos. In some cases, cold wars persist and all efforts of peace-loving committees to unite oppos-

TABLE 3-1
Matings Involving Palomino Horses

		Progeny		
Parents	Ratio Expected	Albino	Palomino	Chestnut
Albino × chestnut	0 : 1 : 0	0	55	0
Palomino × chestnut	0 : 1 : 1	0	57	60
Palomino × Palomino	1 : 2 : 1	17	45	21
Albino × Palomino	1 : 1 : 0	11	3	0

Source: Data of Castle and King, 1951.

ing factions are futile, so that ultimate victory of one set of symbols over another has to await the demise of the combatants and the verdict of general usage.)

Justification for labelling the *Bl* of Andalusians as incompletely dominant, but the c^{cr} of Palominos as recessive lies in the fact that the latter is fully recessive in black horses. Another gene, *D*, dilutes all colors.

Codominance, The M-N Blood Antigens in Man

Among the many different types of blood known in man, none has a simpler genetic basis than those induced by the antigens usually designated M and N. Understanding of that basis might have been easier if these blood types had been labelled M and m, for only a single pair of genes is involved. Neither is dominant over the other, and both are expressed in heterozygotes, a situation sometimes called *codominance*. The genotypes *MM*, *Mm*, and *mm* induce the blood types that the serologist calls M, MN, and N.

The M and N blood antigens differ from some others in that antibodies against them are rare in human blood. For that reason, M and N, unlike some other blood antigens, need not be considered when blood transfusions are made. However, by using antisera containing antibodies induced in other animals, tests can easily be made to determine whether a person is M, N, or MN. Blood cells of people belonging to type M are agglutinated by antiserum containing antibodies against that kind of cell; N cells react with N antiserum and those of type MN react with both M and N antisera.

From such tests applied to 2,734 people and their parents, Wiener (1951) compiled data for all possible parental combinations (Table 3-2). Among these there were remarkably good fits to the 1:2:1 and

TABLE 3-2
Inheritance of the M−N Blood Types

Parental Types	Ratio Expected	Progeny		
		M	MN	N
M × M	1 : 0 : 0	272	1[a]	0
N × N	0 : 0 : 1	0	0	72
M × N	0 : 1 : 0	1[a]	315	0
MN × M	1 : 1 : 0	408	387	1[a]
MN × N	0 : 1 : 1	3[a]	334	300
MN × MN	1 : 2 : 1	163	317	160

[a]Exceptions to expectation.
Source: From Wiener, 1951.

1 : 1 ratios expected, and only six cases that did not conform to expectation from the parental types. From the evidence with respect to those, it seemed fairly clear that what had gone astray was not Mendel's law.

The important thing to note in the foregoing example is that when the two alleles are codominant, all three genotypes (the two homozygotes and the heterozygote) can be identified by an appropriate test. For the MN genotype, it is an agglutination test, but there are others of various kinds.

Nature of Dominance

In cases of incomplete dominance like those cited in this chapter, one should not worry about trying to decide which allele is dominant and which recessive. Is white dominant or recessive to red in roan Shorthorns? Are the albinos that beget Palomino horses homozygous for a dominant gene, as Castle described them, or for a recessive one, as in other interpretations? Debate on such points is futile. The important thing is to recognize that the phenotype of the heterozygote is distinctly different (with rare exceptions) from that of either homozygote, and that effects of both the dominant and recessive alleles are expressed in the individual that carries both. It is as if the heterozygote were a compromise between two forces each of which has tried to attain different ends.

When the phenotype of one homozygote can be considered as the normal or "wild type" for the species, any gene which in the heterozygous state causes a visible modification of that normal state is usually designated as a dominant one. Thus, frizzling is dominant to normal plumage, but, because the heterozygotes are much less frizzled than homozygotes, it is incompletely dominant. Rose comb is completely dominant to single, as far as is now known, but whether the Andalusians are blue because of a dominant gene or a recessive one is a question not worth debating.

Finally, one must not jump to the conclusion that all intermediates between contrasting parental phenotypes result from the action of a single pair of allelic genes. As we shall see later, some crosses between extremes (such as that of large animals with small ones) yield intermediates for which the genetic basis is more complex.

Detection of Heterozygotes

In cases of incomplete dominance considered thus far, the heterozygote is usually so different from both homozygotes that all three genotypes are readily recognized. To be sure, an agglutination test is necessary to identify the MN type of blood, but the same procedure is equally necessary for identification of the M and N homozygotes. Sometimes the range of variation in the heterozygous animal is so great that the extremes are mistaken for homozygotes, as in Shorthorns that are so dark and with so little roan that they are classified as red.

In other cases dominance may appear to be complete, but special tests or minute examination will reveal that some or all of the heterozygotes show some slight effect of the recessive gene.

An example is provided by the modified Thrombotest method which Spurling (1974) devised to detect heterozygotes deficient in Factor VII, one of the factors necessary for clotting of the blood. A deficiency of it causes problems in Beagle colonies that provide dogs to pharmaceutical companies for use in various tests. That deficiency is a hereditary trait.

By use of his Thrombotest method, Spurling (1974) was able to distinguish between normal dogs free of the causative gene and heterozygotes that carried it. For his test he used 30 dogs known to be carriers because they had all produced pups with deficiency of Factor VII, 35 known to be homozygous normal, and four believed (from their pedigrees) to be carriers. Among these 69 dogs, the level of activity of Factor VII on Spurling's logarithmic scale was found to range from 1.4 to 2.2. The known carriers (and the four not proven) were grouped around a mean of 1.73, and for the 35 known to be homozygous normal, the mean was 2.017. A "discrimination threshold" at 1.92 (equivalent to 83.5 per cent activity of Factor VII) would have identified the two genotypes correctly in 64 of the 69 dogs (Fig. 3-6).

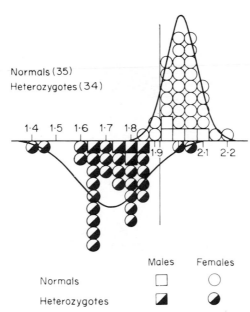

Normals (35)
Heterozygotes (34)

1·4 1·5 1·6 1·7 1·8

1·9   2·1 2·2

	Males	Females
Normals	□	○
Heterozygotes	◲	◑

FIG. 3-6 Activities of factor VII in Beagles, expressed logarithmically, with distributions of assays in known heterozygotes inverted to show better how little they overlap the distributions in dogs known not to carry the causative gene. With a discrimination threshold of 1.92, the test identified the genotype correctly in 64 of the 69 dogs. (After Spurling *et al.*, 1974.)

Another example deals with a mutation that prevents proper functioning of the thrombocytes (platelets) in the blood and ·thus causes inadequate clotting. In man, the disease is called thrombasthenia. Dodds (1976) studied a similar condition in Otter Hounds.

Using a test in which aggregation of the platelets can be stimulated (and measured) by the addition of small amounts of thrombin, she found the aggregation in heterozygotes to be intermediate between that in homozygous normal dogs and that in homozygous recessives afflicted with thrombasthenia (Fig. 3-7).

It is to be hoped that detection of undesirable heterozygotes by suitable laboratory tests will eventually reduce somewhat the number of cases in which the much slower breeding test is still the only way to identify the carriers of unwanted genes.

Dwarf Calves

One of the most intense searches ever made in domestic animals to identify undesirable heterozygotes has been that for carriers of a gene causing undesirable dwarfism in cattle. This defect occurs in

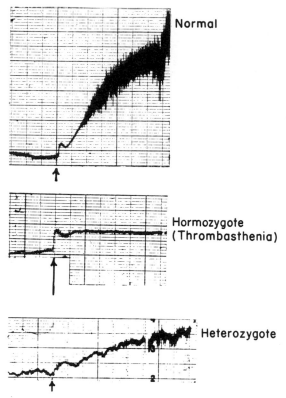

FIG. 3-7 Tracings of aggregations of blood platelets in platelet-rich plasma following addition (arrows) of thrombin. The heterozygote is intermediate between the normal dog and the one with thrombasthenia. (Courtesy of W. Jean Dodds, Division of Laboratories and Research, New York State Department of Health, Albany, New York.)

Herefords and in other breeds of beef cattle. The dwarfs usually show bulging foreheads, disproportionately short broad heads, and malocclusion of the jaws (Fig. 3-8). They are unthrifty, pot-bellied, susceptible to bloating, and low in viability. Their labored breathing has caused them to be known as "snorter" dwarfs. Those that do not survive make undesirable carcasses.

This type of dwarfism is apparently a simple recessive defect. Investigations in three states, as summarized by Pahnish *et al.* (1955), showed that carrier bulls × carrier cows yielded close fits to the 3:1 ratios expected with such a trait (Table 3-3).

In this case, incomplete dominance has caused trouble in two different ways. Heterozygotes tend to be somewhat short-legged and compact, a type which pleased the judges of show-ring Herefords. The

Fig. 3-8 In the foreground: a typical, short-headed, dwarf Hereford calf at 14 months of age, showing short face, bulging forehead, pot belly, and short legs; behind: his twin brother, a normal calf, whose weight, when this photograph was taken, exceeded that of the 362-lb. dwarf by 171 lbs. (From L. M. Julian, W. S. Tyler, and P. W. Gregory in *J. Amer. Vet. Med. Assoc.* **135**: 104–9, 1959.)

resultant corresponding preference by the breeders caused the accumulation of genes for dwarfism in many a fine herd. The problem of eliminating those genes is now recognized by all concerned, but, unfortunately, the heterozygotes cannot be identified with certainty from their appearance. Test-crosses are not very practicable and other progeny tests are slow.

Gregory *et al.* (1953), noting that carriers usually have a slight

TABLE 3-3

Calves from Matings in Which Both Parents Were Known to Carry Dwarfism

Experiment Station	Investigator	Normal Calves	Dwarf Calves
Arizona	Pahnish *et al.*	66	24
California	Gregory *et al.*	46	15
Iowa	Lush and Hazel	197	69

Source: From Pahnish *et al.*, 1955.

prominence in the middle of the forehead, devised a "profilometer" with which to measure it more accurately in young bulls about 15 months old or less. Obviously, the younger that heterozygosity for dwarfing (or for any other defect) can be detected, the better. Other investigators have sought to do this with calves only a week old by studying radiographs of the thoracolumbar vertebrae. While those of dwarfs are then characteristically abnormal, it remains to be seen whether or not carriers can be similarly distinguished from calves free of the dwarfing gene. Some investigators have sought differences in the blood, in the urine, and in pressure of the cerebrospinal fluid, but no diagnostic test has yet proven completely satisfactory.

Apart from the brachycephalic, or short-headed, dwarfs just considered, other types of dwarfism are now recognized. Some of these, particularly the long-headed dwarfism and the "compressed" type, are hereditary, and it seems probable that the genetic bases for these and for the "snorter" dwarfs may be more complex than was first thought.

While most of the serious hereditary defects in domestic animals are apparently completely recessive so far as is now known, it is probable that further study will show some of these to be manifested in heterozygotes to a degree that can be recognized by suitable tests. Neel (1953) listed 20 hereditary abnormalities in man usually considered to be recessive, but now known to be detectable in some or all of the carriers by appropriate examinations or tests. Others have since been added to his list. Since some of these conditions—hemophilia, epilepsy, thalassemia (Cooley's anemia), and others—are diseases that no thoughtful person would like to transmit to his or her descendants, tests for detection of carriers can be very important.

Problems

3-1 Calculate the numbers expected in the various classes in Table 3-1, and the deviations of observed numbers from those expected.

3-2 What numbers of dwarf calves were to be expected in the three groups of calves considered in Table 3-3, if the type of dwarfism studied in all three states was a simple recessive trait?

3-3 On the same basis, when a bull sires a dwarf calf in a large herd, what is the lowest proportion of carriers that one might expect among his normal offspring, apart from chance fluctuations? Under what conditions could that proportion be higher, and (assuming that none of the cows was a dwarf) how high could it go?

3-4 When beautiful Erminette fowls like that shown in Fig. 3-9 were mated *inter se*, their 46 progeny included 20 Erminettes, 16 blacks, and 10 pure white. What genetic basis is indicated for the Erminette pattern?

FIG. 3-9 The heterozygous Erminette. (See Problems 3-4 to 3-7.)

3-5 What numbers of each kind were to be expected on that basis?

3-6 What further breeding tests would you make to verify your hypothesis about these Erminettes, and what results would you expect?

3-7 If the breeder wanted only Erminettes, what kind of birds would he have to mate to get them without any of the undesired kinds? (Careful here; there are several genetically different kinds of white fowls!)

3-8 White Shorthorn heifers are frequently sterile. One writer on this problem reported that a certain white Shorthorn bull had sired 20 roan and red heifers, all normal, and 10 white ones, all sterile. Is there anything here that might make you dubious of that report?

3-9 If a Shorthorn breeder wants no white animals but likes roans, what kinds should he mate together?

3-10 From Scotland, Donald reported the birth of quintuplet Shorthorn calves among which were: 1 red ♂, 2 roan ♀♀, and 2 white ♀♀. Can you tell the colors of their sire and dam?

3-11 For each of the following combinations of M, MN, or N blood types in the parents, determine what blood types could occur in their children:

$$M \times M, \ N \times N, \ M \times N, \ MN \times M, \ MN \times MN.$$

3-12 Blue Andalusian fowls, whether mated to White Wyandottes or to Black Minorcas, produce offspring of the same colors in both cases, and in the same ratio. Why?

3-14 From these crosses, what colors would you expect in the progeny, and in what ratio?

Selected References

ADALSTEINSSON, S. 1974. Inheritance of the Palomino color in Icelandic horses. *J. Heredity* **65:**15–20 (Reviews earlier studies and reports additional evidence.)

DODDS, W.J. 1976. Inherited bleeding disorders. *Purebred Dogs Amer. Kennel Gazette,* **53** (10):31–38.

GREGORY, P.W., C.B. ROUBICEK, F.D. CARROLL, P.O. STRATTON, and N.W. HILSTON. 1953. Inheritance of bovine dwarfism and the detection of heterozygotes. *Hilgardia* **22:**407–50. (Use of the profilometer; the first attempt to identify by special test domestic animals heterozygous for some undesirable defect.)

HUTT, F.B. 1949. *Genetics of the Fowl.* New York. McGraw-Hill Book Co., Inc. (For further information about Frizzles, Blue Andalusians, and other variations.)

JONES, I.C. 1947. The inheritance of red, roan, and white coat colour in dairy Shorthorn cattle. *J. Genetics* **48:** 155–63. (Study of these colors in one large herd for five generations.)

PAHNISH, O.F., E.B. STANLEY, C.E. SAFLEY, and C.B. ROUBICEK. 1955. Dwarfism in beef cattle. *Arizona Agric. Exper. Sta. Bull.* 268. (A useful survey of the problem.)

SPURLING, N. W., L. K. BURTON, and T. PILLING. 1974. Canine factor-VII deficiency: Experience with a modified thrombotest method in distinguishing between the genotypes. *Research in Vet. Science* **16:** 228–39. (Source of Fig. 3–6).

WIENER, A. S. 1951. Heredity of the M–N types. *Amer. J. Human Genet.* **8:** 179–83. (Source of the data in Table 3-2.)

Dihybrids and Independent Assortment; Chi-square Tests

Thus far we have considered only inheritance in which a single pair of allelic genes is involved. When two or more pairs of genes and characters can be studied in one cross, members of each pair usually show up in F_1, F_2, and backcrosses in the same proportions as if they were alone. Exceptions necessitating that word "usually" are considered in later chapters.

However, just as tosses of four coins, two silver and two copper, give us more combinations than tosses of only two of either kind, so do we have more different combinations when two pairs of genes are involved.

Dihybrid Cross

Let us consider a cross between Aberdeen Angus and red Shorthorn cattle. The former are black and polled (*BB PP*), the Shorthorns red, with horns (*bb pp*). All gametes of the Aberdeen Angus must carry the dominant genes *B* and *P*, while those of the Shorthorn have the recessive alleles *b* and *p*. The F_1 progeny have the genotype *Bb Pp*, hence are black and polled, except that some of the males may show scurs (Fig. 4-1).

Because they are heterozygous with respect to two pairs of genes, these F_1 animals are called *dihybrids*. To the zoologist, hybrids are crosses between species or races, but the geneticist's hybrids can be all of one species. They are called monohybrids, dihybrids, or polyhybrids when heterozygous for one, two, or several pairs of genes, respectively. A *dihybrid cross* is one in which two pairs of genes are considered.

When these F_1 dihybrids are mated *inter se* to beget an F_2 generation, each of them will produce gametes of four different kinds: *B P, B p, b P,* and *b p*. This happens because the *B-b* alleles go their separate ways into the germ cells independently of the *P-p* alleles, which are presumably located in a pair of homologous chromosomes other than that carrying *B* or *b*. When the number of chromosomes is reduced in the formation of germ cells, each gamete gets either *B* or *b* from one pair of chromosomes and *P* or *p* from the other. Four different combinations, and only four, are possible. Similarly, when two pairs of coins are tossed—one a silver pair and the other a copper pair—four combinations are possible in any two coins that include one of each kind. These are: both heads, both tails, silver head with copper tail, and copper head with silver tail.

Since each F_1 animal yields four different kinds of gametes, the number of possible combinations resulting from random union of male and female gametes is 4 × 4, or 16. An easy way to see all the

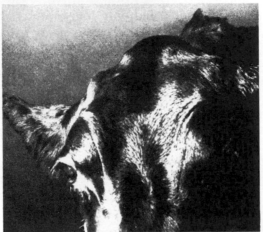

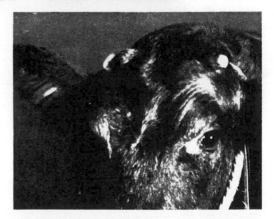

FIG. 4-1   Top: A polled steer from Aberdeen Angus ♂ × F₁ ♀ (from cross Aberdeen Angus ♂ × Shorthorn ♀). No trace of horn in skin or from skull; probably *PP*. Middle: A polled steer from Aberdeen Angus ♂ × F₁ ♀ (from cross Shorthorn ♂ × Aberdeen Angus ♀). There are small knobs of bony growth from the skull, but no horn; probably *Pp*. Bottom: A polled steer from Shorthorn ♂ × Abderdeen Angus ♀, showing scurs; *Pp*. (From J. Hammon in *Endeavour* 9, No. 34, 1950.)

genotypes that could be produced in the F_2 generation is to make a Punnett square, or "checkerboard," like that shown in Fig. 4-2.

From the 16 possible combinations of genes in the F_2 generation, only four different phenotypes are possible. As the checkerboard shows, the ratio of these is:

9 black, polled : 3 black, horned : 3 red, polled : 1 red, horned

Because of the time and expense necessary to produce them, few F_2 generations have been reported from crosses in cattle or in other large domestic mammals. However, with plants, and with mice, fowls, fruit flies, and other animals that reproduce rapidly, it is easy to study the segregation of one, two, or many pairs of genes. From the uniform results obtained in many such experiments, expectations with larger animals can be predicted, so long as the genetic basis for each character is known.

The first dihybrid F_2 generation studied in animals was reported to the Royal Society of London by Bateson in 1901. He had crossed White Leghorns with Indian Game fowls. The former are dominant white, with single (upright) combs. The Indian Games have dark plumage and pea combs. These last show three longitudinal rows of points, the central one being most conspicuous. The F_1 were all predominantly white, with pea combs. Numbers observed and those expected (to the first decimal) with $9:3:3:1$ segregation of two independent pairs of characters were as follows:

	Observed	Expected
White, pea comb	111	106.9
White, single comb	37	35.6
Dark, pea comb	34	35.6
Dark, single comb	8	11.9
	190	190.0

A better fit of observed to expected ratios of phenotypes is not likely to be found.

Probabilities Without Checkerboards

Expectations in the F_2 generation from dihybrids can be determined algebraically by multiplying together the separate probabilities for each pair of genes. For the cross in cattle just considered these are:

(3 black + 1 red) × (3 polled + 1 horned) = 9 black, polled + 3 black, horned + 3 red, polled + 1 red, horned

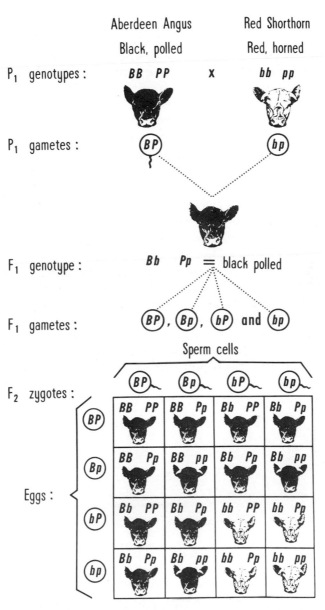

FIG. 4-2 Independent inheritance of the black-red and polled-horned alleles in cattle. The F₂ phenotypic ratio is 9 black polled : 3 black horned : 3 red polled: 1 red horned.

The chance of getting any particular combination from two independent pairs of alternative characters can also be determined from the chances of getting each of these characters by itself. The important principle here is that the chance of getting in combination any two independent events is the product of the separate probabilities that each will occur. In this case the chances of getting the various types in F_2 are: black—¾, red—¼, polled—¾, horned—¼. From these, the probabilities of getting different combinations are as follows:

$$
\begin{array}{lll}
\text{black, polled:} & \text{¾} \times \text{¾} = {}^9/_{16} \\
\text{black, horned:} & \text{¾} \times \text{¼} = {}^3/_{16} \\
\text{red, polled:} & \text{¼} \times \text{¾} = {}^3/_{16} \\
\text{red, horned:} & \text{¼} \times \text{¼} = {}^1/_{16}
\end{array}
$$

It is evident from the checkerboard in Fig. 4-2 that the nine black, polled cattle can have any of four different genotypes. In other words, they look alike, but they would breed differently. The same applies to two of the other phenotypic classes, and there is only one kind for which the genotype is evident from the phenotype. It is the double recessive—the red, horned class.

By adding up the 16 squares in the checkerboard, the frequencies of different genotypes are found to be as follows:

9 black, polled
$$
\left\{
\begin{array}{l}
BB\ PP\dots1 \\
Bb\ PP\dots2 \\
BB\ Pp\dots2 \\
Bb\ Pp\dots4
\end{array}
\right.
$$

3 black, horned
$$
\left\{
\begin{array}{l}
BB\ pp\dots1 \\
Bb\ pp\dots2
\end{array}
\right.
$$

3 red, polled
$$
\left\{
\begin{array}{l}
bb\ PP\dots1 \\
bb\ Pp\dots2
\end{array}
\right.
$$

1 red, horned $bb\ pp\dots1$

These various frequencies in the F_2 generation can be expressed as probabilities. Before that generation is born, we know the probable frequencies of its various types. When the F_2 generation can be classified according to phenotype, the probabilities with respect to genotypes of individual animals can be recognized, as the following samples show. The chance that:

One of the black, polled cattle is true-breeding is	$^1/_9$, or 0.11
One of the black, polled cattle is a dihybrid is	$^4/_9$, or 0.44
A red, polled animal is heterozygous for P is	$^2/_3$, or 0.67
A red, horned animal is $bb\ pp$ is	$1 = 1.00$

Law of Independent Assortment

In concise terms, this principle, often designated as Mendel's second law, is that the genes in each pair of alleles segregate independently of the others. As we shall see later, this rule does not apply to "linked" genes that lie near each other in the same chromosome, but Mendel had no such cases among the seven pairs of contrasting characters which he studied in peas. Such exceptions to the law of independent assortment are relatively infrequent, and few have yet been found in any domestic mammals bigger than a rabbit. It is because of independent assortment that expected ratios can be calculated for crosses in which two, three, or more pairs of independent characters are involved. Those ratios are determined by combining the expectations for each pair by itself.

In the cross shown in Fig. 4-2, all four dominant genes came in from the Aberdeen Angus parent and all four recessive genes from the Shorthorn. Results in the F_1 and F_2 generations would be just the same if each parent contributed dominant alleles of one pair but recessive alleles of the other.

Thus, from a cross of Holstein-Friesian ($BB\ pp$) × Red Poll ($bb\ PP$), the F_1 generation would be black and polled ($Bb\ Pp$). Since the white spotting of Holstein-Friesians is recessive to full color, these F_1 animals would be all black. As before, their gametes would be of four kinds: $B\ P$, $B\ p$, $b\ P$ and $b\ p$; hence the ratio in F_2 would be $9:3:3:1$, just as in the other cross.

Backcross of a Dihybrid

When an F_1 dihybrid bull ($Bb\ Pp$) of the previous example is backcrossed to cows of the double recessive type ($bb\ pp$), the same phenotypes result as in the F_2 generation but the ratio is different. The gametes of the bull are of four different kinds, as before, but the eggs from the cows are all the same, so far as b and p are concerned. The resultant combinations are as follows:

♂ gametes:	$B\ P$		$B\ p$		$b\ P$		$b\ p$
♀ gametes:	$b\ p$		$b\ p$		$b\ p$		$b\ p$
Genotypes:	$Bb\ Pp$		$Bb\ pp$		$bb\ Pp$		$bb\ pp$
Phenotypes:	black, polled		black, horned		red, polled		red, horned
Ratio:	1	:	1	:	1	:	1

It will be noted here that, since the gametes of the double recessive parents contribute no dominant genes, the progeny of this backcross reveal not only the kinds of gametes produced by their sire but also the proportion of each kind. Since these, in turn, depend upon his genotype, this backcross is also a test-cross which shows that sire to be *Bb Pp*. In this case it would also show these two pairs of genes to be independent. Whereas the test-cross ratio is 1 : 1 for a single pair of genes, it is 1 : 1 : 1 : 1 for two pairs of genes segregating independently.

Making New Breeds to Order

In practice, animal breeders are less interested now than formerly in combining various mutations to make new breeds and varieties. Occasionally, however, it may be advantageous to do so. The breeders of Aberdeen Angus, Red Polls, and Polled Herefords must have considered it desirable to get or maintain their preferred color and type, but to eliminate the horns that presumably came with the wild cattle when they were domesticated. If early breeders of red cattle who wished to eliminate the horns could not find the desirable gene *P* in their preferred color, they could easily have introduced it from the Aberdeen Angus, and then eliminated the black. From the backcross just outlined, a quarter of the offspring are red and polled. They are certainly all heterozygous for *P*, but by mating a number of them *inter se*, the desirable *bb PP* genotypes should be found (by breeding tests) in about 25 per cent of the next generation, and these would breed true.

Keeler and Cobb undertook to produce Persian cats (*i.e.*, long-haired) with the color pattern of Siamese. A cross of black Persian × short-haired Siamese yielded in the F_1 generation only black, short-haired cats. Evidently the long hair and the Siamese pattern were recessive. From the F_2 generation, a cat phenotypically black Persian was found by genetic test to be heterozygous for Siamese color. When mated back to her (dihybrid) sire, she produced (presumably among other combinations) some kittens with the desired Siamese color and long hair. A new breed was born.

Trihybrids

When a cross is made between animals that differ in three pairs of independent genes, things may seem a bit complicated, but the results to be expected can easily be calculated. To the two pairs of genes in cattle with which monohybrid and dihybrid crosses have been illustrated, a third pair of alleles can be added—the white face of Hereford cattle, which is almost completely dominant to the colored face of

FIG. 4-3 Polled Hereford bull, Carlos Lamplighter, showing the polled head and Hereford pattern characteristic of the breed. Note pigmentation around the eye, the value of which is discussed in Chapter 23. (Courtesy of American Polled Hereford Association.)

other breeds (Figure 4-3). We shall designate it here by the symbol H.[1] The white-faced black cattle commonly seen in any day's drive in North America result from crosses between Herefords and other breeds. Similar white-faced blacks in Holland are more likely to be purebred Groningens.

Assuming for purposes of illustration that a cross is made between Holstein-Friesians and Polled Herefords, it can be represented as: *BB pp hh* × *bb PP HH*. The resulting F_1 trihybrids, which would look like the animals shown later in Fig. 4-7, can produce eight different kinds of gametes, as can be readily seen from the following dichotomies:

If an F_2 generation were produced, the number of possible different combinations of 8 maternal and 8 paternal gametes would be 64. We could find all of these by making a checkerboard, as before, but

[1]We use H and h here for convenience. Other symbols have been suggested.

it would be a big one. The number and ratio of F_2 phenotypes can easily be determined by the algebraic method. Thus $(3\ B + 1\ b) \times (3\ P + 1\ p) \times (3\ H + 1\ h)$ give the combinations that follow. To simplify matters, a single symbol is used for the dominant character, but it must be remembered that a black animal B can be either BB or Bb; a polled one, PP or Pp; and a white-faced one, HH or Hh.

27	B	P	H:	black, polled, white face
9	B	P	hh:	black, polled, colored face
9	B	pp	H:	black, horned, white face
9	bb	P	H:	red, polled, white face
3	B	pp	hh:	black, horned, colored face
3	bb	P	hh:	red, polled, colored face
3	bb	pp	H:	red, horned, white face
1	bb	pp	hh:	red, horned, colored face

In only one of these eight phenotypic classes will these F_2 animals all have the same genotype. The triple recessives are all $bb\ pp\ hh$. Frequencies of different genotypes in the other seven classes are easily calculated as the product of the separate probabilities for each pair of genes. To illustrate, let us consider the 27 animals showing all three dominant alleles.

With respect to B and b the F_2 ratio is $1\ BB : 2\ Bb : 1\ bb$. Among the three that show the dominant allele, the chance of getting BB is $\frac{1}{3}$, and that of getting Bb is $\frac{2}{3}$. The same applies to the P-p and H-h pairs. Combining these, the expected frequencies of all possible genotypes among the cattle showing B, P, and H are as follows:

$$BB\ \ PP\ \ HH: \quad \frac{1}{3} \times \frac{1}{3} \times \frac{1}{3} = 1 \text{ in } 27$$
$$BB\ \ PP\ \ Hh: \quad \frac{1}{3} \times \frac{1}{3} \times \frac{2}{3} = 2 \text{ in } 27$$
$$BB\ \ Pp\ \ HH: \quad \frac{1}{3} \times \frac{2}{3} \times \frac{1}{3} = 2 \text{ in } 27$$
$$Bb\ \ PP\ \ HH: \quad \frac{2}{3} \times \frac{1}{3} \times \frac{1}{3} = 2 \text{ in } 27$$
$$BB\ \ Pp\ \ Hh: \quad \frac{1}{3} \times \frac{2}{3} \times \frac{2}{3} = 4 \text{ in } 27$$
$$Bb\ \ PP\ \ Hh: \quad \frac{2}{3} \times \frac{1}{3} \times \frac{2}{3} = 4 \text{ in } 27$$
$$Bb\ \ Pp\ \ HH: \quad \frac{2}{3} \times \frac{2}{3} \times \frac{1}{3} = 4 \text{ in } 27$$
$$Bb\ \ Pp\ \ Hh: \quad \frac{2}{3} \times \frac{2}{3} \times \frac{2}{3} = 8 \text{ in } 27$$

This same kind of analysis will give the expected frequency of genotypes within any of the eight phenotypic classes. Correspondingly, by using probabilities of $\frac{1}{4} : \frac{2}{4} : \frac{1}{4}$ for each pair of genes in the F_2, the expected frequencies of genotypes among the whole 64 zygotes in the F_2 population are easily determined. Altogether there are 27 different genotypes in the trihybrid F_2 population. The number of phenotypes (eight) is the same as the number of different F_1 gametes.

When a trihybrid is backcrossed to the triple recessive type, the number of different phenotypic classes expected is the same as the number of different F_1 gametes, *i.e.*, eight. All eight should be found in approximately equal numbers.

Polyhybrids

From what happens when 1, 2, or 3 pairs of genes are involved in a cross, it is evident that when polyhybrids with still more pairs of genes have to be considered, one might need either bigger and better checkerboards or some skill in computing complex probabilities to determine the expectations. Fortunately, there are some general relations which tell us what to expect when the number of pairs of genes in a cross is 1, 2, 3, 4, or any number, n (Table 4-1).

A point worth remembering is that the number of individuals necessary in the F_2 generation to get either the homozygous recessive or the homozygous dominant type is increased four-fold with each additional pair of genes. For a dihybrid cross that number is 16, for trihybrids it is 64, and with four pairs of genes it becomes 256. Furthermore, since the laws of chance can be capricious within these relatively small numbers, any breeder hoping to extract the multiple recessive would need considerably more than these numbers to be sure of getting what he seeks.

Anyone curious to know why there are so many different shapes and sizes, such differing abilities and dispositions within flocks and families, might devote a few moments to calculations with 2^n. The number of chromosomes in man is 23 pairs. If two parents are both heterozygous in only one specific pair of genes in each chromosome,

TABLE 4-1
Expectations in Crosses Involving Different Numbers of Pairs of Genes

Pairs of Genes (number)	Kinds of F_1 Gametes (number)	Possible Combinations of F_1 Gametes (number)	Phenotypes in F_2 When Dominance is Complete (number)	Genotypes in F_2 (number)
1	2	4	2	3
2	4	16	4	9
3	8	64	8	27
4	16	256	16	81
n	2^n	4^n	2^n	3^n

the number of possible different phenotypes in their children (assuming complete dominance in all 23 pairs) would be 2^{23}, or 8,388,608. With incomplete dominance in some pairs, the number of phenotypes would be still higher. The number of genotypes possible in such a family would be 3^n, which works out to a 12-digit figure resembling closely that for the national debt of the United States. Since most parents probably differ in many more than 23 pairs of genes, it is scarcely surprising that no two people are alike except identical twins.

A Tetrahybrid F_2 in Rabbits

Four pairs of independent genes were involved in a cross made by Parkhurst and Wilson (1933) who wished to make a "rex lilac" rabbit. The recessive rex gene, r, eliminates the long, coarse, guard hairs and leaves a coat containing only short, fine hairs, so that the fur is short and soft like that of a mole. Lilac rabbits are homozygous for three recessive genes a, b, and d, the alleles of which are present in the common cottontails of North America, in other "wild-type" rabbits, and in breeds like the Belgian Hare which have the same phenotype as the wild ones. The four pairs of genes concerned are:

A = agouti	D = intense pigmentation
a = non-agouti	d = dilute pigmentation
B = black	R = normal fur
b = chocolate	r = rex fur

Belgian Hares are basically black (BB), but the agouti gene A causes in most hairs a sub-apical band of yellow-brown and this gives the coat a grizzled, grey appearance. It also eliminates much of the pigment on the ventral surface of the animal, which appears light-bellied. The agouti pattern is common in wild animals and gets it name from the agouti, a rodent found in South and Central America. Familiar examples of the pattern are provided in North America not only by the wild rabbits and hares, but also by field mice, red squirrels, ground hogs, and other rodents.

When the rex mutation appeared, rabbits with the agouti pattern that would otherwise have been called Belgian Hares became "Castorrex" when made homozygous for r. (Students who have difficulty in getting any precise definition of a breed, and of a variety, might remember this as one of several cases in domestic animals in which one breed differs from another in only a single pair of genes.) Parkhurst and Wilson had Castorrex but wanted to transfer the rex mutation to their lilac rabbits. These last were non-agouti, chocolate, and dilute, but with normal fur.

The cross was thus *AA BB DD rr* × *aa bb dd RR*. The 38 rabbits of the F₁ generation showed all four dominant genes (Fig. 4-4), and from these an F₂ generation numbering 524 was raised. Since the F₁ animals were heterozygous in four pairs of genes, they could each produce 16 kinds of gametes, and the number of possible combinations of these was 256. The number of different phenotypes expected was 16, and all of these were found. The actual numbers obtained (Table 4-2) agreed fairly well with those expected, but there were significant discrepancies in some classes. The table shows only the recognizable

FIG. 4-4 A tetrahybrid F₁ rabbit (above), *Aa Bb Dd Rr*, from the cross described in the text. Among 524 F₂ offspring from such rabbits were two quadruple recessives: non-agouti, chocolate, dilute, and rex—the desired lilac rex (below). (From R. T. Parkhurst and W. K. Wilson in *J. Heredity.*)

TABLE 4-2

Segregation in F_2 Generation of Four Pairs of Genes from the Cross of
Castorrex × Lilac Rabbits

				Expected		
	Phenotypes			In 256	In 524	Observed
A B D R	Agouti (black)	normal		81	165.8	190
A B dd R	"	blue	"	27	55.3	62
A bb D R	"	chocolate	"	27	55.3	52
A bb dd R	"	lilac	"	9	18.4	31
aa B D R	Non-agouti black		"	27	55.3	42
aa B dd R	"	blue	"	9	18.4	15
aa bb D R	"	chocolate	"	9	18.4	21
aa bb dd R	"	lilac	"	3	6.1	7
	Totals,	normal		192	393.0	420
A B D rr	Agouti (black)	rex		27	55.3	44
A B dd rr	"	blue	"	9	18.4	16
A bb D rr	"	chocolate	"	9	18.4	11
A bb dd rr	"	lilac	"	3	6.1	7
aa B D rr	Non-agouti black		"	9	18.4	16
aa B dd rr	"	blue	"	3	6.1	5
aa bb D rr	"	chocolate	"	3	6.1	3
aa bb dd rr	"	lilac	"	1	2.1	2
	Totals, rex			64	130.9	104
	Totals, all kinds			256	523.9	524

Source: Data of Parkhurst and Wilson, 1933.

genes for the different phenotypes; hence those designated as *A* (for
example) could have been *AA* or *Aa*, but those shown as *aa* must have
been *aa*.

The most consistent disagreement between observed and expected numbers in this population was the shortage of rex rabbits. A
similar deficiency of rexes is generally found in matings from which
that mutation segregates, and reasons for it are considered later. The
chance of getting the much desired lilac rex combination was only 1 in
256, but fortunately two such rabbits were obtained. Considering the
purpose of the cross and the fact that about 575 other rabbits co-
operated valiantly to achieve it, it is to be hoped that the two lilac rexes
were of opposite sexes, but on that crucial point, the record, alas, was
silent.

A Tetrahybrid Test-Cross in Fowls

The fact that independent pairs of alleles will segregate in conformity with expectation when several such pairs are involved in one cross is well illustrated in a test-cross made by the senior author. A cross of White Leghorns × Mottled Houdans yielded F_1 birds heterozygous for the following pairs of genes:

From the Houdan		*From the Leghorn*	
i	colored	*I*	dominant white
D	duplex comb	*d*	simplex comb
Mb	muffs and beard	*mb*	normal
W	white skin	*w*	yellow skin

Duplex combs are bifurcated. Muffs and beard result from elongated feathers at the sides of the face and on the throat (Fig. 4-5). F_1 males of the genotype *Ii Dd Mbmb Ww* were backcrossed to Anconas, which carry only recessive alleles of these four pairs of genes. The Ancona gametes were therefore all *i d mb w*. The distribution of their

FIG. 4-5 A Mottled Houdan male showing the white-mottled, black plumage, crest, muffs and beard, duplex comb, and extra toe (left foot) that are characteristic of this French breed.

TABLE 4-3

Distribution of 389 Chickens Among 16 Classes Resulting from the Backcross
of Quadruple Heterozygotes to Multiple Recessives

Gametes Formed by the Heterozygous Males	Resultant Phenotypes	Numbers Found
I — D — Mb — W	Dominant white, duplex, muffs, white skin	22
w	" " " " yellow skin	24
mb — W	" " "no muffs, white skin	23
w	" " " " yellow skin	25
d — Mb — W	" " single, muffs, white skin	28
w	" " " " yellow skin	17
mb — W	" " "no muffs, white skin	25
w	" " " " yellow skin	27
i — D — Mb — W	Colored, duplex, muffs, white skin	30
w	" " " yellow skin	24
mb — W	" " no muffs, white skin	21
w	" " " yellow skin	19
d — Mb — W	" single, muffs, white skin	26
w	" " " yellow skin	26
mb — W	" " no muffs, white skin	28
w	" " " yellow skin	24

389 daughters among the 16 different phenotypic classes expected
and found in this backcross (Table 4-3) shows a remarkably good fit of
observed numbers to the slightly over 24.3 expected in each class.

These classifications were made at about five months of age.
Sometimes animals homozygous for several recessive genes do not
fare as well in the struggle for existence as do those carrying more
dominant genes. The fact that all 16 types were found at five months
in approximately equal numbers shows not only that the four pairs of
alleles segregated independently, but also that all 16 phenotypes were
equally viable.

The Chi-Square Test for Closeness of Fit

In some of the examples of Mendelian segregation used in the foregoing pages, it was obvious at a glance that the numbers observed in the
different classes fitted closely to those expected. In others there were
discrepancies large enough to suggest that the ratios expected were
not realized. For help in interpreting such cases, statisticians use the
convenient *chi-square* test. Deviations of observed from expected

numbers in each class are squared, and these figures are then divided by the expected numbers. The resulting figures are then added to get the constant known as χ^2 which is named for the Greek letter chi. From a table provided in convenient form by Fisher (1946), one can determine the probability of getting by chance deviations as great as those observed, or greater (Table 4-4).

Chi-square Test Applied to Mendelian Ratios

A few examples will illustrate the procedure. As we saw in Table 3-1, Castle and King found that Palomino horses mated *inter se* yielded 17 albino, 45 Palomino, and 21 chestnut offspring. The steps taken to determine whether or not these numbers fit well those expected with a $1:2:1$ segregation of these three phenotypes are shown in Table 4-5.

To see if the value for χ^2 of 0.89 thus determined represents significant deviations from expectations, the table of probabilities must be consulted. In doing so, the number of *degrees of freedom (n)* is important. When Mendelian ratios are being tested, n is always one less than the number of phenotypic classes. In this example there are just three kinds of horses. After some of them have been classified as of any one specific color, all the rest must be fitted into the remaining two classes; hence the number of degrees of freedom is two. Table 4-4 tells us that, when $n = 2$, a χ^2 of 0.89 lies somewhere between the two that indicate corresponding probabilities (P) of .50 and .70. It is apparently closer to .70. The exact probability could be determined from a more extensive table of χ^2, or by interpolation, but in this case that is unnecessary. The table shows that deviations as great as those observed, or greater, would be expected by chance in 50 to 70 per cent of similar trials. In other words, the fit of observed to expected ratios in the Palominos of Castle and King is a good one. It supports their conclusion that Palominos are monohybrids.

In a test with fowls made to determine whether or not duplex comb (D) and polydactyly (extra toe, Po) segregate independently, F_1 dihybrids from the cross *DD PoPo × dd popo* were backcrossed to double recessives, with results as shown in Table 4-6. If these two pairs of alleles were independent, a ratio of $1:1:1:1$ was to be expected, but there were deficiencies of *D po* and *d Po* chickens, with corresponding excesses of the double-dominant and double-recessive classes.

With four classes, the number of degrees of freedom is three. Table 4-4 tells us that when $n = 3$ the probability of getting by chance a value for χ^2 as high as 11.341 is only .01; hence the χ^2 of 13.69 actually found must correspond to a value for P of $<.01$. In other

TABLE 4-4
Table of Chi-Square

n = \ P	.99	.98	.95	.90	.80	.70	.50	.30	.20	.10	.05	.02	.01
1	.0002	.0006	.004	.016	.064	.148	.455	1.074	1.642	2.706	3.841	5.412	6.635
2	.0201	.0404	.103	.211	.446	.713	1.386	2.408	3.219	4.605	5.991	7.824	9.210
3	.115	.185	.352	.584	1.005	1.424	2.366	3.665	4.642	6.251	7.815	9.837	11.341
4	.297	.429	.711	1.064	1.649	2.195	3.357	4.878	5.989	7.779	9.488	11.668	13.277
5	.554	.752	1.145	1.610	2.343	3.000	4.351	6.064	7.289	9.236	11.070	13.388	15.086
6	.872	1.134	1.635	2.204	3.070	3.828	5.348	7.231	8.558	10.645	12.592	15.033	16.812
7	1.239	1.564	2.167	2.833	3.822	4.671	6.346	8.383	9.803	12.017	14.067	16.622	18.475
8	1.646	2.032	2.733	3.490	4.594	5.527	7.344	9.524	11.030	13.362	15.507	18.168	20.090
9	2.088	2.532	3.325	4.168	5.380	6.393	8.343	10.656	12.242	14.684	16.919	19.679	21.666
10	2.558	3.059	3.940	4.865	6.179	7.267	9.342	11.781	13.442	15.987	18.307	21.161	23.209
15	5.229	5.985	7.261	8.547	10.307	11.721	14.339	17.322	19.311	22.307	24.996	28.259	30.578

Note: The values for P (top line) are the probabilities of getting by chance figures for χ^2 as great as those given in the columns below. The values of χ^2 increase as do the degrees of freedom (n), which are given in the left column. For use of the table, see context.

Source: From R. A. Fisher's *Statistical Methods for Research Workers*, by permission of the author and publishers. Oliver and Boyd, Edinburgh. That book gives values for χ^2 with degrees of freedom up to 30.

TABLE 4-5

χ^2 Test Applied to Three Classes Expected in Ratio of 1 : 2 : 1

Phenotypic Classes	Observed Numbers (o)	Expected Numbers (e)	Deviation (o−e)	(o−e)²	$\dfrac{(o-e)^2}{e}$
Albino	17	20.75	+3.75	14.0625	0.678
Palomino	45	41.50	−3.50	12.2500	0.209
Chestnut	21	20.75	−0.25	0.0625	0.003
	83	83	0.0		0.890

$$\chi^2 = 0.89 \quad n = 2 \quad P = .5-.7$$

words, the probability that the observed deviations from expected numbers could have happened by chance is less than 1 in 100. It matters little how much smaller than .01 the value for P actually is. Since deviations as great as those found are to be expected by chance less frequently than once in a hundred trials, it was clear that the excess of *D-Po* and *d-po* combinations was statistically significant. In other words, unlike every other example considered earlier in this chapter, the two pairs of alleles concerned had not shown independent assortment.

Results like these pose the question: How great can the deviations from expectation be without rendering untenable the hypothesis under test? It is common usage among biologists to consider that when χ^2 (or any other measure of deviation from expectation) yields a probability less than .05, the deviations are significant. When P = <.01, they are "highly significant." Students sometimes recognize the import of these probabilities better when these are expressed in other ways. A probability of .05 means one chance in twenty that deviations as great as those observed (or greater) could have resulted by chance. It also means odds of 19:1 against that same possibility.

We must remember than the χ^2 test is used to determine the validity of some hypothesis. When it indicates that the results observed are not compatible with the original hypothesis, all is not lost. We are merely warned that some other interpretation of the data must be sought. In the case cited in Table 4-6, the investigators were delighted to find that the hypothesis tested (that the *D-d* and *Po-po* alleles segregate independently) was not sustained. The most likely alternative explanation was that the two loci were *linked* in the same chromosome, and this was confirmed by further breeding tests. We shall return to *linkage* in a later chapter.

TABLE 4-6
χ^2 Test Applied to Four Classes Expected in Equal Numbers

Phenotypic Classes	Observed Numbers (o)	Expected Numbers (e)	Deviation (o − e)	(o − e)²	$\dfrac{(o - e)^2}{e}$
D Po	143	129	+14	196	Since e is 129
D po	108	129	−21	441	for all four
d Po	109	129	−20	400	classes, a
d po	156	129	+27	729	single divi-
	516	516	0	1766	sion suffices:
					1766/129 =
					13.69

$\chi^2 = 13.69 \qquad n = 3 \qquad P = <.01$

In other χ^2 tests for closeness of fit of observed numbers to Mendelian ratios, significant deviations may mean that the genetic basis postulated to explain the results is not the correct one, or, as we shall see in the next section, that deficiencies in one class, or in more, have resulted from subnormal viability of certain genotypes.

In determining χ^2, it is immaterial whether the deviations of observed numbers from expected ones are plus or minus, since both kinds become plus after squaring. Table 4-6 illustrates a short-cut possible for test-crosses and for all cases in which the expected number is the same in all classes. The squared deviations in each class need not be divided separately by the expected number for that class, but can be added up and their sum then divided by the expected number. The χ^2 test should not be used when the number in any class is less than five unless one applies Yates' correction, details of which can be found in books on statistical methods.

Tests of Fit in Two-Term Ratios

The χ^2 test can be used to test closeness of observed to expected numbers in simple 3:1, 7:1, and 1:1 ratios, and others which have only two classes.

Among the 524 rabbits considered in Table 4-2, there were 420 with normal fur and 104 of the rex type. The expected numbers were 393:131, and the deviation (in each class) 27. In this case χ^2 is 729/393+729/131=7.42. The number of degrees of freedom is one (because if a rabbit did not have normal fur, there was only one other kind that it could have), and the table shows P to be <.01.

The deviations from expectation are clearly so great that the original hypothesis (that normal and rex would segregate in the ratio of 3:1) was not sustained. This does not necessarily refute the hypothesis that rex is a simple recessive. It does mean that there was a significant shortage of rex rabbits below expectations on that hypothesis. This, in turn, suggests that at some stage, either before birth or after it, but before classification, mortality could have been higher in rexes than in rabbits with normal fur.

The significance of deviations in two-term ratios can also be determined in terms of the *standard error* of the ratio, but this method is not so simple as the χ^2 test, nor is it any more accurate. A set of tables prepared by Warwick (1932) shows probabilities of getting all possible numbers of dominant and recessive types in any total number of animals up to 50 for any of nine different Mendelian ratios. Investigators who work with domestic animals will find it useful.

Contingency Tables

Apart from its usefulness in determining how well observed numbers fit Mendelian ratios, the chi-square test can be a convenient help to biologists who want to know whether or not two populations differ significantly in some trait. We may wonder whether the pink pills have saved more pups than the blue ones, or whether mortality is significantly lower in the chicks given Smith's vaccine than in those that got Brown's. For such purposes we set up a *four-fold contingency table* in which the numbers observed are compared with the numbers expected and the significance of the differences is determined by χ^2.

In doing this, we usually test the hypothesis that the two groups compared do *not* differ in response to the treatment. Even though we are secretly hoping that the new pink pills are better than the old blue ones, and mortality is obviously lower in the pink-pilled pups, the hypothesis tested is that survival of the pups is *not* related to the color of the pill. It is a test of independence. Accordingly, the expected numbers are calculated on the basis that survival is independent of the treatment. The following example illustrates the usefulness of the simple contingency table.

In New Zealand, Ward (1945) found in 20 herds of dairy cattle in the Canterbury district that among 195 dam-and-daughter pairs there were 86 dams susceptible to mastitis and 109 that were resistant. Assuming that each cow had one daughter, and converting his percentages to daughters, the number of susceptible daughters was 77 in the former group and 61 in the latter. The question is: Do resistant cows tend to beget resistant daughters, and do susceptible cows tend to beget susceptible ones?

Assuming (for purposes of our test) that they do not, the observed numbers are arranged as in Table 4-7, and the numbers expected if susceptibility of the daughter were independent of the kind of dam are computed from the marginal totals.

Let us begin with the upper left cell of the table.

Any one daughter in the whole 195 has the following probabilities:

Of being susceptible: 138/195

Of having a susceptible dam: 86/195

Of having both these events befall her: 138/195 × 86/195

For all 195 daughters, the number which would be themselves susceptible and have a susceptible dam is thus

$$195 \times \frac{138}{195} \times \frac{86}{195} = 60.86$$

Expected numbers in the other three cells can be computed similarly, but this is unnecessary because one can determine them by subtraction of 60.86 from the marginal totals. Our deviation of observed from expected numbers (o − e) is found to be 16.14 in all four cells, and, proceeding as in Table 4-6, χ^2 is found to be 26.20.

With a four-fold contingency table, the number of degrees of freedom, n, is always one. We can see why that is so if we remember that, in the example just worked out, as soon as the expected number was calculated for *one* cell, those for the other three cells were automatically fixed by the marginal totals. They could not be anything other than the numbers shown in Table 4-7.

With $\chi^2 = 26.2$ and $n = 1$, Table 4-4 tells us that the probability of getting by chance deviations as great as those shown in our contingency table must be far less than .01. (It is also less than .001.) In other words, our hypothesis that susceptibility of the daughter is not related to susceptibility of the dam is invalid. The alternative interpretation, that resistant cows do tend to beget resistant daughers (*i.e.*, that resistance to mastitis is to some extent genetically determined), was confirmed by similar data from 15 herds in the Manawatu area. We shall return to this subject in a later chapter. One should remember that chi-square tests are worked out from numbers—not from percentages—and that they are more reliable for large numbers than for small ones. For information about the use of the chi-square test

TABLE 4-7
Four-fold Contingency Table, with Expected Numbers in Italics

| | Daughters | | Totals |
	Susceptible	Resistant	
Dams:			
Susceptible	77 60.86	9 25.14	86
Resistant	61 77.14	48 31.86	109
Totals	138	57	195

with larger contingency tables and in other ways, the student should consult books on statistical methods, two of which are cited at the end of this chapter.

Problems

4-1 Detlefsen and Carmichael (1921) found that a black Mule-foot (solid hoof) boar mated with Duroc-Jersey (red) sows produced only black, mule-footed offspring. A backcross of these to Duroc-Jerseys yielded 8 black Mule-foot, 9 red Mule-foot, 11 black cloven, and 14 red cloven. What is the genetic basis for the mule-foot condition?

4-2 For black and red?

4-3 What numbers were to be expected in the four classes?

4-4 How many of these 42 animals were homozygous with respect to the four genes considered in this cross? How many were heterozygous in one pair only?

4-5 If the F_1 black, mule-footed pigs had been mated *inter se* and produced the same number of offspring as the backcross, how many red mule-foots would be expected? How many red cloven?

4-6 How would you proceed to get a stock of red mule-foots that would breed true?

4-7 A cross of red, polled (*NN PP*) × roan, horned (*Nn pp*) cattle yields offspring half of which are roan, half red, but all polled. When the roan, polled animals are mated together, what proportion of their calves should also be roan, polled? What proportion should be white and horned?

4-8 What kinds of offspring would be expected, and in what proportions from the following matings: (a) *Nn Pp* × *Nn PP*; (b) *nn Pp* × *Nn pp*; (c) *Nn Pp* × *nn pp*?

4-9 Pure Rose-combed Black Bantams are homozygous for the completely dominant genes *R* and *C* determining their distinctive comb and color. Single-combed White Bantams carry the recessive alleles of both genes. From results in the following matings, determine the genotypes of the parents in each case.

| | Progeny | | | |
Parents	Rose, Black	Rose, White	Single, Black	Single, White
Rose, black × single, white	12	8	11	13
Rose, black × single, white	9	13	0	0
Rose, black × rose, black	30	17	13	5
Rose, black × rose, white	27	0	13	0

4-10 How many kinds of gametes can be produced by an animal heterozygous in five pairs of genes?

4-11 When such animals are mated *inter se*, what proportion of their offspring might be expected to be quintuple recessives?

4-12 When fowls heterozygous for the recessive mutant, snow-white down (*sw*), and for pea comb (*P*) were backcrossed to double recessives, their progeny consisted of 123 cream (normal) and pea-combed, 94 cream single-combed, 106 snow and pea-combed, and 117 snow and single.

To determine whether or not these figures fit the expected numbers, find χ^2, n, P, the value for χ^2 at the .05 point, and the same at the .01 point.

4-13 How often in 100 trials would you expect to get by chance as poor a fit, if the genes segregate independently?

4-14 Mohr and Wriedt found in Oplandske cattle in Norway that a bull, Amor, mated to 27 of his own daughters sired 55 calves among which 11 had very short spines and other defects (see Fig. 9-12). None of these survived. What Mendelian ratio is most closely approximated here?

4-15 What is the probability (P) of getting by chance the observed deviation from it?

4-16 If the defect were a simple recessive, what conditions would have to apply to yield that ratio?

4-17 Considering that the defect is rare and had not previously been reported, were those conditions likely?

4-18 From matings of this sort what other ratio would be expected? Why?

4-19 Does the observed ratio differ significantly from that other ratio? What is χ^2? What is P?

4-20 What explanation seems most likely for the excess of short-spined calves?

4-21 Hafez *et al.* (1958) reported that in a small herd of purebred Herefords in the State of Washington a bull, Prince Larry S. Regal, sired 2 dwarfs and 3 albinos among 6 calves, one from each of 6 cows. These cows—listed below as A to F, with their calves—included two full sisters and two half-sisters of the bull.

A: Albino, dwarf (Fig. 4-6)   D: Colored, dwarf (Fig. 4-6)
B. Albino, not dwarf   E. Colored, not dwarf
C. Albino, not dwarf   F: Colored, not dwarf

What seems the most likely basis for the kind of albinism found in this herd? For the dwarfism?

4-22 Did these two characters segregate independently?

4-23 For which of the 7 parental animals was the genotype revealed with respect to albinism? To dwarfism? To both?

4-24 Since two pairs of alleles were segregating, was any dihybrid Mendelian ratio to be expected among the 6 calves? Why not?

4-25 Could Prince Larry ever be used in any herd of cattle with little risk of his siring more dwarfs or albinos? If so, what kind of a herd?

4-26 In that event, what is the minimum proportion of his calves you would expect to carry the gene for dwarfism? For albinism? For both?

4-27 Africander cattle are normally red and horned, but in South West Africa yellows are popular. An experimental farm there obtained a yellow bull without horns, the latter condition being apparently a mutation. When used

FIG. 4-6   Dwarf half-sisters; one albinotic, the other not, but both showing the typical bulging forehead of the "snorter" dwarf. (From E.S.E. Hafez, C. C. O'Mary, and M. E. Ensminger in *J. Heredity.*)

with red, horned cows, he sired calves as follows: yellow, polled, 7; yellow, horned, 6; red, polled, 7; red, horned, 7.

What do these results tell about the inheritance of yellow and red, and of hornlessness in this breed?

4-28 What does his progeny test reveal about the genotype of the bull with respect to these characters?

4-29 Which of his progeny should be bred together as a start toward establishing a yellow, polled breed?

4-30 What proportion of their calves should have (a) the phenotype desired? (b) the genotype desired?

4-31 A cross between Aberdeen Angus and Herefords produces the trihybrids (*Bb Pp Hh*) shown in Fig. 4-7. If such F_1 animals are mated *inter se,* how many squares would you need in a checkerboard that shows all possible combinations of F_1 gametes in the F_2 generation?

4-32 What proportion of that F_2 generation should be homozygous for the three dominant genes concerned? For the three recessive genes?

4-33 If the breeder wants to make a red, polled, white-faced breed, what proportion of the F_2 generation should have (a) the phenotype desired? (b) the genotype desired?

4-34 What proportion of the red, polled, white-faced animals in the F_2 generation should be homozygous for *P*? For *H*? For both?

FIG. 4-7 White-faced, polled, black calves, trihybrids from the cross: Aberdeen Angus × Hereford. (Courtesy of American Angus Association.)

4-35 How should the breeder proceed to develop his herd of polled cattle that will breed true for red coat and white face?

4-36 For the fit of observed to expected numbers in Table 4-3, determine χ^2, n, and P.

4-37 Seeking to determine whether or not he could transmit leukosis from one chicken to another, a veterinarian inoculated 273 of them with material from leukotic birds and kept 215 uninoculated chickens as controls. Subsequent cases of leukosis numbered 86 in the former group, and 42 in the controls. Was transmission accomplished, *i.e.*, was there a significant difference between the two groups? With a four-fold contingency table, find χ^2, n, P, and the answer.

Selected References

DETLEFSEN, J. A., and W. J. CARMICHAEL. 1921. Inheritance of syndactylism, black and dilution in swine. *J. Agric. Res.* **20:** 595–604. (Source of data used in Problems 4-1 to 4-6.)

FISHER, R. A. 1958. *Statistical Methods for Research Workers.* 13th ed. Edinburgh: Oliver and Boyd. New York: Hafner. (A standard reference book.)

HAFEZ, E. S. E., C. C. O'MARY, and M. E. ENSMINGER. 1958. Albino-dwarfism in Hereford cattle. *J. Heredity* **49:** 111–16. (The remarkable case considered in Problems 4-21 to 4-26.)

PARKHURST, R. T., and W. K. WILSON. 1933. Rexing the lilac rabbit. *J. Heredity* **24:** 35–39. (Attempt to produce quadruple recessives, and source of data in Table 4-2.)

SNEDECOR, G. W., and W. G. COCHRANE. 1980. *Statistical Methods. Agriculture and Biology.* 7th ed. Ames, Iowa State Univ. Press. (Particularly useful because of its many examples of the use of statistics in interpreting experiments with plants and animals.)

WARD, A. H. 1945. Inheritance of susceptibility to mastitis. In *21st Ann. Report New Zealand Dairy Board*, Wellington, N.Z.: 59–61. (Source of data in Table 4-7.)

WARWICK, B. L. 1932. Probability tables for Mendelian ratios with small numbers. *Texas Agric. Exper. Sta. Bull.* 463. (Useful probabilities of getting various combinations in numbers up to 50 for any of nine different ratios.)

Interaction of Genes; Modified Ratios

Sometimes two different genes (not alleles) produce by their combined action a phenotype different from that induced by either gene alone. Just as a dominant gene can suppress its recessive allele at the same locus, so can a gene at one locus interact with others at different loci. In other interactions, one gene may prevent the expression of others; hence combinations of characters that would be expected if each pair of alleles exerted separate effects are not found or cannot be recognized.

With such interactions, results to be expected from simple Mendelian segregation will differ somewhat from the normal dihybrid and polyhybrid ratios considered earlier. A few examples of modified ratios are given in this chapter to illustrate different kinds of interactions.

Complementary Genes; Walnut Combs

Previously we have considered four types of combs that are breed characteristics in the fowl; now we come to a fifth—the walnut comb. The differences among these are shown in Fig. 5-1, and the following summary lists some of the commoner breeds that show these combs:

Single: Leghorns, Plymouth Rocks, Rhode Island Reds
Rose: Wyandottes, some Leghorns, Hamburgs
Pea: Brahmas, Cornish or Indian Games
Duplex: Houdans, Polish, Buttercups
Walnut: Malays, Orloffs, Chantecler

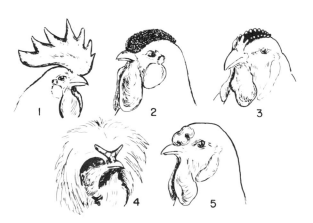

FIG. 5-1 Types of comb in the fowl: 1. single; 2. rose; 3. pea; 4. duplex; 5. walnut.

The walnut comb is small, with an uneven surface, often show-ing irregular furrows. Its name comes from its resemblance to half a (shelled) walnut. It was adopted as a breed characteristic for Chantec-lers in Canada and for Orloffs in Russia because, being small and close to the head, it is less susceptible to freezing in cold weather than some of the other combs.

When Bateson and Punnett were studying the inheritance of different combs, they found that sometimes walnuts mated *inter se* threw pea-combed and rose-combed offspring. It was eventually found

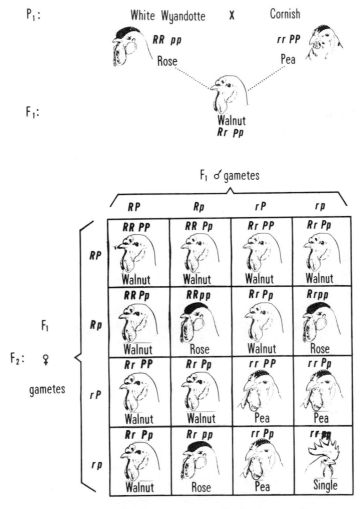

FIG. 5-2 Interaction of complementary genes in the cross of pure rose-combed and pure pea-combed fowls produces walnut combs in F_1 and yields a ratio in F_2 of 9 walnut : 3 pea : 3 rose : 1 single.

that the walnut type results from the combined action of the genes P and R, which separately induce pea and rose combs, respectively. Thus pure Wyandottes are $RR\ pp$ and pure Cornish are $rr\ PP$. A cross between such fowls yields the dihybrid $Rr\ Pp$ which has a walnut comb. When these are mated *inter se*, the resultant F_2 generation shows approximately 9 walnut:3 rose:3 pea:1 single (Fig. 5-2).

The differences between this cross and the simple dihybrid ones considered earlier are that (a) the F_1 generation shows a new phenotype, different from that of either parent, and (b) the F_2, while it contains four classes in the usual ratio, also shows a new type—the single comb—that crops out from two generations of ancestors, none of which had single combs. In earlier chapters, single comb was treated as a simple recessive to rose, and also to pea. To simplify matters, we can still do so whenever either rose or pea is involved without the other. In segregations involving both R and P, we should remember that single is a double recessive, $rr\ pp$. Similarly, when duplex combs are concerned, the single comb is dd.

In cases like this, when two genes at different loci working together produce something different from the effect of either one alone, they are said to be *complementary* genes.

The 9:7 Ratio

There are three different recessive mutations that cause the short-haired coat of rex rabbits. When any one of these is crossed with normal animals, there is the usual segregation in F_2 of 3 normal:1 rex. Phenotypically, the three types are indistinguishable. When Castle mated together rexes from two different sources, all the F_1 animals had normal coats, and the F_2 ratio was 55 normal:33 rex.

These results are understandable if we remember that one of the P_1 rexes must have had one rex mutation, and the other parent another. Following the usual practice of designating different genes that induce the same phenotype by a common letter with different numbers appended, the original cross was:

$$R\text{-}1\ R\text{-}1\ r\text{-}3\ r\text{-}3 \times r\text{-}1\ r\text{-}1\ R\text{-}3\ R\text{-}3$$

The F_1 dihybrid $R\text{-}1\ r\text{-}1\ R\text{-}3\ r\text{-}3$, although carrying two genes for rex, showed only a normal coat because the dominant alleles of those genes were also present. We should by now be able to dispense with the usual checkerboard and to write down the expected combinations of the four genes in the F_2 generation as follows:

$$R\text{-}1\ R\text{-}1\ R\text{-}3\ R\text{-}3\ldots1$$
$$R\text{-}1\ R\text{-}1\ R\text{-}3\ r\text{-}3\ \ldots2$$
$$R\text{-}1\ r\text{-}1\ R\text{-}3\ R\text{-}3\ \ldots2$$
$$R\text{-}1\ r\text{-}1\ R\text{-}3\ r\text{-}3\ \ldots4$$

normal: 9

$$R\text{-}1\ R\text{-}1\ r\text{-}3\ r\text{-}3\ldots1$$
$$R\text{-}1\ r\text{-}1\ r\text{-}3\ r\text{-}3\ldots2$$

rex:3

$$r\text{-}1\ r\text{-}1\ R\text{-}3\ R\text{-}3\ldots1$$
$$r\text{-}1\ r\text{-}1\ R\text{-}3\ r\text{-}3\ldots2$$

rex: 3

$$r\text{-}1\ r\text{-}1\ r\text{-}3\ r\text{-}3\ldots1$$

rex: 1

rex: 7

On working out Problem 5-12, we find that the numbers of normal and rex rabbits found by Castle in his F_2 generation fitted very well the 9:7 ratio expected.

A very similar situation applies to the two different mutations, *wa-1* and *wa-2*, which cause a wavy coat in mice. One of these cropped up in a mouse colony at Edinburgh University and the other at Harvard University. Both are completely recessive and, so far as the wave is concerned, are phenotypically indistinguishable. The Harvard mice go in for curled whiskers somewhat more than those of the Edinburgh strain. When Keeler crossed these two stocks, the 27 mice of the F_1 generation were all perfectly normal (Fig. 5-3). This proved that the two mutations were genetically quite distinct, as were the two

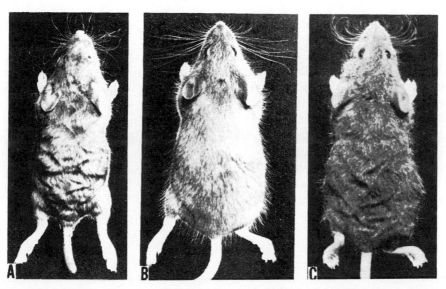

FIG. 5-3 Crossing waved-1 (A) and waved-2 (C), similar recessive mutations found in mouse colonies in Edinburgh and Boston, respectively, yielded in the F_1 generation only mice with normal coats (B), and showed thus that the two similar phenotypes were caused by two different mutations. (From C. E. Keeler in *J. Heredity.*)

kinds of rex in rabbits. It has since been found that *wa-1* and *wa-2* are in different chromosomes.

Keeler did not report an F_2 generation. Had one been produced, the expectation for it would have been 9 normal : 7 waved, unless the double recessives showed some cumulative action of the four recessive genes which would distinguish that class from the others. In that event, the expected ratio would be 9 normal : 6 waved : 1 super-waved or otherwise exceptional. None of this offers any comfort to straight-haired daughters of wavy-haired fathers in *Homo sapiens*, but it does show that they are not alone.

A classical case of complementary genes, and still the standard example of the 9 : 7 ratio in many texts, was provided by two white varieties of sweet peas studied by Bateson and Punnett. When crossed, these yielded purple flowers in the F_1 and a ratio of 9 purple : 7 white in the F_2 generation. Two different dominant genes were necessary for formation of color. Each of the white varieties had one but lacked the other. The dihybrid F_1 plants had both genes, hence could produce color, but in their progeny only 9 in 16 got both dominant genes, while 7 in 16 got only one or neither. These last could have only white flowers.

Test for Independence or Identity of Similar Mutations

From the rex rabbits, waved mice, and white sweet peas, we should remember that identical mutant phenotypes can be caused by entirely different genes. When in some breed or stock a recessive mutation crops out that is apparently the same as one already studied elsewhere, there is always the question whether the new one is really genetically identical with the old, or whether it might be a *mimic, i.e.,* the result of a change in a gene at a different locus. The standard test to answer this question is to mate together the mutants of different origin. If they are genetically identical, all the progeny should show the same mutation as the parents, but if they are different, as with the waved mice and the rex rabbits, then all of the offspring should be normal.

This applies also to dominant mutations, but, since any new dominant mutation is almost certain to occur in only one chromosome, the first animal to show it will be a heterozygote and will breed accordingly. In contrast, recessive mutations can be transmitted (unknown) through several generations of heterozygotes and will be detected only when a chance mating of two such carriers results in the appearance of the homozygous recessive mutation.

Doubt is inevitable about the identity of mutations that are similar but somewhat different in different breeds. Here the question is whether different gene changes are responsible, or whether one and the same gene mutation is expressed differently because of the influence of the many genes that make one breed different from another. For example, epithelial defect, which is a recessive mutation found thus far in at least three breeds of cattle in the United States, affects comparatively small areas of skin in Ayrshires (Fig. 2-9), more in Holstein-Friesians, and still more in Jerseys (see Fig. 9-3). We cannot yet be sure that the mutation is at the same locus of the same chromosome in all three breeds. In this case, as in so many others, the homozygous recessive animals die at early ages, so any breeding tests would have to be made with heterozygotes.

A 9:3:4 Ratio in Mice

In the common house mouse, many different color varieties have been preserved by fanciers. Wild mice, like most other wild rodents, have the agouti pattern described (in rabbits) in the previous chapter. One fancy variety lacks the gene A, which causes the agouti banding of the hairs, and is therefore solid black. Another common mutation is albinism, c. Since albinotic mice can be genetically agoutis, but unable to show that pattern because they cannot form color, crosses between such albinos and black (non-agouti) mice produce in the F_1 generation mice that are all agoutis:

P_1 Black ($CC\ aa$) × Albino ($cc\ AA$)
F_1 Agouti ($Cc\ Aa$)
F_2 9 C—A— :3 C—aa : 3 $cc\ A$—: 1 $cc\ aa$
         Agouti   Black   Albino

(Dashes in the F_2 genotypes indicate that the animal may be homozygous or heterozygous for the gene preceding the dash.)

As with other dihybrids, the two pairs of genes carried in the heterozygous state by the F_1 agoutis should yield in the F_2 generation a phenotypic ratio of 9:3:3:1, but, in this case, the last two classes are indistinguishable. Mice of the genotypes $cc\ AA$, $cc\ Aa$, and $cc\ aa$ are all albinos. Accordingly, the visible phenotypic ratio is 9 agouti: 3 black: 4 albino.

The albinotic mice, like the white sweet peas, illustrate interactions in which one pair of genes completely prevents the manifestation of others. Similarly, recessive white fowls (cc) may be genetically mottled, barred, Columbian, or other patterns, but their genotypes with respect to these can be determined only when they are crossed with other birds that bring in the gene (C) for color.

Complementary Genes and Reversion

In every example considered thus far in this chapter except the walnut combs, the crossing of different breeds or mutant strains of animals, and of different varieties of white sweet peas, has resulted in an F_1 generation that showed the "wild type" of the species concerned. The rabbits had normal coats, not rex; the mice from the waved parents had smooth hair, and those from blacks × albinos were agoutis; and the sweet peas were purple as are the wild ones from which horticultural varieties have arisen. Such *reversion* to the wild or ancestral form is not uncommon when cultivated varieties, domesticated breeds, or geneticists' mutant stocks are crossed.

Darwin noticed that different breeds of the domestic fowl, when allowed to intercross at random, tended to produce offspring resembling the Red Jungle Fowl. He was led from that fact (and others) to believe that jungle fowl to be the ancestor of all domestic breeds of chickens. In most similar cases it seems probable that complementary genes originally occurring together have become separated under domestication, a process in which breeders preserve unusual mutations and use them to differentiate breeds and varieties. When some of these last are crossed, former combinations of complementary genes are restored, original phenotypes reappear, and the cross is said to show reversion to the wild or ancestral type.

Broodiness in fowls provides a good example of reversion in a familiar domestic animal. Broody hens want to sit on their eggs, not to lay more of them. They want to hatch chicks, to become mothers. Under natural conditions in the jungles of Asia, this process is indispensable for the reproduction of the species. From the standpoint of the hen, it is one of the tragedies of advancing civilization that, while motherhood provides the basis for the whole dairy industry, is actively encouraged in sheep, pigs, and other domestic animals, and, so long as it occurs under certain approved conditions, is universally acclaimed in at least one other species, motherhood in a modern poultry flock is almost as unpopular as it would be in a boarding school for young ladies. In most parts of the world, big incubators have made broodiness obsolete. Accordingly, by constant selection that trait has been eliminated almost entirely from Leghorns and reduced to a minimum in other breeds.

When two different breeds are crossed, however, the incidence of broodiness is usually much higher in the F_1 progeny than in either parent breed. This suggests that its expression is dependent upon complementary genes. If there were only two pairs of these, A-a and C-c, elimination of A from one strain and C from another could make both strains non-broody, but a cross between them would restore an Aa Cc combination and thus permit broodiness. Evidence that this happens was found by Goodale, who also demonstrated that the trait could be greatly reduced by the proper kind of selection.

Later evidence has shown that other genes are probably involved, in addition to those with complementary action. The manifestation of broodiness is also influenced by the environment.

Paralysis in the Cross: Great Dane × St. Bernard

Complementary genes are apparently responsible for a peculiar kind of paralysis of the hind limbs in F_1 dogs from this cross studied by Stockard (1936). Onset is sudden, at about three months of age, seldom later. There is no sign of pain, and the symptoms vary from slight paralysis to inability to walk without help. In mild cases the hind quarters sag. Muscles of the thigh are paralyzed, but not the sartorius, which flexes the hip joint. The basic abnormality is a loss of motor neurones in the lumbar region of the spinal cord. Compensation for this paralysis leads to rotation of the hind legs and an ambling gait. The condition is not lethal, and some affected dogs live to eight or ten years.

In Stockard's pure Great Danes and St. Bernards, this kind of paralysis was uncommon, but from reciprocal crosses between them, among 57 F_1 dogs alive at three months, only three or four were unaffected. In an F_2 generation of 66 animals, at three months of age about one-third were paralyzed. The condition was found in a few crosses from Bloodhounds, but not among 2,000 other puppies in the same kennels at the same time.

Altogether, these facts suggest that complementary genes contributed by the Great Danes and St. Bernards interact to cause the paralysis. Since dog breeders are more anxious to keep their breeds pure than to cross them, the condition may not be serious within either breed, but breeders should watch for any signs of paralysis in prospective breeding stock.

Epistasis and a 13:3 Ratio in the Fowl

In any pair of alleles, the dominant gene prevents (either completely or partially) the expression of its recessive partner. Sometimes the

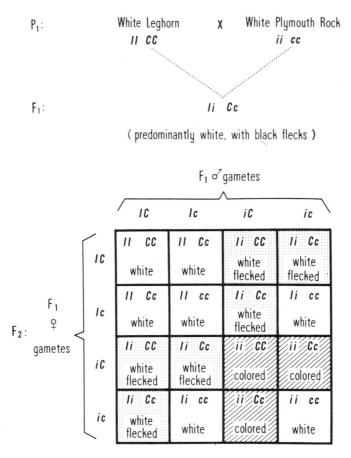

FIG. 5-4   Results in the F_1 and F_2 generations from the cross of dominant white × recessive white fowls.

expression of a dominant gene is suppressed by the action of another gene at a different locus, perhaps in a different chromosome. Such a masking of the influence (and presence) of one gene by another is known as *epistasis* (standing upon). The all-powerful gene that blocks out another is said to be epistatic to it, and the one suppressed is *hypostatic*.

How such interactions can modify expected ratios is illustrated by a well-known case in the domestic fowl. White Wyandottes, White Plymouth Rocks, and several other white breeds have white plumage because they lack the gene C necessary for formation of melanin in feathers. They are not true albinos because they do have melanic pigment in the eye. With respect to plumage color their genotype is *cc*.

In White Leghorns the situation is quite different. Their plumage is just as white as that of the other white breeds, but genetic analyses have shown that they do carry C. Its action is suppressed in White Leghorns by an inhibiting gene, I, which, in the homozygous state, completely prevents the formation of melanic pigment in the plumage. White Leghorns are said to be *dominant whites* because when they are crossed with colored breeds the F_1 plumage is predominantly white. In contrast, crosses of White Plymouth Rocks and White Wyandottes with colored breeds yield only colored offspring; hence such white breeds are called recessive whites.

When White Leghorns are crossed with a recessive white breed, the F_1 progeny are usually not pure white because I is not completely epistatic to C in heterozygotes. Most chicks and adult birds of the genotype Ii show some black flecking in the plumage, but they are predominantly white and a few are entirely so. As Fig. 5-4 shows, the interaction of I, C, and their alleles in such a cross yields an F_2 ratio of 13 predominantly white:3 colored. If one were able to recognize every bird heterozygous for I, the ratio would be 7 pure white:6 white, with black flecking:3 colored. Of the pure white, 3 would be homozygous for dominant white only, 3 for recessive white only, and one for both kinds, but only breeding tests could distinguish these three genotypes.

Students perplexed about the difference between dominance and epistasis might remember that dominance and recessivity apply to the outcome of a private conflict within one pair of alleles at one locus, whereas epistasis involves always two pairs of alleles, or more.

Interaction of Genes Affecting Coat Color in Dogs

Two pairs of genes are known to determine certain basic coat colors in dogs. Thus:

BB or Bb = black coat, when E is also present
BB or Bb = red coat, black eyes and nose, in dogs ee
    bb = brown, chocolate, or liver, when E is present
    bb = yellow, or pale red with brown nose, in dogs ee

The interaction of these genes is such that the dominant alleles of both pairs are necessary to cause a black coat. B and E are thus complementary genes. Some investigators consider E as a gene for the extension of black pigment, which is restricted in red dogs to their eyes and noses. The recessive gene b reduces the intensity of black, and the double recessive, $bb\ ee$, showing the cumulative effects of genes that reduce or restrict black, is yellow or pale red, the lightest of the four phenotypes.

Because of the interaction of these two pairs of genes, the colors to be expected in a litter of puppies will depend, not on the color of the parents, but on their genotypes. Two parents, both black, may have a litter showing any of four different phenotypic ratios (Table 5-1). Furthermore, the ratio of colors in the litter does not always reveal the genotypes of the parents. Among the four phenotypic ratios possible in litters from two black parents, only one reveals the genotypes of both parents (see Problem 5-7).

In red and brown dogs, more of the genotype is revealed by the color than in black ones, and, correspondingly, the colors appearing in a litter give more indication of the genotype of the parents. Red dogs must be either *BB ee* or *Bb ee*. Any yellow pup from two red parents shows that both of them are *Bb ee*. However, if such a mating produces only red offspring, one of the parents could carry *b*, but only matings to a yellow dog or to a red one known to be *Bb ee* could reveal which of the parents is heterozygous.

TABLE 5-1

Colors Expected in Puppies from Matings of Black and Red Dogs of Differing Genotypes

Phenotypes and Genotypes of Parents	Phenotypic Ratio in progeny
A. Black × black	
1. *BB EE* × any genotype	all black
2. *BB Ee* × *BB Ee* or *Bb Ee*	3 black : 1 red
3. *BB Ee* × *Bb EE*	all black
4. *Bb EE* × *Bb EE* or *Bb Ee*	3 black : 1 brown
5. *Bb Ee* × *Bb Ee*	9 black : 3 red : 3 brown : 1 yellow
B. Black × red	
1. *BB EE* × *BB ee* or *Bb ee*	all black
2. *BB Ee* × *BB ee* or *Bb ee*	1 black : 1 red
3. *Bb EE* × *BB ee*	all black
4. *Bb EE* × *Bb ee*	3 black : 1 brown
5. Bb Ee × *BB ee*	1 black : 1 red
6. *Bb Ee* × *Bb ee*	3 black : 3 red : 1 brown : 1 yellow
C. Red × red	
1. *BB ee* × *BB ee* or *Bb Ee*	all red
2. *Bb ee* × *Bb ee*	3 red : 1 yellow

Diallel Crosses

With respect to these coat colors, the double recessive genotype of the yellow dog is the only one that is evident from the color without any breeding tests. If it were desired to know the full genotype of any of the other kinds, the fastest procedure would be to make test-crosses to that double recessive. This might be difficult with some breeds or in some places where preference of the breeder for red dogs rather than brown ones has eliminated the gene *b*, or reduced it to such a low frequency that yellow dogs are rare.

In such circumstances, *diallel tests* are useful. These are crosses in which the genotype of the unknown is determined by mating it to two or more animals, preferably of different types. Diallel tests are particularly useful for determining genotypes with respect to complex characters like egg production and viability, but the example of coat colors in dogs will serve to illustrate in simple terms how diallel tests work.

If a black male dog mated with black females throws only black offspring, his genotype is unknown except that he must carry B and E. If, when mated with a red dog, he sires any red pups, it is clear that he is Ee, but he could still be either BB or Bb. Similarly, if in any mating with a red, brown, or yellow female he throws any brown or yellow pups, he must be Bb. If two or three litters from the black male × red females yield eight or ten pups all of which are black, the chances are good that the black sire is EE.

The 15:1 Ratio; Duplicate Genes, and Herniated Pigs

Sometimes either of two dominant genes can cause the same effect. When that happens, the recessive character can be manifested only by individuals homozygous for both recessive alleles. As a result, when dihybrids are mated *inter se*, the usual 9:3:3:1 ratio expected in an F_2 generation is modified to 15:1.

Examples of such *duplicate* genes are more common in plants than in animals, but Warwick (1931) concluded after several years of study that inguinal hernia in young male pigs most likely results from homozygosity for two pairs of duplicate recessive genes, h and h-1. In this kind of hernia, which is a common defect in swine, loops of the intestine pass through the inguinal canal into the scrotum. The condition is not present at birth but may appear at any time thereafter up to one month of age. Genetic study of the abnormality is complicated by its expression in only one sex, so that the genotype of the female can be determined only by progeny tests.

If duplicate genes are responsible for it, normal males could be homozygous for the dominant alleles, or heterozygous in one pair or both, and only males of the genotype *hh h-1 h-1* would show hernia. Pigs are not the best material for complex genetic analyses. Warwick was unable to get conclusive evidence that only two pairs of duplicate genes are responsible, although he had seven sows that, when mated to herniated boars, produced from three to five herniated sons each, and no normal ones. However, his objective was less to demonstrate Mendelian ratios than to find out whether or not the abnormality is influenced by heredity. This he did by deliberately using herniated boars and selecting to *increase* the incidence of scrotal hernia in his herd. With this procedure the proportion of male pigs that showed hernia was raised from 1.7 per cent in the unselected stock to 91 per cent in his sixth generation. Obviously, selection in the opposite direction should help to keep the incidence of the defect at a low level. Similar selection and genetic studies have shown that there is a genetic basis for both inguinal and abdominal hernia in other animals.

Problems

5-1 Using the genetic basis given in this chapter for certain coat colors in dogs, and the symbols *B* and *E* for the dominant genes involved, assign the most probable genotypes to the five females and two males that produced in diallel crosses the litters described below.

	Black ♀ A	Black ♀ B	Black ♀ C	Red ♀ D	Red ♀ E
By red ♂ X	3 black 2 red	all black	all black	all red	all red
By red ♂ Y	2 black 3 red 1 brown 1 yellow	5 black 1 brown	all black	4 red 1 yellow	all red

5-2 If we assume that the genotypes assigned in 5-1 are correct, could ♀♀ B, C, or E have a yellow pup in other matings? If so, which ones? What color would the sire have to be?

5-3 could ♀ ♀ C, D, or E have a brown pup? If so, which ones? What color would the sire have to be?

5-4 If the breeder makes a test-cross of a black male suspected of being a dihybrid for these two pairs of genes, what phenotypic ratio would you expect in the progeny if his suspicion is correct?

5-5 Under what conditions could two black parents have a yellow pup? What proportion of their pups would you expect to be yellow?

5-6 What would be the phenotypes of dogs which, when mated together, could produce all four colors in approximately equal numbers? (Two different matings will do it.)

5-7 In Table 5-1, when the colors of both parents are known, which phenotypic ratios in the progeny reveal the genotypes of both parents?

5-8 From Castle's cross of rex rabbits that yielded no rexes in the F_1 generation, what kinds and numbers were to be expected in his 88 F_2 rabbits according to the genetic basis explained in this chapter?

5-9 How many of the 33 rex rabbits in Castle's F_2 population should have been homozygous for one rex gene or the other, but not for both?

5-10 If we suppose that, as occasionally happens, some disaster had wiped out both the parental rex stocks, how could they have been re-established from the F_2 generation just considered, with assurance that neither rex race carried any of the genes that distinguished the other?

5-11 Aleutian blue minks differ from wild ones in being homozygous for the recessive gene *al*. Imperial platinums, also bluish, are *Al Al* but homozygous for *ip*. The double recessive, a very light blue called sapphire, was for some years more valuable than any of the other three kinds mentioned.

From matings of minks all believed to be *Al al Ip ip*, as reported to Shackelford, four mink raisers obtained 53 dark, wild type; 17 imperial platinum; 24 Aleutian; and 8 sapphires. What numbers were to be expected of each kind if the genotypes of the parents were as supposed?

5-12 To obtain sapphires, a breeder in the State of Washington was expecting to mate *inter se* some minks that were heterozygous for *Al* and *Ip*, but a carelessly driven truck crashed into the pens and killed two of the dihybrid males; the others escaped. The breeder then mated his 20 dihybrid females to imperial platinum males that did not carry the gene *al*.

What phenotypes were to be expected in the resultant 96 kits, and in what ratio?

5-13 What ratio of phenotypes could have been expected if these 96 had been sired by dihybrids as originally planned?

5-14 The breeder sued the truck driver. If you had been the judge, would you have awarded him any damages? How could they be assessed? (The first verdict allowed him $8,125, but it was later reversed by a higher court.)

5-15 Could the minks raised (from the matings that were made) have been used to produce sapphires in the next generation? If so, what phenotypes should have been mated together to yield the most sapphires?

5-16 What proportion of such matings could (theoretically) have produced some sapphires?

5-17 In the litters from such matings that yielded sapphires, what phenotypic ratio was to be expected if enough kits were born to show it?

5-18 What was to be expected in the litters that contained no sapphires?

5-19 What kind of breeding test would be necessary to prove the validity of Warwick's tentative hypothesis that scrotal hernia in swine is not a simple recessive defect, but is probably caused by the action of duplicate recessive genes? (Remember that only boars are affected, and that these can be used for breeding, but that the search for 15 : 1 ratios under such conditions is scarcely feasible in animals as large as pigs.)

Selected References

CASTLE, W. E. 1940. *Mammalian Genetics*. Cambridge, Mass. Harvard Univ. Press. (Reports studies of mutants in rabbits and other animals, with good illustrations of breeds.)

DARWIN, C. 1890. *The Variation of Animals and Plants Under Domestication*. 2nd ed. New York. D. Appleton & Co. Inc. (Apart entirely from its many references to reversion, this book, which first appeared in 1868, is a treasury of information about variation in familiar species, and the reasons considered responsible for it a century ago.)

KEELER, C. E. 1935. A second rexoid character in the house mouse. *J. Heredity* **26:** 189–91. (With evidence that it is genetically different from the similar type first known.)

LITTLE, C. C. 1957. *The Inheritance of Coat Color in Dogs*. Ithaca, N.Y. Cornell Univ. Press. (With interpretations different from Winge's.)

SHACKELFORD, R. M. 1949. Six mutations affecting coat color in ranch-bred mink. *Amer. Naturalist* **83:** 49–68. (Behavior in crosses of various mutants from Aleutians to sapphires, including even "goofus" and green-eyed pastels.)

STOCKARD, C.R. 1936. An hereditary lethal for localized motor and pre-ganglionic neurones with a resulting paralysis in the dog. *Amer. J. Anat.* **59:** 1–53. (Extensive details of the defect discussed in this chapter.)

WARWICK, B. L. 1931. Breeding experiments with sheep and swine. *Ohio Agric. Exper. Sta. Bull.* 480. (Evidence that inguinal hernia is hereditary in swine.)

WINGE, Ö. 1950. *Inheritance in Dogs*. Ithaca, N.Y. Cornell Univ. Press. (Mostly about colors, and genetic bases therefor in various breeds.)

CHAPTER

6

Chromosomes and Genes

In the previous five chapters we have learned quite a bit about heredity without any consideration whatever of the mechanism whereby variations in combs, horns, color, and other characters are transmitted from one generation to the next. Mendel did the same. The structures that we now call chromosomes were scarcely known when he crossed his famous peas in the little garden at Brünn, and were not given their present name until 1888. Although they were much studied by cytologists during the last quarter of the nineteenth century, it was not until 1902 that Sutton showed how the behavior of chromosomes during the formation of germ cells (a process which had been elucidated in great detail by Boveri) provided a mechanism for the segregation and inheritance of Mendelian characters.

We have since learned that those characters and all inheritance (except for substances transmitted in the cytoplasm, such as plastids in plants and antibodies in animals) are induced by the little units of complex protein that we now call genes. These are borne in the chromosomes. It is appropriate, therefore, that we should consider the nature and behavior of chromosomes and the genes they carry.

Chromosomes

These all-important little structures get their names from their special affinity for basic stains, a characteristic that greatly facilitates the study of them. The word *chromosome* means "colored body." Chromosomes have been described as deeply staining bodies into which the nucleus resolves itself during cell division.

Cytologists find that chromosomes, which may seem at certain stages and at low magnification to be dark little rods of various sizes, are much more complex when seen at different stages of cell division and at higher magnification. In many species each chromosome has distinctive differences from all the rest, so that it can be identified, and designated at least by a number and sometimes even by a name.

The chromosome is often seen to be coiled, and the degree of coiling is responsible (in large part) for changes in the length of any one chromosome during cell division. Some have a little bit of chromosome—a *satellite*—attached to the main body only by a slender thread of chromatin. The central thread or string of the chromosome is referred to as the *chromonema*, meaning colored thread. Along its length are little thickened regions called *chromomeres*.

In some cases, one or more parts of a chromosome accept basic dyes faster than others and retain those dyes longer. Such parts are said to contain *heterochromatin*, in contrast to the *euchromatin* in the remainder of the chromosome. By the term euchromatin (*i.e.*, good chromatin), the cytologists, who coined the term, probably meant merely that its staining reactions were normal. The geneticist, also, may well label it as good chromatin, but for a different reason. There is evidence that the euchromatic regions carry the genes, which are his chief stock-in-trade, whereas the heterochromatic ones are genetically inert.

After differences in lengths, the characteristic most helpful in identifying chromosomes is probably the location of the *centromere*. This is best seen in *anaphase*, the stage of cell division in which the chromosomes have separated and are on their ways to opposite poles of the cell. Each chromosome then seems to be pulled either to one pole or the other by a very delicate strand or thread called a *spindle-fibre*, and the point of attachment of that fibre is the centromere.

If the centromere is right at the midpoint of the chromosome, then, as that chromosome proceeds (or is pulled) toward the pole, there are two equal arms on either side of that centromere, and the chromosome appears V-shaped. However, if the centromere is nearer one end that the other, one arm will be correspondingly shorter than the other, and the chromosome will appear J-shaped. Thus the ratio of the longer arm to the shorter one, which varies from one chromosome to another, helps the cytologist to identify the different chromosomes. (See Figs. 6-1 and 6-2.) Sometimes the spindle-fibre is attached right at the end

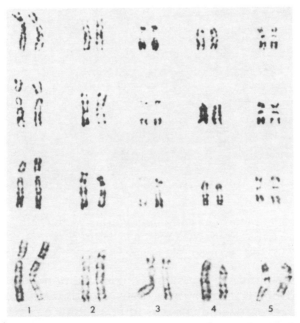

FIG. 6-1 G-banding patterns of the five largest pairs of chromosomes of the pig. The top two rows were pre-treated with saline-citrate solution, the lower two with trypsin. (From Pace *et al.* in *J. Heredity* as cited. Copyright © 1975 by the Amer. Genetic Association.)

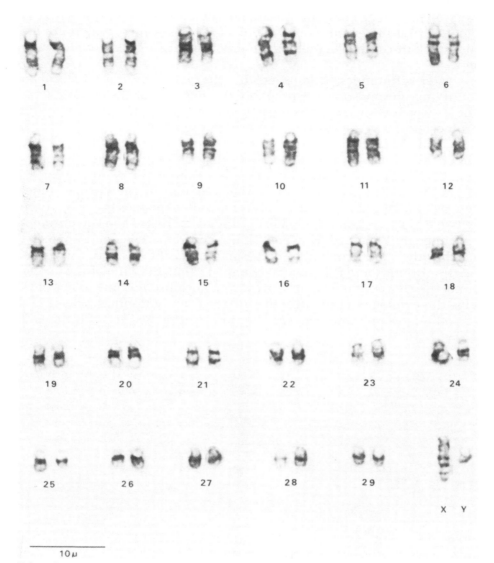

10μ

FIG. 6-2   Mitotic chromosomes of a bull, stained with Giemsa stain to reveal the transverse bands. (From Gustavsson *et al.* in *Hereditas* as cited.)

so that the travelling chromosome is as straight as a log in a stream.

The centromere differs from the rest of the chromosome and can frequently be recognized by its less intense staining, even without any directing spindle-fibre. Because it is considered a focal point for movement of the chromosome, it is often designated as the *kinetochore*, rather than the centromere.

Finally, the most remarkable thing about a chromosome is its ability to reproduce another just like itself. This reproduction is essential for the process of cell division by which organisms grow.

Numbers of Chromosomes

Apart from some exceptions which will be considered in a later chapter, the number of chromosomes per cell is constant for all individuals of the same sex in any one species.

In somatic cells, as distinct from germ cells, there are two chromosomes of each kind—a pair, one member of which was received by the individual from its dam, the other from its sire. The total number is referred to as the *diploid* number or $2n$. In germ cells, each of which contains only one member of each pair of chromosomes, the number is half that in somatic cells, and is called the *haploid* (or n) number. For example, man has 23 pairs of chromosomes; hence his diploid count is 46, and his haploid number is 23.

Diploid chromosome numbers for some familiar animals are given in Table 6-1. Most of the figures shown are comparatively high, a fact that makes the chromosomes and genes of those species somewhat more difficult to study than they would be if there were fewer chromosomes. Obviously it is harder for the cytologist to distinguish and identify the 32 different kinds of chromosomes in a horse than the four pairs in the fruit fly, *Drosophila melanogaster*. As we shall see later, it is also more difficult for geneticists to locate genes in the particular chromosomes that carry them when a mutation may belong in any one of 32 chromosomes than when it must belong to one of four.

This does not mean that genetic studies are impossible with animals having many chromosomes. The chief factors limiting anyone who wishes to study heredity in horses are not their 32 pairs of chromosomes, but uniparity and the long interval between generations. In the big ascarid worm, *Parascaris equorum* (formerly longknown as *Ascaris megalocephala*) which specializes in horses, the haploid number of chromosomes is only 1 in the type called *univalens* and 2 in the *bivalens* form. Cytologists have given that worm much study, but geneticists have not. We know a little about the genes of horses, but nothing about those of the worms inside them.

Sex Chromosomes

Among the chromosomes of most animals (if not all) there is one pair of chromosomes which determines sex. They are usually recognizable in suitable material by some distinction of size, or by location of the

TABLE 6-1
Diploid Chromosome Numbers in Some Familiar Animals
(Mostly from Makino)

Common Name	Scientific Name	2n
Honey-bee	Apis mellifera	32, 16
Pigeon	Columbia livia	80
Domestic duck	Anas platyrhyncha	80
Muscovy duck	Cairina moschata	80
Fowl	Gallus gallus	78
Turkey	Meleagris gallopavo	82
Rabbit	Oryctolagus cuniculus	44
House mouse	Mus musculus	40
Rat	Rattus norvegicus	42
Guinea pig	Cavia cobaya	64
Horse	Equus caballus	64[a]
Ass	Equus asinus	62[a]
Goat	Capra hircus	60
Sheep	Ovis aries	54
Cattle	Bos taurus	60
Zebu	Bos indicus	60
Pig	Sus scrofa	38
Dog	Canis familiaris	78
Cat	Felis catus	38
Fox	Vulpes fulva	38
Mink	Mustela vison	30

[a]The horse and the ass, both believed for many years to be 2n = 66, have recently been shown by Makino and his associates to have the numbers here given.

Note: For explanation of the paired number, see text.

centromere, and often by the fact that the two members of the pair are remarkably different. It is customary to refer to these as the X-chromosome and the Y-chromosome. In mammals and in many other animals, males have one of each kind, but females carry two X's, no Y. Males are said to be XY and females XX. All the rest of the chromosomes are called *autosomes*. The relation of these sex chromosomes to the determination of sex is considered in the next chapter. They are introduced here chiefly to permit better understanding of the illustrations.

In birds, reptiles, Lepidoptera, and some other animals, this situation is reversed. The male has two sex chromosomes alike, but the female has only one X. Sometimes the Y-chromosome is smaller than the X; sometimes it seems to be missing, so that one sex is XX and the other X—, or, as is sometimes written, XO. To distinguish more clearly between such animals and those in which the males have only one X-chromosome, the former are now usually said to be ZZ-ZW and the latter XY-XX.

It was long believed that in birds the female carried only one sex chromosome, V-shaped, with its centromere metacentric (at the middle). In 1961, however, Frederic identified a smaller, straight (acrocentric) chromosome which was unpaired in female cells of the domestic fowl. He called it a Y-chromosome (*i.e.*, what many cytologists now call a W-chromosome, the dissimilar partner of the Z-chromosome). Subsequent research has revealed that the same situation occurs in over 100 species of birds (Bloom, 1974).

The honey-bee is a special case. The drones develop by parthenogenesis from unfertilized eggs, the females from fertilized ones. The former are truly fatherless waifs, carrying only chromosomes from their mother. They are haploids.

Study of Chromosomes

Thanks to advances in cytological techniques during recent years, we are now seeing chromosomes better than ever before.

In earlier years, to study chromosomes one took little bits of some tissue in which cells were likely to be dividing rapidly, such as an amnion, an embryo, or a testis, "fixed" their cells with some killing fluid, got the tissue infiltrated with paraffin, cut it in very thin slices with a microtome, and eventually, after several other steps, including the all-important staining, studied the sections under the microscope. Since the sectioning had sliced away parts of some cells, a great deal of searching and counting of chromosomes was necessary to be sure that the microscopist was seeing the entire *chromosome complement*.

With a newer technique, the cytologist grew (in tissue-culture) cells from bone marrow, from the blood (monocytes), or even from skin or other parts of the body. These were treated with a plant hormone (phytohemagglutinin) which induced cell division, then with a drug (colchicine) which arrested the same process. The cells were then studied under a cover-slip in what the cytologist called a "squash preparation." The light pressure exerted to flatten the cells caused the chromosomes to be beautifully spread out. As there had been no cutting of the cells, good preparations showed every chromosome and exact, definitive counts were not so difficult to make as in sectioned material.

When a cell divides, each *daughter-cell* gets one element, or strand, of every chromosome. Recalling that diploid chromosomes represent pairs of homologous chromosomes, one member of each pair having come from each parent, we should be able to match up the scattered chromosomes in pairs, each pair having both members identical in length, in staining, and in location of the centromere. That is exactly what cytologists do, often cutting out the separate chromosomes (of one cell) from photographs or drawings, and then matching them in pairs to make what is called a *karyotype* or *karyogram* (Figs. 6-1, 6-2).

Even better procedures followed rapidly, and squash preparations were replaced by a technique in which cells suitable for study (usually from tissue cultures) are dropped on a glass slide, spread out, and, to arrest division of any dividing cells, are killed with a suitable fixative. After drying for several days (without any cover slip), they are treated with a trypsin preparation, which, by digesting some of the chromatin, causes what remains to show in bands transverse to the long axes of the chromosomes. Differences in width of these bands are rendered more visible by staining the preparation with Giemsa stain. The technique is now generally known to cytologists as G-banding (Fig. 6-1).

This technique is sometimes used in combination with another called "quinacrine fluorescence," or Q-banding, which accentuates differences in the width and density of the transverse bands. Different intensities of banding are observed from one end of a chromosome to the other, and these differences can be recorded photoelectrically. Using both techniques, Gustavsson and his co-workers were able to differentiate the 30 pairs of chromosomes in cattle and to describe their banding patterns so minutely that every one can be identified (Figs. 6-2, 6-3). The X-chromosome of cattle is easy to identify by its submedian centromere as well as by the fact that it appears twice in female cells but only once in males. The Y-chromosome, which also has a submedian centromere, is very small and appears as a small cross.

Idiograms

Differences among chromosomes are somewhat easier to see when they can be compared with an *idiogram*. This is a diagram, a formalized karyotype, with each chromosome straightened out to show its full length, and the position of the centromere marked by a constriction or otherwise. A karyotype (or karyogram) shows the chromosomes of a single cell and both members of each pair, but an idiogram shows the ideal chromosome, the average type derived from study of many cells, and shows only one of each pair.

In some species, many of the individual chromosomes can be identified fairly well by differences in the location of the centromere

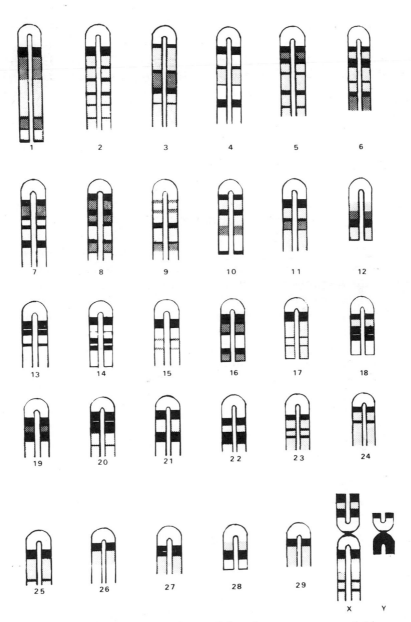

FIG. 6-3 Diagram showing distinctive G-banding patterns and intermediate shades (dark grey and light grey) by which the individual chromosomes can be identified. The intermediate shades (densities) were recorded photoelectrically after Q-banding (quinacrine mustard) which reveals differing degrees of fluorescence. (From Gustavsson *et al.* in *Hereditas* as cited.)

and hence in the relative lengths of the two arms which it separates. Moreover, some chromosomes carry tiny satellites. These conditions are found in human chromosomes (Fig. 6-4). By special techniques and staining, even those that are much alike (Fig. 6-4: 8 and 9, 10–12) can now be differentiated.

The G-banding and Q-banding techniques (and variations thereof) are now being applied to the chromosomes of domestic animals. As we shall see later, they make possible the identification of component parts of translocations—aberrations in which one chromosome (or a part thereof) becomes attached to another.

Mitosis

In the process of cell division, each dividing somatic cell passes to the two daughter-cells equal shares of the chromosomes, together with approximately equal shares of the cytoplasm and the contents thereof. Such a division of the chromosomes is called *mitosis,* or *equational division.* It contrasts with the unequal division of nuclear material that occurs when germ cells are formed—a process known as *meiosis,* or *reductional division.* The all-important feature of mitosis is that from each chromosome two identical strands of chromatin and genes are distributed. One of these strands goes to one daughter-cell, and the other to the other daughter-cell.

The process of cell division is a continuous one, but special names have been given to the several recognizably different stages. In the so-called *interphase* or *"resting stage"* (not a good name, because the

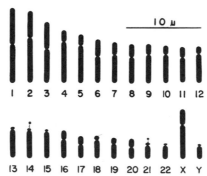

FIG. 6-4 Idiogram of human chromosomes showing one member of each autosomal pair and the two sex-determining chromosomes, X and Y. Positions of the centromeres are shown as constrictions. Since this figure was prepared, Chromosome 13 has been found to carry a satellite, as do Chromosomes 14 and 21. (From E.H.Y. Chu and N. H. Giles in *Amer. J. Human Genetics.*)

nucleus is then busily duplicating each chromosome, and hence is far more active than it seems), the cell shows a *nucleus* that is recognizable (in stained preparations) as a dark, rounded body (Fig. 6-5). Its stainable material is dispersed in a fine network, in which chromosomes can seldom be recognized. Darker spots within it are the *nucleoli*, or there may be only one of these, a *nucleolus*. In the somewhat fluid cytoplasm surrounding the nucleus, there are watery droplets called *vacuoles*, and little bodies of assorted sizes and shapes known as *mitochondria* and *Golgi bodies*. The functions of these are not fully understood, but they (like the chromosomes) apparently reproduce themselves and, when the cell divides, are passed in about equal amounts to the two daughter-cells. The mitochondria are known to be centres of cellular metabolism. Under the electron microscope, one can see still smaller bodies, called ribosomes, which are believed

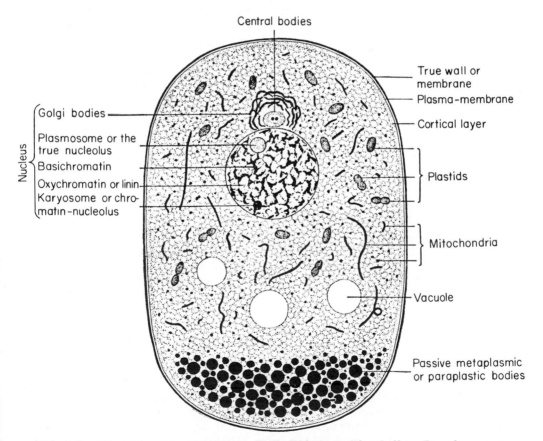

FIG. 6-5 Diagram of a cell. (From E. B. Wilson in *The Cell in Development and Heredity*, 3rd ed., courtesy of The Macmillan Company.)

to be involved in the synthesis of proteins. An important item in the cytoplasm (though often not visible until later stages of cell division) is the *centrosome,* located beside the nucleus. Its two *central bodies,* or *centrioles* seem to become foci, or centres of attraction, for the chromosomes as they separate into two groups in mitosis.

Animal cells differ from plant cells in having no cellulose in the cell wall, which is a semi-permeable membrane, and no plastids. These last are self-reproducing bodies, often carrying a pigment (*e.g.,* chloroplasts) and believed to store energy and to manage their own reproduction. Cells of higher plants lack centrioles but seem to manage very well without them. To see easily the various stages of mitosis, we should examine under the microscope some tissue that is growing at a rapid rate and hence has many cells in the process of dividing. Tips of onion roots are standard material for botanists, but those whose aversion to the pungent *Allium cepa* extends to its very root tips can do almost as well with developing eggs of the whitefish. A cross-section through a well-prepared blastula of that species shows—even at low magnification—cells in all stages of mitosis.

In the *prophase,* when division is getting under way, the chromosomes become distinguishable, usually in double strands. These, in turn, show that, while the cell was apparently resting, the nuclear material was actively engaged in reproducing itself. The centrioles move to opposite poles of the cell, but are sometimes difficult to see because of the *aster,* the radiating, star-like structure that surrounds them (Fig. 6-6). The nuclear membrane disappears.

In the next stage, the *metaphase,* the spindle, consisting of delicate fibres, is formed, and the double chromosomes become aligned along the *equatorial plate* at the centre of the cell. In side views the spindle is usually distinct, and the chromosomes appear to be tightly packed along its equator. In polar view, however, they are often seen to be nicely spread out in the circle that is made by a cross-section through the spindle (Fig. 6-6). In sectioned preparations this is the view most favorable for making counts of chromosomes in somatic cells.

One must not think of the chromosomes at this stage as stationary bodies, like soldiers toeing a line and standing at attention. Moving pictures of living cells (such as the remarkable ones made by the Bajers of the Jagellonian University in Cracow) show that during metaphase the chromosomes are in constant motion, almost as if they were competing and jockeying for position like restive race-horses at the starting line. The simile is better if we can imagine that they are not single horses, but pairs of identical twins, and that the two members of each pair stick together, side by side, until the starting gun is sounded. At that signal the twins separate; one runs in one direction, the other in the opposite one.

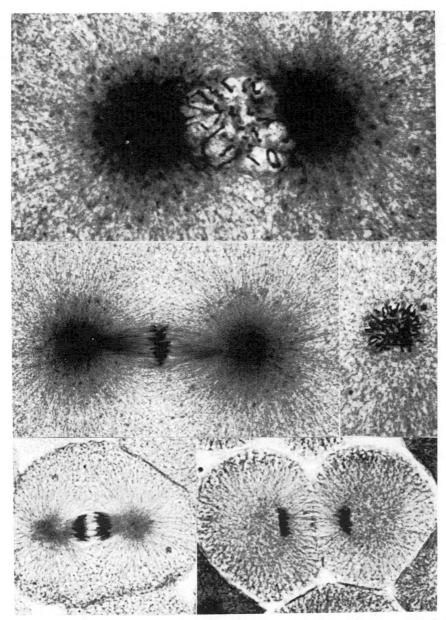

FIG. 6-6 Mitosis in dividing cells of fertilized eggs of the whitefish, *Coregonus clupeiformis*. Upper: prophase, centrosomes obscured by the dense asters. Middle, left: metaphase, side-view, spindle very distinct; right: metaphase, polar view. Lower, left: early anaphase; right: telophase, with daughter-cells partly separated. Magnification of the prophase is much greater than for the others. (Courtesy of General Biological Supply House, Inc., Chicago.)

Although each chromosome has a partner, or *homologous* chromosome, in the same cell, the two members of that pair are "on their own" in mitotic divisions. They are distributed at random on the equatorial plate. This stand-offishness is quite different from events in meiosis, where two homologous chromosomes *do* come together and stay together through part of that process.

In the *anaphase* the chromosomes move toward the opposite poles, and in good material one may recognize both early and late stages of this process. It is during the anaphase that one sees best the point of attachment of the spindle-fibre, *i.e.*, the centromere. The cell is usually pinched in around the middle as cleavage begins.

Finally, in the *telophase* the chromosomes of each daughter-cell clump together to form the nucleus of that cell (Fig. 6-6). As they do so, the spindle disappears, a new nuclear membrane is formed, and the original cell is divided into two by a furrow that begins at the periphery of the cell, at its equator, and moves inward. In plant cells, a cell wall is laid down, but in animal cells it is almost as if the two daughter-cells were pinched apart by constriction of the equator of the parent cell.

By the process just described, one cell becomes two. Each of the new cells carries the same number of chromosomes as the parent cell, and (except for occasional accidents or mutations) every somatic cell thus gets the same collection of genes.

Maturation of Germ Cells

As the number of chromosomes in somatic cells remains the same for all members in successive generations of a species, and since each new individual arises from the union of two gametes, it is evident that there must be some mechanism whereby the number of chromosomes is reduced in gametes but restored to the diploid number when male and female gametes unite to start a new individual. That mechanism is provided by a process consisting of two cell divisions, once referred to as *reduction division*, but now more commonly known as *meiosis*. This process is the essential feature of the somewhat longer procedure leading to the formation or *maturation* of definitive germ cells (Fig. 6-7).

The development of reproductive cells really begins in the embryo, soon after fertilization, when certain cells become differentiated as *primordial germ cells* and eventually concentrate at the site of the developing gonads. There they undergo the repeated mitotic divisions that accompany growth of the testis or ovary. At that immature stage these cells are known as spermatogonia in males, and oögonia in females.

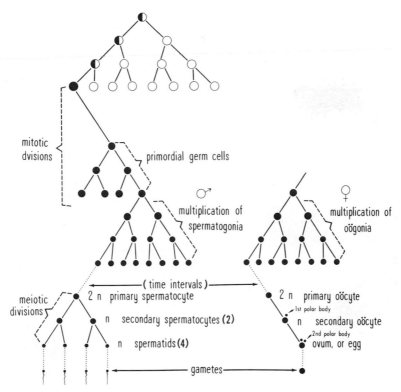

Within the figure:

mitotic
dvisions

primordial germ cells

♂ multiplication of
spermatogonia

♀ multiplication of
oögonia

—— (time intervals)——

meiotic
divisions

2 n primary spermatocyte

2 n primary oöcyte
1st polar body
n secondary oöcyte
2nd polar body
ovum, or egg

n secondary spermatocytes (2)

n spermatids (4)

——————— gametes ———————

FIG. 6-7 Diagrammatic representation of the maturation of germ cells in animals. (After E. B. Wilson in *The Cell in Development and Heredity*, 3rd ed., courtesy of The Macmillan Company.)

In a cross-section of a testis, one can see under the microscope large numbers of cells, most of which are spermatogonia, just inside the basement membranes of the tubules. At sexual maturity, many of these develop successively into *primary spermatocytes*, which may be recognized by their larger size and by indications that their nuclei are more active. In the first meiotic division, each primary spermatocyte divides to form two *secondary spermatocytes*. These, in turn, divide equally to produce two *spermatids*, which become transformed into mature male germ cells, the *spermatozoa*. During the two meiotic divisions the homologous chromosomes of each pair disjoin, so the number of chromosomes in the spermatids is half that in the primary spermatocytes. Every primary spermatocyte has the diploid, $2n$, number of chromosomes, but in the four functional gametes to which it gives rise, the chromosomes are haploid, n.

The process just reviewed whereby spermatozoa are produced is called *spermatogenesis*. As is evident in Fig. 6-8, a cross-section

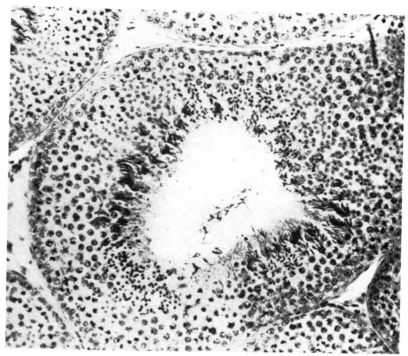

FIG. 6-8   Photomicrograph of a section through an active seminiferous tubule of a cock, showing various stages of spermatogenesis from primary spermatocytes near the outer wall to free spermatozoa in the lumen. (From R. A. Miller in *Anat. Record* **70**: 155–89.)

through an active seminiferous tubule shows cells in various stages of transformation from primary spermatocytes just inside the periphery to fully formed spermatozoa near the lumen.

In *oögenesis*, the formation of female gametes, the process is similar in some respects to gametogenesis in males, but quite different in another. One *primary oöcyte* gives rise only to one functional gamete, not four. At the first meiotic division, the cytoplasm does not divide, and half the chromosomes are extruded from the cell in a little *polar body* almost devoid of cytoplasm. The same process is repeated when the *secondary oöcyte* divides. Sometimes the first polar body divides also. The result of two meiotic divisions of a primary oöcyte is thus one functional ovum and two or three tiny, non-functioning polar bodies (Fig. 6-7). The immediate product of the second meiotic division is sometimes called an *oötid*, to correspond with the spermatid— the immature germ cell of the male before it grows a tail. As the oötid is not immature very long, if at all, and requires no transformation, it is simpler (and not incorrect) to forget about oötids, and to think of the product of two meiotic divisions as the ovum, or egg.

Meiosis

As just related, the maturation of the germ cells would appear to be a rather simple process, but study of the chromosomes during the meiotic divisions has shown their activities to be very complex. The pattern of their behavior is much the same in oöcytes as in spermato-cytes. It is also essentially the same in grasshoppers as in man, in all animals in-between, and in plants. The important feature differentiat-ing meiotic divisions from mitotic ones is that in the former the two homologous members of each pair of chromosomes come together side by side in *synapsis*, and then separate, with one member going to one of the two new cells, the other member to the other. This process, which usually occurs in the first meiotic division, results in the formation of two cells, each having one member of every pair of chromosomes, *i.e.*, the haploid number.

In the prophase of the first meiotic division, the chromosomes appear as thin threads, a stage referred to as *leptotene* (Fig. 6-9b). Homologous chromosomes then come together side by side in pairs (Fig. 6-9c and d), in synapsis, but it is very difficult at this time to distinguish individual chromosomes. In this stage, which is called *zygotene*, each apparent chromosome is really a *bivalent* element comprising both members of a homologous pair.

The chromosomes then contract and thicken in what is called the *pachytene*, or thick-thread stage (Fig. 6-9c). By this time it is usually evident that each chromosome has split lengthwise (except at the centromere) and that the bivalent element has thus been transformed to a *tetrad* containing four strands called *chromatids*. It should be remembered that two of these must have come from the maternal parent and the other two from the father.

At some time after a four-strand stage has been reached, one strand may lie across another chromatid at an angle, the point of contact being called a *chiasma*. Appearance of these *chiasmata* marks the *diplotene* stage (Fig. 6-9g) in which homologous chromosomes (formerly attracted to each other) now repel each other, *i.e.*, they disjoin. The chiasmata, which become prominent at this time, mark points where interchanges have occurred between chromatids. The exact mechanism of these interchanges is unknown.

Remembering that homologous maternal and paternal chromo-somes can carry different alleles at any locus, we see that if any interchange occurred between chromatids of maternal and paternal

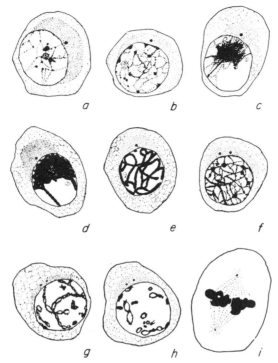

FIG. 6-9 Stages preceding the first meiotic division of spermatocytes of the fowl. a: prophase; b. leptotene; c and d: zygotene; e. pachytene; f: chromosomes diffused before breaking up into tetrads; g: diplotene, showing chiasmata; h: diakinesis showing tetrads; i: metaphase of first meiotic division. (From R. A. Miller in *Anat. Record* **70:** 155–89.)

origins, the two resulting new strands could carry new combinations of genes, different from those in chromatids that have had no such interchange. We shall pursue this intriguing situation in a later chapter.

In the next stage, *diakinesis*, the chromosomes become shorter, denser, and more clearly recognizable as separate bivalent units (Fig. 6-9h). As there is only the haploid number of tetrads, diakinesis is a favorite stage at which to count chromosomes.

In the metaphase of this first meiotic division (Fig. 6-9i), the tetrads line up at the equator of the cell, much as in mitotic metaphases. Almost in unison they break apart into *dyads*, each consisting of two chromatids, and these move to opposite poles of the spindle. At this stage the two strands of the dyad are still held together by a centromere and, except for any interchanges that may have occurred, both are of either maternal or paternal origin.

An important point to remember is that in this metaphase it is entirely a matter of chance within each tetrad whether the dyad of maternal origin lies on the side of the equator whence chromosomes go to (let us say) the north pole, or on the opposite side whence they go to the south pole. The same applies to the dyads of paternal origin. We thus have the mechanism for independent segregation of chromosomes and their genes to the gametes.

The *second meiotic division*, which usually follows after a short interphase, unlike the first, is more comparable to the equation divisions of mitosis, and the chromosomes again go through prophase, metaphase, anaphase, and telophase. In the metaphase, the centromere of each dyad divides, and the two chromatids of that dyad separate. In the anaphase they proceed to opposite poles of the spindle, each no longer merely a part of a bigger organization, but now entitled to full status as an independent chromosome of its own. The resultant germ cells thus have one representative—and only one—of each pair. They are haploid.

Although sexes do not differ in the essential features of meiosis, there can be great differences in the timing of some stages and in the intervals between them. For example, the first meiotic divisions in cockerels do not begin until those males become sexually mature, but in female chicks primary oöcytes are recognizable in embryos at 14 days of incubation. Three days later, the chromosomes in those cells have gone as far as the synaptene stage. Thereafter there is a long resting period, and the first polar body is not extruded until several months later, shortly before the ovum is released from the follicle when the pullet begins to lay. The second meiotic division of the egg is completed after fertilization. It is believed that in birds, as in some other species, entrance of the spermatozoön is a stimulus necessary for completion of that second division.

Fertilization

After penetration of the egg by a spermatozoön, the nuclei of these two cells (then referred to as *pronuclei*) fuse in the *fertilization spindle* which results in the first cleavage of the fertilized egg.

In mammals, amphibia, and many other animals, that division is *holoblastic*, or complete, but in eggs with much yolk, like those of birds and reptiles, the cleavage is *meroblastic, i.e.*, confined to a very shallow layer on the surface of the egg.

In remarkable studies of these stages in the eggs of cows, Hamilton and Laing were able to find both male and female pronuclei just prior to their fusion, a fertilization spindle and an extruded polar body, and the two-celled stage resulting from the first cleavage (Fig. 6-10).

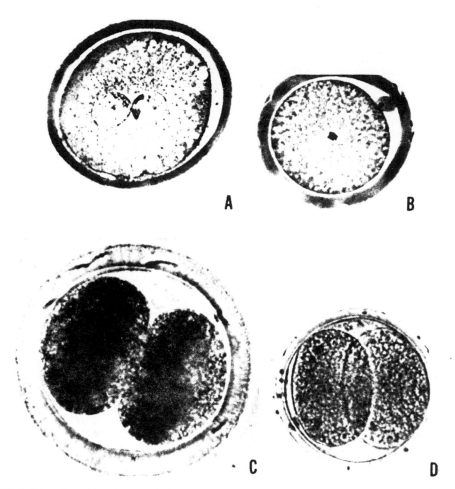

FIG. 6-10 Fertilization and cleavage in eggs of the cow. A: ♂ and ♀ pronuclei just prior to fusion. B: Fertilized egg showing first cleavage spindle and polar body. C: Two-cell stage, photographed alive. D: Two-cell stage after fixation and clearing. Cells taken at 40 to 55 hours post-oestrum. (From W. J. Hamilton and J. A. Laing in *J. Anatomy*.)

The Nature of the Gene

A detailed account of events leading to our present concept of the chemical and physical structure of genes is beyond the scope of this book. Suffice it to say that, after being limited for many years to the belief that genes must be comparable to large molecules of protein, that vast field was eventually narrowed down (about 1945 to 1950) to *nucleoproteins*. These compounds consist of a nucleic acid combined with a protein. The two known types of nucleic acid, deoxyribonucleic acid (DNA) and ribonucleic acid (RNA), get their names from the sugars that they contain—deoxyribose and ribose, respectively.

DNA occurs in the nucleus, more specifically in the chromosomes, and is now known to be the material that transmits the inheritance from one generation to the next, and from one cell to all the rest. In modern genetic parlance it "carries the genetic information." For those who (like these authors) find it difficult to conceive of some one chemical entity confiding its wishes to another, a good way out is to think of the DNA (or the genes) as being able to direct the chemical processes of the cell along certain lines.

The Structure of DNA

According to Watson and Crick, whose concept of DNA and model of its structure were advanced in 1953 and have since been generally accepted, a molecule of DNA consists of two long chains of nucleotides in a configuration resembling somewhat a spiral staircase. More accurately, it is a double helix (Fig. 6-11). A nucleotide consists of a phosphate group, deoxyribose sugar, and a base. That base may be either of two purines—adenine or guanine (A or G); or two pyrimidines—thymine or cytosine (T or C). Each base in a nucleotide of one chain is tied to its opposite number in the corresponding nucleotide of the other chain by a hydrogen bond. Both bases are held to their respective chains by the deoxyribose part of the nucleotide.

Each cross-link (or step on the staircase) can thus be visualized as a purine held by hydrogen bonds to a pyrimidine, or as these same three in reverse order (Fig. 6-12). Adenine pairs only with thymine, and guanine only with cytosine; hence the following four combinations—and only these—are possible: A-T or T-A, G-C or C-G.

One might well question how these four pairs of bases (plus the deoxyribose and phosphate) can produce the innumerable kinds of genes that cause the hundreds and thousands of variations within one species, and the far greater numbers that result in the marked differences between species. The answer lies in the very great numbers of different possible sequences of the four pairs along the two chains that make up a molecule of DNA. Watson (1960) believes that such a molecule has a length of at least 10,000 nucleotides. One can visualize the possibilities by thinking of the genes as long words made up from an alphabet of only four letters, but with those letters repeated in a multiplicity of different sequences.

It is to be hoped that no reader will get an impression from the

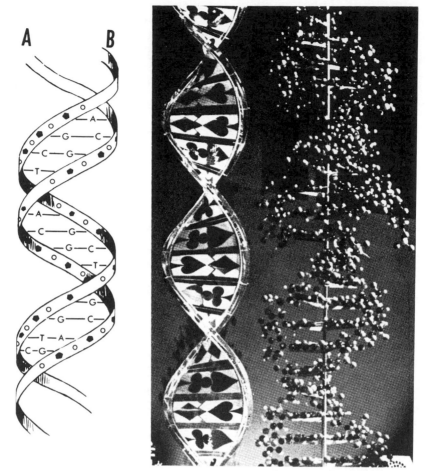

FIG. 6-11 Left: Diagrammatic model showing a short segment of a DNA molecule according to the Watson-Crick configuration. In the two chains of polynucleotides, A and B, the black pentagons represent deoxyribose sugar, to which the bases, G and C or T and A, are attached to form cross-links between the chains. Open circles in A and B represent phosphates between the sugars. Right: Parts of models of a short segment of a DNA molecule in which G. Gamow has shown (left) the four bases, guanine, cytosine, adenine, and thymine as four card suits, rather than as chemical formulae. Hearts can pair only with clubs, and diamonds only with spades. A molecular model is on the right. (Courtesy of American Association for Advancement of Science, from photograph of its exhibit at the Seattle World's Fair.)

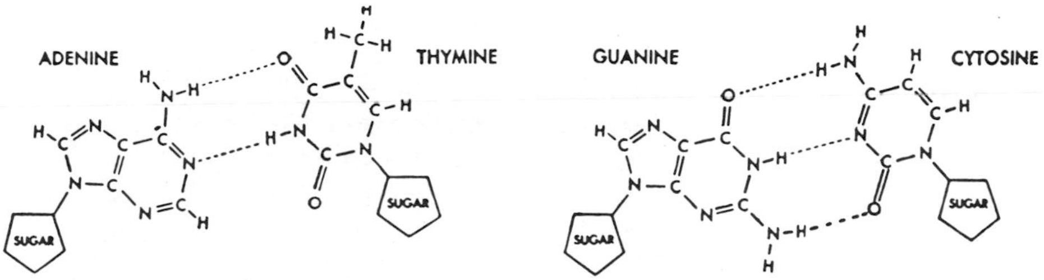

FIG. 6-12 Cross-links in the DNA molecule. The purine, adenine, is tied by hydrogen bonds to the pyrimidine, thymine, in one kind of link. In the other, the purine, guanine, is similarly linked to cytosine. Both kinds of cross -links are attached at each end to the deoxyribose sugar of the nucleotide. (From A. Kornberg in *Science* **131:** 1503–8, 1960.)

diagram in Fig. 6-11 that a gene resembles a spiral fire-escape at the back of an apartment building. Molecular models made by biochemists suggest that DNA is more like a long, narrow, and twisted bunch of grapes, with constrictions at regular intervals.

Genes, Mutational Sites, and Cistrons

Genetic studies with bacteria, yeasts, fungi, bacteriophages, and viruses have proven (by techniques which breeders of cattle or chickens are unable to use) that big genes have little genes inside. In other words, a big gene may consist of many subgenes, or *mutational sites*, in linear order. According to Demerec, mutations at sites within one gene-locus usually affect related phenotypes. The largest number known at any one locus he puts at 400, but calculations give an estimate of 2,000 at another. It is considered that a site of this kind corresponds to one nucleotide pair of the Watson-Crick model of DNA.

A series of such mutational sites is called by some geneticists a *cistron*, but others hold that the term *gene* is still adequate. We shall use it for the "everyday" genetics to which this book is confined. Readers who might like to look beyond that horizon could well begin with the writings cited at the end of the chapter. The smallest length of DNA capable of mutation (which may be a single pair of bases) is often referred to as a *muton*.

The Size of a Gene

From what we have just seen about the structure of genes, it is clear that the question "How big is a gene?" which has intrigued biologists for many years, must now be revised to read, "How long is a gene?" Presumably the length of a gene depends on the number of pairs of nucleotides in the molecule of DNA. Watson tells us that these are spaced about 3.4 Ångstrom units apart. From his estimate that the number of pairs of nucleotides in a molecule of DNA is at least 10,000, but can go to 20,000, it would appear that a gene should be at least 3.4 microns in length (or 0.0034 millimeter) but might be twice that figure.

Obviously it is much narrower. From the size of the purine and pyrimidine bases that make the cross-ties between the two strands of the helix, Watson estimates the diameter of the double helix to be only 20 Ångstrom units, or 0.002 micron. It seems clear from these estimates that none of us is likely to see any genes when looking at ordinary material, even down an extraordinary microscope.

Coding, or How Genes Work

One of the most dramatic advances in genetics since the rediscovery of Mendel's laws is probably the elucidation of the mechanisms whereby genes induce the synthesis of different proteins within the cell. Much of that work is done by RNA, or ribonucleic acid, which differs from DNA in that its sugar component is ribose, instead of deoxyribose as in DNA. Also unlike DNA, RNA occurs in a single strand. Without going into details the procedure for the synthesis is believed to be as follows.

One strand of the DNA molecule serves as a template along the length of which RNA is formed by enzymatic action from elements in the cytoplasm. This RNA, which (for reasons soon to be seen) is called *messenger RNA*, contains a series of bases which are complementary to those in the DNA. In other words if three successive bases in the strand of DNA were C-G-T, their opposite numbers in the messenger RNA would be G-C-A. This complementary pairing is the same as that shown in Figs. 6-11 and 6-12, except that A in the DNA induces in RNA the formation of uracil (U) rather than thymine. Uracil differs from thymine only in lacking a CH_3 group.

When completely formed, the strand of messenger RNA moves into the cytoplasm and becomes attached to the surface of minute cellular structures called *ribosomes*. Starting at the end of the strand of messenger RNA, the bases therein, working in groups of three, or as *triplets*, attract complementary triplet bases in molecules of a different kind of RNA called *transfer RNA*. To each molecule of this transfer RNA there is attached one specific amino acid. Such a molecule thus has two important functions. One of these is to catch from the cytoplasm either the particular amino acid called for by its bases or the enzymes which form that amino acid. The other function is to provide a sequence of three bases that is complementary (and hence attractive) to some corresponding triplet of bases in the messenger RNA.

As the process of accumulating molecules of transfer RNA continues along the strand of messenger RNA, one amino acid is added after another and these join to form a polypeptide chain, or protein. Thus, to recapitulate, the bases in the DNA, acting in triplets, induce formation of complementary triplet nucleotides in messenger RNA, and these, in turn, waylay on the ribosomes corresponding molecules of transfer RNA. These last carry with them the amino acids called for by the combination of bases in the original DNA, and, as those amino acids are linked up in succession from one end of the strand of messenger RNA to the other, so are they formed into the specific proteins needed for the genes to do their work.

The triplet bases of DNA and RNA which carry the information for protein synthesis are referred to as the genetic code. The set of three bases in messenger RNA which specify an amino acid may be referred to as a *codon*, and the corresponding set of three bases in the transfer RNA as an *anticodon*. The first base in any triplet can be C, G, A, or U, and the second and third bases can also be any of these four, so the number of different combinations of bases possible in groups of three is $4 \times 4 \times 4$, making 64 combinations possible to specify 20 amino acids. The genetic code is described as a *degenerate code* because there is more than one codon for most of the amino acids specified. Three of the codons, UAA, UAG, and UGA, have been referred to as nonsense codons since they do not code for an amino acid. They do serve the purpose of terminating the amino acid chain of a polypeptide when a protein is being synthesized, so they can also be referred to as termination codons.

While determination of the genetic code whereby genes direct the synthesis of proteins is a remarkable advance, there will still be much work for geneticists, embryologists, and biochemists to do before we learn how the genes build some proteins into Pekingese puppies or Palominos, and others into Suffolk sheep or Shorthorns.

Reproduction of the Gene

Earlier in our description of mitosis, the two-strand chromosomes seen in prophase were said to show that during the so-called "resting stage," or interphase, the chromosome is actually very busy in the process of reproducing itself. We can now understand how that is done. Within each strand of the DNA every base acquires from the cytoplasm another base that is its proper pairing partner (*i.e.*, C-G, G-C, A-T, T-A), so the result is the formation of two new complementary half-units of the DNA molecule. Together these two units make a new gene just like the older one. Such new genes, with their associated proteins, make up the new chromosome.

It is not difficult to realize that, in this process of replication of the genes, mishaps may occasionally occur. One pair of bases might get reversed, so that it is (for example) T-A instead of A-T. Perhaps a pair might somehow get out of its proper place in the new molecule, or even be left out altogether. Any of these events would result in a new sequence of bases, a new codon, and a resultant change in cellular amino acids. If the cell is still viable, the geneticist might find some change in color of skin, hair, or flower, some change in blood chemistry or metabolism—in short, a mutation. We shall consider such events further in Chapter 18.

Problems

Understanding of this chapter is not easily tested with problems capable of mathematical solution, as in some of the earlier ones. The important thing is to understand the processes of mitosis, meiosis, and the maturation of germ cells, and their functions in heredity and in reproduction of the species. You can drill yourself on the meaning of the new terms by compiling a list of the words in italics and then testing your understanding of those terms. If you think some of these questions are from the clouds, just take a look at those on the same subject in some other texts.

6-1 Normal gametes must be haploid, but must all haploid cells in animals be germ cells? (Remember *Apis mellifera*.)

6-2 For what purpose is colchicine used (a) by cytologists? (b) in medicine? (You'll have to look up that last on your own; you might want to use it some day—for yourself or for a dog.)

6-3 If you asked your grocer for a dozen haploid gametes of *Gallus gallus* with adherent parts, but with no zygotes among them, he would probably send for the men in the white coats. What would he call them? How could the producer guarantee that there would be no zygotes?

6-4 Hybrids between the domestic duck and the Muscovy are sterile; mules are the same, but the fully fertile Santa Gertrudis cattle came from equally fertile hybrids between Shorthorns and Brahmans (zebus). After looking over Table 6-1, can you draw any conclusions about the relation of chromosome numbers to the fertility or sterility of inter-specific or intergeneric hybrids?

6-5 Can hybrids be fertile if their parents have the same number of chromosomes? Must such hybrids always be so? What other genetic influences could affect their ability to produce functional gametes?

6-6 In the dog, how many X-chromosomes should there be in a random sample of 100 cells of each of the following kinds:

Primary spermatocytes   Primary oöcytes

Secondary spermatocytes   Secondary oöcytes

Spermatozoa   Unfertilized ova

6-7 In what cells of cattle should the autosomes number 29? What cells should have 58?

6-8 How many functioning gametes should normally be produced from 100 primary spermatocytes? From 100 spermatids? From 100 secondary oöcytes?

6-9 Are you clear on the differences among tetrads, dyads, and chromatids?

6-10 Recalling that drones develop from unfertilized eggs, that they are haploid, but that fertilized eggs of the honey-bee are diploid, how must the maturation of germ cells in drones differ from that in males of other species?

6-11 Would you expect a parthenogenetic turkey to be haploid or diploid? (See Chapter 19.)

Selected References

AVERS, C. J. 1976. *Cell Biology.* New York. Van Nostrand. VIII + 568 pp. (Full details, up-to-date, about what goes on in cells.)

BLOOM, S. E. 1974. Current knowledge about the avian W chromosome. *BioScience* **24:** 340–43. (A good review.)

CHU, E. H. Y., and N. H. GILES. 1959. Human chromosome complements in normal somatic cells in culture. *Amer. J. Human Genet.* **11:** 63–79. (Well illustrated; source of Fig. 6-4.)

GUSTAVSSON, I., M. HAGELTOORN, and L. ZECH. 1976. Recognition of the cattle chromosomes by the Q- and G-banding techniques. *Hereditas* **82:** 157–66. (Full details of how they did it. Source of two figures (Figs. 6-2 and 6-3) shown in this chapter.)

HAMILTON, W. J., and J. A. LAING. 1946. Development of the egg of the cow up to the stage of blastocyst formation. *J. Anat.* **80:** 194–204. (In addition to the cells shown in Fig. 6-10, this paper shows fine photographs of unfertilized ova and of early stages of development.)

HSU, T. C., and K. BENIRSCHKE. 1976–1977. *An Atlas of Mammalian Chromosomes.* Vols I–X. New York. Springer-Verlag. (Ten volumes, profusely illustrated, chromosomes of common animals and rare ones.)

MAKINO, S. 1951. *An Atlas of the Chromosome Numbers in Animals.* Ames. Iowa State Coll. Press. (Covers 3,317 species, including 563 vertebrates; best place to find reference to literature on chromosomes of domestic animals.)

PACE, J., P. K. SRIVASTAVA, and J. F. LASLEY. 1975. G-band patterns of swine chromosomes. *J. Heredity* **66:** 344–48. (Source of Fig. 6-1.)

SHARP, L. W. 1943. *Fundamentals of Cytology.* New York. McGraw-Hill Book Co., Inc. (Details about chromosomes, cell divisions, gametogenesis in animals, all easily read and well illustrated.)

WATSON, J. D. 1968. *The Double Helix.* New York. Atheneum. XVI + 226 pp. (A delightful account of events leading to his discovery, with Crick, of the structure of DNA.)

WATSON, J. D., and F. H. C. CRICK. 1953. Genetical implications of the structure of deoxyribonucleic acid. *Nature* **171:** 964–67. Also in *Cold Spring Harbor Symposia Quant. Biol.* **18:** 123–31, 1953. (The foundation of current concepts.)

CHAPTER
7

Sex Determination and Sex-linked Inheritance

Sex Determination; Some Early Theories

Anticipatory speculation whether "it" would be a boy or a girl was undoubtedly just as popular a pastime in all the long previous centuries as in this twentieth one. In addition, earlier generations could also speculate more than is now justified about what makes "it" one sex or the other. That they certainly did. Surveying what was known about sex determination in 1889, Geddes and Thomson stated that in the seventeenth century Drelincourt had compiled no fewer than 262 theories on that subject, that Blumenbach had "quaintly remarked that nothing was more certain than that Drelincourt's own theory formed the two hundred and sixty-third," and that in their opinion (G. and T.), the number had been "well-nigh doubled" since Drelincourt's time.

In an age when plausible theories on how to run the world are so freely bandied about it, it is encouraging to reflect that every last one of those 500-odd theories about what determines sex, each of which must have seemed plausible to its proponent, was wrong. Sex-determining chromosomes were not even identified as such until some fifteen years after Geddes and Thomson wrote their first edition. We cannot take space here to match Drelincourt's compilation, but a few samples of earlier views will serve to illustrate the remarkable range of imagination with which some of our ancestors were blessed.

Aristotle thought that males were more likely to be conceived when north winds were blowing; the gentler sex was naturally associated with southern zephyrs. By some the offspring were believed to be the same as (or opposite to) the older (or younger) parent. An ingenious idea with unlimited scope for precipitating domestic strife was Starkweather's theory (as late as 1883!) that the superior parent settles the issue and produces the opposite sex. A simple theory that was advocated from the time of Anaxagoras (*circa* 450 B.C.) down to 1921, and is still held by some dog breeders, is that the right ovary produces males, while the left one, being on the weaker side, must produce females. Here, at least, was a theory that could be subjected to experimental tests, but when Doncaster and Marshall surgically removed a single ovary from rats, they found that the remaining one, whether right or left, yielded litters with equal numbers of both sexes.

Enough of this. Lest the reader be too scornful about the would-be scientists of yesteryear, he should remember that any theory, even a blind guess, is likely to be right, by chance alone, in half the cases to which it is applied. With small numbers, it could easily be right more often than not.

Heterogametic Males

It was not until about 1905 that cytologists showed beyond any further doubt that certain insects produce two kinds of sperm cells, which differ by having either one or the other of an unequal pair of chromosomes. These we now know as the sex-determining chromosomes—X and Y. Even earlier in other insects, half the spermatocytes had been found to carry an unpaired chromosomal element, or accessory chromosome, which the other half lacked. McClung had suggested (in 1901) that these two kinds of spermatozoa occurring in approximately equal numbers might be responsible for the similarly equal numbers of the two sexes that result from fertilization. When it was further established that female germ cells of these various insects did not differ as did the spermatocytes, but all carried one X-chromosome, the role of that chromosome in determining sex (in these insects, at least) was clear.

Obviously males carried one X-chromosome and either an unequal partner—the Y-chromosome—or none at all. As we now say, the males are either XY or XO. In either case they are *heterogametic, i.e.,* producing different kinds of gametes, half of which are male-determining, the other half female-determining. In species having such heterogametic males, the females are XX, and *homogametic,* producing gametes all of which are equal so far as their sex-determining potency is concerned.

From many cytological and genetic studies, it has now been shown that, so far as species studied thus far can be a guide, males are heterogametic in Nematoda, Mollusca, Echinodermata, and most Arthropoda. In the last-named group, all insects studied thus far seem to have heterogametic males, except the Lepidoptera and Trichoptera. In some fishes the males are heterogametic; in others they are homogametic. Mammals apparently all have heterogametic males.

Heterogametic Females

To distinguish heterogamety in females from that of males, species with the former are sometimes said to have a ZZ-ZW (or ZO) type of sex determination in contradistinction to the XX-XY (or XO) type of mammals and of other animals listed above. For most of us it is simpler to think of the sex chromosomes as X and Y, and to remember that the males are heterogametic in most species, but are homogametic in some groups.

Animals having heterogametic females include all the Aves, apparently all Reptilia, and some, at least, of the Amphibia. In Lepidoptera (moths and butterflies) the females are heterogametic. One must be careful about generalization for whole groups based on studies of a small number of species. In the Poeciliidae, a family of viviparous Cyprinodont fishes, some species have heterogamety in males, others in females. In one of them, *Platypoecilus maculatus,* Gordon found that the males are heterogametic in some Mexican rivers, while in the Belize River of British Honduras (now Belize) the females are heterogametic. Verily, exceptions that "prove the rule" are not wanting in biology!

Identifying the Heterogametic Sex

This can be done in two ways. When cytologists can recognize some chromosome as being paired in one sex, but matched in the opposite sex with an unequal partner, or having no partner at all (XO), that chromosome must be the sex chromosome; but all this is much more easily said than done. In mammals that have been examined thus far, the X- and Y-chromosomes are distinctly unequal in size and shape, but sometimes, as in some fishes, these two are so much like all the autosomes that they cannot be readily identified.

Another way of determining which sex is heterogametic is to make reciprocal crosses involving dominant and recessive alleles of *sex-linked characters*. These are characters induced by genes located in the sex chromosomes. As we shall see shortly, such reciprocal crosses yield differing results in the F_1 and F_2 generations. For that reason they provide a standard test to determine whether some previously unstudied mutation is in the sex chromosome, or in an autosome. At the same time, reciprocal crosses also serve to show which parent has only one X-chromosome and which has two.

Usually the cytological evidence of heterogamety is found before the genetic evidence (because of the scarcity of known sex-linked characters), but sometimes this situation is reversed. One such case served the useful purpose of demonstrating that, when interpreting what is seen through the lenses of a microscope, we must not jump to hasty conclusions. It was shown in 1908 that barred plumage in the fowl must be caused by a dominant sex-linked gene, and that hens must be heterogametic. Cocks should therefore have two sex chromosomes, females only one. Between 1908 and 1933, no fewer than seven different cytologists identified the sex chromosomes in the fowl as the largest one, and all found it to be paired in males, and unpaired in females. Unknown to some of them, the true situation was first recognized by the Japanese cytologist, Suzuki, in 1930. He found that the fowl's sex chromosome is not the largest, but the fifth in total length, and V-shaped. This was independently confirmed three years later by Russian cytologists who did not know of Suzuki's work, and by Yamashina and others since. A karyotype of the chromosomes of the fowl is shown in Fig. 7-1.

One difficulty about identifying the sex chromosome in birds is that it should be most evident in the anaphase of the first meiotic division of the heterogametic sex. However, maturation divisions in primary oöcytes of birds are difficult to find, because there is only one nucleus in each fully formed cell (*i.e.,* yolk of egg). Miller (1938) overcame this difficulty by making a female undergo spermatogenesis. This was done by removal of the (single) left ovary from a female chick at four days of age, with resultant formation of a testis-like organ on the right side. In sections through it, Miller could find many primary spermatocytes dividing, and, in some of these, could clearly see the V-shaped sex chromosome passing to one pole, without any partner going to the other (Fig. 7-2).

The first evidence that in some species the females could be heterogametic came from crosses made by Doncaster in 1906 between

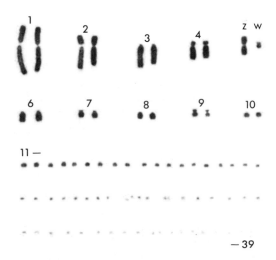

FIG. 7-1 Mitotic chromosomes of a female fowl. The Z-chromosome is fifth in size. Its partner (W) falls in size between chromosomes 8 and 9, but can be identified by banding. (Courtesy of S. E. Bloom, Cornell University.)

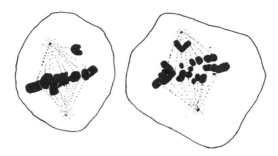

FIG. 7-2 Camera lucida drawings of dividing primary spermatocytes of an ovariotomized female fowl showing the unpaired, V-shaped sex chromosome moving ahead of the autosomes to one pole of the spindle. (From R. A. Miller in *Anat. Rec.*)

two types of the currant moth, *Abraxas grossulariata*, one of which, the *lacticolor* mutant, was pale in color. His findings were explicable only on the basis that the female moths were heterogametic and the *lacticolor* type was caused by a recessive gene in the sex chromosome. As a result of these first experimental crosses with sex-linked genes, species in which females are heterogametic are sometimes said to have the Abraxas type of sex determination.

Sex Determination

The simplest way to understand how gametes of two different kinds determine sex and yield a sex ratio of 1 : 1 is to consider the process as a backcross of a heterozygote to a homozygote. Fig. 7-3 shows how this works in mammals or in any other species in which the males are heterogametic. The male produces spermatozoa of which half carry an X-chromosome, and half carry a Y; all the female gametes carry one X; none carries a Y.

Times at which Sex Is Determined

An important point to note at this stage is the time at which sex is determined. In species with heterogametic males, sex must depend on whether the spermatozoön that effects fertilization carries an X-chromosome or not. Obviously, sex is determined in such cases at fertilization. It is not thus in species with the Abraxas type of sex determination, for in them the spermatozoa are equipotential. Obviously sex in such species is determined at reduction division in

Parents: X X ♀   X   X Y ♂

Sex chromosomes:

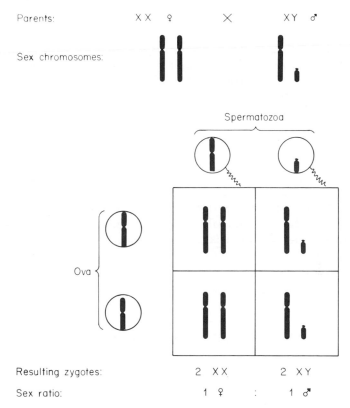

FIG. 7-3   Diagram showing how two kinds of spermatozoa and one kind of ova yield a sex ratio of 1 : 1.

oögenesis, presumably the first meiotic division of the primary oöcyte. Those that retain the X-chromosome in the first polar body will eventually form ova that can only become females.

Thus sex is determined at fertilization in mammals, but in birds it is determined at the first meiotic division of the oöcyte. In the fowl this occurs just before the ovarian follicle ruptures to release what we commonly call the yolk of the egg.

Complications

Before concluding happily that sex determination is simple, with X = ♂ and XX = ♀ (except for those non-conforming groups in which this situation is reversed), readers should be informed of these authors' innate propensity for hedging. Thus far we have ignored the autosomes in this chapter, but as we shall see later, they too are involved in sex determination, at least in some species, if not in all. In Drosophila, it is the ratio of X-chromosomes to autosomes that determines sex— not just the presence of one X or two. In some species the balance between male-determining and female-determining forces is so delicate that it can be upset by environmental influences. In others, genetically determined sex can be modified in part by hormones.

Then there is the Y-chromosome to consider. In most species it seems to be inert, but in Drosophila it carries one gene, at least. Why not say that the Y is what determines maleness, since (in species with heterogametic males) it does not occur in females? One reason is that there are some species with the XO type, having no Y, in which the solitary X clearly is the sex determinant. Such flies have even been found occasionally in Drosophila. These, too, are males but they are sterile. In that species the Y is apparently necessary for fertility, but in some others (of the XO type) it is not. Finally, in recent years, somewhat to our surprise, it has been shown that the Y-chromosome *is* necessary for complete differentiation as males in mice and men.

Control over Determination of Sex

When it became clear that in mammals two kinds of spermatozoa are produced, it was natural that biologists should try to find some way of separating the male-determining ones from the female-determining kind, and thus be able to produce, by artificial insemination, males or females as desired. Such an advance would undoubtedly be welcomed by the producers of dairy cattle, most of whom would much rather have heifers than bull calves. Unfortunately, as this is written, no effective scheme for controlling sex has yet been proven practicable.

First attempts were based on the possibility that, as the Y-chromosome is generally much smaller than the X, the two kinds of spermatozoa might differ slightly in size. If so, perhaps they could be separated by centrifugation. It was a nice theory, but, like many another, it didn't work out when careful experiments were made to test it. Another possibility was that the X-bearing and Y-bearing spermatozoa might differ in ability to survive in slightly alkaline or slightly acid media. If so, by an appropriate douche of the female reproductive tract before mating, matters might be arranged as desired. But when this scheme was tried with pigs at three different experiment stations, it didn't work either.

One possibility seemed that the two kinds of sperm cells might be separated by putting them in an electrophoretic current. Gordon found that under such conditions some spermatozoa (of rabbits) swim to the anode and some to the cathode. The latter, when tested by artificial insemination, produced an excess of males, while those from the anode yielded a predominance of females. Unfortunately, these results have not been confirmed. Matters can not yet be arranged to get the much-wanted excess of heifers in one species, or the oft-desired male heir in another.

Sex-linked Inheritance

A moment's reflection will tell us that genes in the X-chromosome will not segregate in exactly the same way as those we have considered in earlier chapters. All previous illustrations of Mendelian segregation were deliberately chosen because the genes involved were all autosomal. The ratios considered, therefore, applied equally to males, to females, and hence to populations including both sexes.

Since (with the usual rare exception) the Y-chromosome does not carry genes, it is obvious that any gene in the sex chromosome of a male mammal will go to his daughters, not to his sons. Those daughters need not show any effect if that gene is recessive and if they get its normal allele from their mothers. Conversely, when a female carries a recessive gene in one of her X-chromosomes, but not in the other, she can pass that *sex-linked* gene not only to half her sons, but also to half her daughters. In the latter it will produce no visible effect if it is completely recessive, and if the daughter gets from her sire the normal, dominant allele of that recessive gene. In the son getting such a gene from his dam, it is a different matter. He cannot get the protecting, dominant allele in the Y-chromosome from his sire. Accordingly, he shows the *sex-linked recessive character*. If all this sounds a bit confusing at first glance, a good example should clarify matters somewhat. Hemophilia provides just such a case.

Hemophilia in the Dog

Hemophilia prevents normal clotting of the blood by interfering with one of the several essential steps in that complex process. A specialist would say that victims of hemophilia A lack a substance variously known as the antihemophilic factor (AHF), antihemophilic globulin, or Factor VIII, which, in normal blood, interacts with blood platelets to accelerate the transformation of prothrombin to thrombin. This applies only to the classical hemophilia, now designated as hemophilia A to distinguish it from other similar conditions. It was known for over a hundred years in man before recognition of the same disease in dogs provided excellent material for experimental study of the defect.

Afflicted puppies usually show first symptoms of hemophilia at six weeks to three months of age. Common signs include lameness (from bleeding into the joints), large subcutaneous swellings (see Fig. 7-4) and, eventually, paralysis of one or more legs. Minor injuries resulting from normal play, which would cause only a slight bruise in most dogs, can prove fatal to hemophilic puppies. One died from bleeding that followed eruption of a tooth. Another suffered a fatal rupture of a blood vessel when picked up by its owner. In laboratory tests with hemophilic dogs, the time required for blood clotting (in very small samples) varied from 22 to 40 minutes, far longer than the range of 2.8 to 6.3 minutes found in normal dogs.

Hemophilia A in dogs is caused, as in man, by a sex-linked recessive gene. Since hemophilic puppies seldom survive to the age of reproduction, the usual kind of mating that yields hemophiliacs is that of a heterozygous female with a normal male. If we designate the causative gene as h and its dominant allele as H, the behavior of these

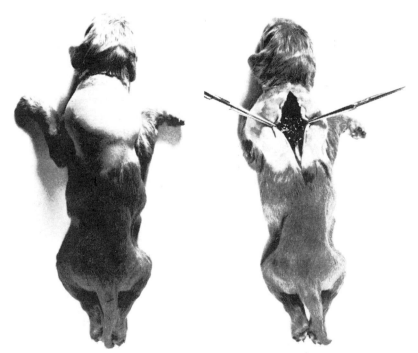

FIG. 7-4 Hemophilic puppy showing large swelling between the shoulders caused by internal bleeding. The skin was shaved after death. (From F. B. Hutt *et al.* in *J. Heredity.*)

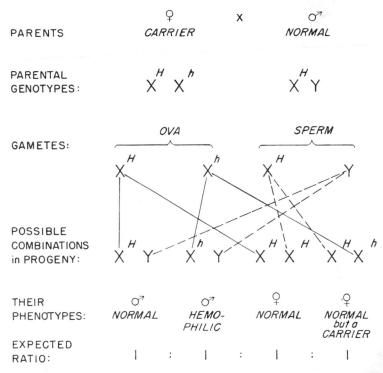

FIG. 7-5 Chart showing sex-linked transmission of hemophilia by a heterozygous female mated to a normal male. X^h = sex chromosome carrying gene for hemophilia; X^H = sex chromosome carrying the normal allele.

genes and resultant segregation of hemophilia in that sort of mating are as shown in Fig. 7-5.

As is evident in that illustration, litters from carrier females contain males among which half are normal, and half hemophilic. Half of their sisters carry the gene, but show no effect of it because they have also the protecting dominant allele H. The remaining females lack h. In eight such litters, the senior author and his associates found the following:

	Normal	Hemophilic
In females	32	none
In males		
Observed	23	17
Expected	20+	20−

These figures show a good fit of observed to expected ratios. A point to remember in genetic studies with dogs is that the excess of males over females at birth is higher in that species than in other domestic animals.

From such litters, the owner may wish to salvage some of the pups for further breeding, provided that he can avoid further trouble from the gene h. It is not necessary to discard the whole litter. Any full brother of a hemophilic pup that does not himself develop the disease must be $X^H Y$ and, hence, as safe for breeding as any unrelated male. It is impossible to say which females are HH (safe) and which Hh (unsafe) without a breeding test. That test, which would consist simply of mating the suspect to any male at all, and raising perhaps six or seven male offspring, would probably not be justified except with particularly valuable stock. The easiest solution would be to discard all sisters of hemophilic pups, but to use normal brothers, if desired.

An important point is that females in the dog and in man do not escape hemophilia just because they are females. In other words, that disease, unlike some others, is not one that can occur only in one sex. In both species hemophilic females have been known. In the dog, they were first produced by Brinkhous and Graham, who managed by repeated blood transfusions to raise hemophilic males to maturity, and then mated them with females known to be heterozygous for hemophilia. As expected, about half the puppies were hemophilic in both sexes. (See also Problem 7-2.)

Some of the authors' students, few of whom, if any, are supporters of monarchism, have a hazy impression from what they have read that hemophilia (in man) occurs chiefly in certain royal families of Europe and, therefore, can hardly be considered a serious disease! They are wrong on both counts. The Hemophilia Foundation estimated the number of human cases in the United States in 1950 at 40,000. From her studies of the disease in Illinois, Birch (1937) estimated that about 57 per cent of those afflicted die within 5 years of birth and that only about 5 per cent can expect to live more than 40 years. In many human pedigrees the condition disappears after three or four generations; hence, geneticists believe that new mutations (from H to h) are responsible for its persistence in human populations.

Christmas Disease in the Dog

One of the several different kinds of hemophilia that have been identified in comparatively recent years is Christmas disease, or hemophilia B. Because one team of investigators found it in a human family whose surname was Christmas, and published their results during Christmas week, it is not surprising that their name for the condition should persist, or that when the identical condition appeared in Cairn Terriers, it should still be called Christmas disease. Hemophilia B is less severe than the classical hemophilia A, which we have just considered. It is attributed to a diminished activity of Factor IX, a substance in the serum that is necessary for formation of plasma thromboplastin. Rowsell and his associates found Christmas disease in Cairn Terriers to be identical with that in man, and, like the latter, to be transmitted as a sex-linked recessive character. Their pedigree of the condition shows typical transmission of such a trait from carrier females to half their sons, with 3 affected females, from matings of hemophilic ♂ × carrier ♀ (Fig. 7-6).

The Tortoise-Shell Cat

As with autosomal characters considered in Chapter 3, sex-linked genes sometimes show incomplete dominance or recessivity, so that the heterozygote is clearly distinguishable from both homozygotes. Tortoise-shell cats that show both black and orange, sometimes with white spotting as well, are examples of such cases. They are commonly called calico cats when partly white.

Black and orange coats in cats are determined by the sex-linked alleles, o and O, and heterozygotes show both colors.[1] Table 7-1 shows the kinds of offspring that are possible among black, orange, and tortoise-shell cats.

This is a good point at which to introduce a new term: *hemizygous*. It is applied to sex-linked genes in the genotype of the heterogametic sex. In these cats, the females can be homozygous or heterozygous, but the males, having only one X-chromosome (and presumably no genes in their Y-chromosome) can be neither. They are hemizygous with respect to O and o or any other sex-linked genes. In species having heterogametic females, that sex can be similarly hemizygous for sex-linked genes.

As the matings in Table 7-1 show, there should never be any tortoise-shell male cats. How could the males be heterozygous for O and o? Yet there *are* tortoise-shell males. Geneticists tried for fifty years to account for them, and we shall see in Chapter 19 how they now do so.

[1]These alleles were designated as B and b in the first edition of this book, but some committee has since decided that o and O are better.

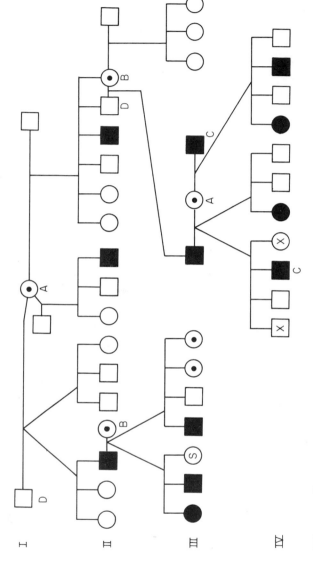

FIG. 7-6 Pedigree of Christmas disease (hemophilia B) in Cairn Terriers showing typical sex-linked, recessive inheritance. The letters A, B, C, and D identify four animals, all of which appear twice in the pedigree. (Courtesy of H. C. Rowsell.)

Male ☐ ○ Female
● Bleeders
⊙ Carriers
S Stillborn
⊠ Died in first week

144

TABLE 7-1

Crosses Involving Orange, Black, or Tortoise-Shell Cats

		Offspring	
Sire	Dam	♀♀	♂♂
Orange, O— × black, oo		Oo tortoise	o— black
Black, o— × orange, OO		Oo tortoise	O— orange
Orange, O— × tortoise, Oo		⎰OO orange ⎱Oo tortoise	O— orange⎱ o— black ⎰
Black, o— × tortoise, Oo		⎰oo black ⎱Oo tortoise	O— orange⎱ o— black ⎰

Hypotrichosis and Anadontia in Calves

The hairless calf shown in Fig. 2-7 owed its plight to a recessive gene which, because it caused the condition in a female as well as a male (Fig. 2-8), must have been autosomal.

By contrast, the hairless calf shown in Fig. 7-7 probably resulted from the action of a recessive, sex-linked gene. Some reviewers of this text may disagree with that statement, because it is based only on the three such calves, all males, that Drieux and his associates found in France. However, a good reason for inviting just such trouble is the hope that a little publicity here may eventually lead to further information about the character when, as is entirely probable, the same mutation occurs again in cattle, or in other mammals, elsewhere.

Drieux's calves were completely hairless at birth, even without eyelashes. One of them got a somewhat downy coat by 15 days; in another, bristly hairs appeared at 2½ months, and were sparsely distributed over the body by four months of age. Teeth were absent (anadontia) except in one calf which at six months showed two upper first molars. In these hairless calves, the tongue seemed longer than normal, possibly because the lack of teeth permitted it to loll out of the mouth. The epidermis was thin, with the outer layer keratinized, and hair follicles rudimentary. Because the calves could eat only soft food, attempts to rear them were abandoned at six months of age.

Evidence that this condition is caused by a sex-linked recessive gene came from four facts:

1. All three were males.
2. The mother of one was a daughter of the cow that produced the other two.

FIG. 7-7 Calf with hypotrichosis and anadontia, apparently sex-linked. (From H. Drieux *et al.* in *Rec. Méd. Vét.*)

3. These cows were crossbreds of Maine-Anjou-Normandy stock, while the sire was a completely unrelated Charolais.
4. His 180-odd calves by other cows in the same district were all normal.

The mother of one hairless calf also produced (by the same sire) two other bull calves, both normal.

Unknown to Drieux, who concluded that the abnormality is sex-linked, some additional support for his belief is provided by the fact that a very similar hypotrichosis combined with anadontia is caused in man by a recessive sex-linked gene. It has been reported from areas as far apart as Sind, in Pakistan, and Illinois. Further evidence is desirable, but it would appear that, with this case, and the two kinds of hemophilia considered earlier, we have at least three sex-linked mutations that have occurred both in man and in other mammals.

Sex-linked Barring in the Fowl

The examples considered hitherto in this chapter have all dealt with sex-linked characters of mammals, *i.e.*, of species having heterogametic males. To reverse our mental processes and to find what happens in crosses involving heterogametic females, let us consider reciprocal crosses between two breeds of fowls: Barred Plymouth Rocks and Rhode Island Reds. These have long been popular breeds in North America, and, in spite of competition from Leghorns, females from the crossing of these two breeds are still favorites for egg production in some Northeastern states.

The kind of barred plumage characteristic of Barred Plymouth Rocks, Cuckoo Leghorns, Coucou de Malines, and some other breeds results from white bands lacking in melanin across feathers that would otherwise be solid black. This barring is caused by a sex-linked gene, *B*, which is incompletely dominant. Two of these genes cause a wider white bar than one. As a result, purebred, homozygous Barred Rock males (*BB*) have lighter plumage than females of the same breed, because the latter must all be hemizygous (*B—*). Heterozygous males (*Bb*) are barred, but have narrow bars and darker plumage, like those of hens.

To produce good birds for egg production, the usual cross is that of Rhode Island Red ♂♂ × Barred Plymouth Rocks ♀♀ (Fig. 7-8). Strains bred for high egg production are readily available in both breeds, and the egg production of the crossbred F_1 progeny is raised still further by the hybrid vigor resulting from the inter-breed cross. Another big advantage is that this cross yields in the F_1 generation males that are genetically barred, and females that are not. The chicks do not show barring in their downy plumage, but those that will later develop barred feathers have a whitish spot in the occipital region of the head (Fig. 7-9). The non-barred, pullet chicks lack that spot. It follows that the chicks from this cross can be sorted according to sex as soon as they come out of the incubator, and as fast as they can be handled. Pullet chicks go to poultrymen who raise them for egg production and want no males. The latter are raised for meat.

It will be noted in Fig. 7-8 that the F_1 males show the phenotype of their dam, while the F_1 females have that of their sire. This is characteristic of one of the two ways in which crosses can be made between parents differing in sex-linked alleles. Such crosses are sometimes said, therefore, to show *criss-cross inheritance*.

The reciprocal cross (Barred Plymouth Rock ♂♂ × Rhode Island Red ♀♀) yields different results. All the F_1 progeny are barred (Fig. 7-10).

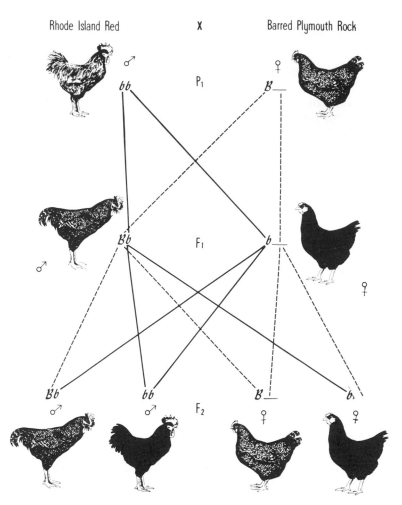

FIG. 7-8 Sex-linked transmission of barring in the cross of Rhode Island Red ♂ × Barred Plymouth Rock ♀.

FIG. 7-9 Chicks from the cross of Fig. 7-8, with two showing the head-spots which tell that they will be barred, and hence males; the unspotted, non-barred chicks must be females.

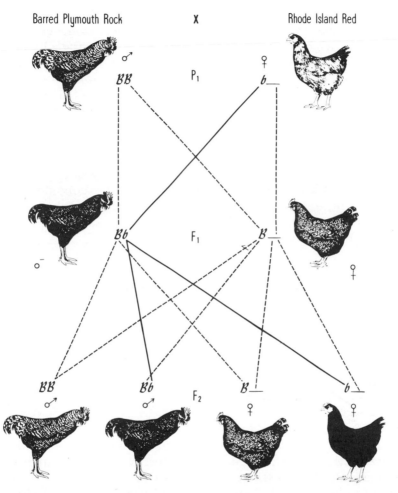

FIG. 7-10 The reciprocal cross to that shown in Fig. 7-8, yielding different results in the F_1 and F_2 generations.

In practice, poultry breeders seldom raise F_2 generations of these crosses, but geneticists have done so, and have found that, as should be expected, the two kinds of crosses yield different results in F_2. One has barred and non-barred birds in equal numbers in both sexes (Fig. 7-8); the other yields 3 barred : 1 non-barred, and all of the latter are females.

A sex-linked cross utilizing a different pair of sex-linked alleles is popular in Britain. Like the one just considered, it utilizes Rhode Island Red males—not because they are non-barred, but because they are gold (*ss*). When these are mated with Light Sussex females, which are silver (S—), criss-cross inheritance is again the rule and the F_1 males are silver (Ss), the females gold. These can be sorted at hatching just as readily as the non-barred and barred chicks of the cross considered above.

Differences between F_1 generations from reciprocal crosses provide a standard test for sex-linked inheritance in species with which it is possible to make such experimental matings. Obviously the student of human heredity might envy his colleagues who work with mice, fowls, or fruit flies; and no one has yet made sex-linked crosses with elephants. Such experimental matings would ordinarily not be feasible with mutations that cause death or disability before the age of reproduction, as do the defects of dogs and calves considered in this chapter. Fortunately, it is not too difficult to recognize what mutations are sex-linked by their incidence in the two sexes and by their transmission in pedigrees.

Autosexing Breeds in the Fowl

Many poultrymen would like to be able to identify sex of their chicks in pure breeds at hatching without the necessity of having to cross two different breeds. This is possible in certain *autosexing* breeds, the first of which was the Cambar, produced by Punnett in the late 1920's.

Most autosexing breeds utilize the gene for barring, and they are based on the fact that the difference in color between BB♂♂ and B— ♀♀ is accentuated when other genes are present that reduce the pigmentation of the plumage. In such cases the difference between the sexes can even be recognized in the down of newly hatched chicks. Recognizing the potential value of such a means of identifying sex, Punnett added B from Barred Plymouth Rocks to Silver Campines, and eventually eliminated the b allele of that breed to produce pure Silver Cambars. In this autosexing breed the BB males are easily distinguished from the B— females (Fig. 7-11).

Other autosexing breeds—Legbars, Dorbars, Ancobars, etc.— were soon produced, but none has yet been utilized on a large scale by the poultry industry. One reason is that they have not yet been bred to lay as well as the good strains of other breeds. Another is that a fast and accurate method of determining sex in newly hatched chicks by examination of the cloaca came into general use soon after the advent of the Cambars. If that procedure had not been discovered, the autosexing breeds would undoubtedly have been improved in economic traits, and would have made a very important contribution to the poultry industry.

FIG. 7-11   Cambar chicks, with down colors telling that they are (from left to right): ♀, ♂, ♀, ♂. (Courtesy of Michael Pease.)

The *K-k* Alleles of the Fowl

One factor preventing the use on a large scale of sex-linked genes for identification of sex of chicks at hatching is that none has yet been found that can be used with complete satisfaction in crosses involving White Leghorns. This is so because the dominant white of that breed (*I*) is epistatic to *B* and *S*, and because other sex-linked mutations of the species are not desirable. A helpful exception, and one with potential promise, is provided by the *K-k* alleles, which were first shown to be sex-linked by the Russian geneticist, Serebrovsky.

These determine the rate of feathering during the first few weeks after hatching. By eight or ten days of age, chicks of the genotype *kk* (♂) or *k*— (♀) show little tails perhaps half an inch long, and wing feathers extending to the end of the body. By contrast, chicks that are *KK*, *Kk* (♂), or *K*— (♀) have no tails at that age, and very short wings (Fig. 7-12).

The cross of rapidly feathering ♂ × slowly feathering ♀ yields slow males (*Kk*) and rapid females *k*—, and such chicks can be sorted with an accuracy exceeding 95 per cent at eight days of age. Hatcherymen cannot afford to hold many chicks to that age. However, the slow and rapid types can be differentiated with the same degree of accuracy right at hatching (by a difference in their first wing feathers), and it may be possible to raise that degree still higher by breeding to accentuate the difference between them.

FIG. 7-12 Chicks showing at nine days slow feathering (left) and rapid feathering (right).

The predominant egg-producing hen in North America today is a cross derived from two or more strains of White Leghorns. Most Leghorns have rapid feathering, but strains with slow feathering exist, and sex-linked crosses involving K and k are now utilized to sex chicks at hatching. Apart entirely from that use, the gene k is already making an important contribution to the poultry industry. It has been incorporated into many of the parent strains that produce the world's broilers because it provides some insurance against the unwanted "bare-backs" at market age. To geneticists the K-k alleles are of interest as examples of single genes that affect function (growth) rather than form.

Other Sex-linked Characters

In his great compendium, McKusick (1978) needs over 100 pages to review the conditions known (or reported) to be sex-linked in man. Almost a third of those pages deal with variant forms of the enzyme, *glucose-6-phosphate dehydrogenase*, conveniently dubbed G6PD. About 30 sex-linked genes are known in the mouse (see Map, Figure 10-5), and four in the dog, but, as this is written, only one specific sex-linked gene has been demonstrated in any domestic animal bigger than a dog. It is considered in Chapter 9.

The pink-eyed cinnamon canaries studied by Durham and Marryat provided (in 1908) the first evidence of sex-linked inheritance in birds. Differing degrees of sex-linked albinism have since been demonstrated in budgerigars, turkeys, and the fowl. Sex-linked genes are also known in pigeons. Some of those known in the fowl are considered in Chapter 10.

Problems

7-1 Attempting to separate ♂-determining and ♀-determining spermatozoa of the rabbit by electrophoresis, M. J. Gordon found that among 87 offspring from sperm that had migrated to the anode there were 25 ♂♂, and among 80 rabbits from cathode sperm there were 51 ♂♂. Using a four-fold contingency table, apply the χ^2 test to these data and determine from the value found for P whether or not further research in this field looks promising. (Then see p. 138).

7-2 In a suspect litter of eight puppies, four of which are ♂♂, when one pup shows hemophilia, how many others would be expected to do likewise eventually?

7-3 How many of those that should not become hemophilic could be identified at once? How?

7-4 How many of the litter that are not hemophilic at maturity might be expected to transmit the disease to their offspring?

7-5 Could these last be identified? Wholly? Partly?

7-6 How many of the non-hemophilic dogs could be identified for certain at maturity as unable to transmit hemophilia, and how?

7-7 Would the transmitters show it in their own pups, or would it skip a generation?

7-8 What is the probability that the father of the hemophilic pups would sire more of them in three large litters from the same bitch? From unrelated bitches of a different kennel?

7-9 What is the probability that the mother of this litter would produce other hemophilic pups in three large litters by the same sire? By an unrelated sire of another breed?

7-10 Would you expect recessive autosomal mutations to show up (as "sports") more quickly after their occurrence than sex-linked mutations, or *vice versa*? In either case, why?

7-11 In which sex would you expect sex-linked mutations to appear first in turkeys? In rabbits?

7-12 When a black (♀) cat produces a litter containing one tortoise-shell kitten and three black ones, what would you know about (a) the color of the sire? (b) the sex of the black kittens?

7-13 When a tortoise-shell cat brings home her litter of tortoise-shell, black, and orange kittens, by what criteria might it be possible to incriminate the orange tom across the street as the probable sire?

7-14 By what criterion could the suspect orange tom in Problem 7-13 be exonerated?[2]

7-15 White eye (w) is a sex-linked mutation, recessive to the normal red eye of the fruit fly, Drosophila, in which males are heterogametic. Make a diagram of

---

[2]The answer to this question is not, as one bright lad wrote "If he had been castrated".

the kind of cross you would have to make with this mutation to demonstrate criss-cross inheritance in the F_1, and compare your expectations in the F_2 generation with those shown in Fig. 7-8.

7-16 If Marquise and Fauvette, the mothers of Drieux's hairless and toothless calves, had been available for further breeding, by what kinds of matings and with what results might one have proven beyond doubt that the mutation was sex-linked, and not autosomal?

7-17 Coleman *et al.* (*J. Heredity* **51**, p. 158, 1960) studied a peculiar tremor in Bronze turkeys which proved to be hereditary and was called "vibrator." Affected birds had normal viability, and, when mated *inter se*, produced progeny all of which were vibrators. However, when vibrator toms were crossed with normal hens, the female offspring were all vibrators, but the males were normal. How would you explain that?

7-18 In North Carolina, Graham and his associates investigated the incidence in certain families of a deficiency of phosphorous in the blood, which is associated with a type of rickets resistant to therapy with vitamin D. Fourteen affected fathers married to unaffected wives sired 21 daughters, all of whom had hypophosphatemia, and 16 sons, none of whom was affected. What is the genetic basis for the condition? How does it differ from that for hemophilia?

7-19 When an undesirable kind of hereditary albinism appeared in his Mammoth Bronze turkeys, the breeder appealed to this author for help in eliminating the defect. Of five suspect toms provided for testing, three were found to carry the causative gene, and when mated with unrelated normal females, they sired 229 offspring. Among these were 45 albinotic ones, all females. The owner could test his toms in single-male matings, but did not want to discard any more of his breeding stock than was necessary to eliminate the gene.

What would you tell him? What should he discard? What could he safely keep?

Selected References

DRIEUX, H., M. PRIOUZEAU, G. THIÉRY, and M.-L PRIOUZEAU. 1950. Hypotrichose congénitale avec anodontie, acérie et macroglossie chez le veau. *Recueil de Méd Vét.* **126**: 385–99. (Details of the naked calves.)

GEDDES, P., and J. A. THOMPSON. 1889. *The Evolution of Sex.* London: Walter Scott. New York: Scribner's. (Good browsing to learn how much and how little was known of the subject before sex chromosomes were discovered.)

GORDON, M. J. 1957. Control of sex ratio in rabbits by electrophoresis of spermatozoa. *Proc. Nat. Acad. Sci.* (Washington) **43**: 913–18. (Source of material discussed in text and problems.)

GRAHAM, J. B., V. W. McFALLS, and R. W. WINTERS. 1959. Familial hypophosphatemia with vitamin D-resistant rickets. II: Three additional kindreds of the sex-linked dominant type with a genetic analysis of four such families. *Amer. J. Human Genet.* **11**: 311–32. (Source of data in Problem 7-18.)

HUTT, F. B., C. G. RICKARD, and R. A. FIELD. 1948. Sex-linked hemophilia in dogs. *J. Heredity* **39:** 2–9. (Descriptions and genetic evidence.)

McKUSICK, V. A. 1978. *Mendelian Inheritance in Man.* 5th ed. Baltimore. Johns Hopkins Univ. Press. XCI + 975 pp. (Indispensable; hereditary diseases in man, with references thereon.)

MILLER, R. A. 1938. Spermatogenesis in a sex-reversed female and in normal males of the domestic fowl, *Gallus domesticus. Anat. Rec.* **70:** 155–89. (Successful quest for cytological evidence that the female is heterogametic.)

PUNNETT, R. C., and M. S. PEASE. 1930. Genetic studies in poultry. VIII. On a case of sex-linkage within a breed. *J. Genetics* **22:** 395–97. (The autosexing Cambars.)

ROWSELL, H. C., H. G. DOWNIE, J. F. MUSTARD, J. E. LEESON, and J. A. ARCHIBALD. 1960. A disorder resembling hemophilia B (Christmas disease) in dogs. *J. Amer. Vet. Med. Assoc.* **137:** 247–50. (Account with pedigree of condition discussed in this chapter.)

STERN, C. 1973. *Principles of Human Genetics.* 3rd ed. San Francisco. W. H. Freeman & Co. (A good source of information about sex-linked traits and others in man.)

CHAPTER

Sex-limited and Sex-influenced Traits

After reviewing in the previous chapter examples of differential manifestations of inherited characters caused by genes in the X-chromosome, it seems desirable to point out that, apart entirely from sex-linked traits, some inherited characters can be shown only by one sex. They are thus *sex-limited*. Others appear in both sexes, but more often, or with more complete expression, in one sex than in the other. These are *sex-influenced*. Sex-limited and sex-influenced characters can be caused by autosomal genes, but sex-linked traits cannot.

Y-borne Genes

Any gene carried in the Y-chromosome of a heterogametic male would be transmitted to all his sons, but only to his sons. The trait induced by such a gene would be sex-limited. A few such cases have been demonstrated in certain fishes, and some are known in insects, but none has been incontrovertibly demonstrated in any species above the fishes.

For many years it was believed that certain pedigrees of peculiarities in man showed Y-borne characters, but as a result of a thorough examination of these by Stern (1957), it is clear that further evidence must be adduced before we can consider any human trait to be caused by a Y-borne gene. Even the celebrated *ichthyosis hystrix* of John Lambert, whose scaly skin, said to have "looked and rustled like the bristles or quills of an hedge-hog, shorn off within an inch of the skin," was exhibited before the Royal Society of London in 1731, must now be banished from the texts on genetics in which for many years it had pride of place as the outstanding example of a Y-borne, or *holandric*, trait in man. Investigations by Penrose and Stern have shown that, while there were undoubtedly four males thus affected, it is not clear that the scaly skin was transmitted only to males, and to all males in the family concerned.

All of this does not mean that there can be no genes on mammalian Y-chromosomes, but only that, if such genes exist, they are likely to be rare, and proofs will have to be good.

Some evidence suggests that a locus in the Y-chromosome known as a histocompatibility locus may determine, not only the compatibility of different tissues, but also maleness. There is also the possibility that the Y-chromosome carries genes that are (so far) unrecognized because they occur in only one form, *i.e.*, have not mutated.

Sex-limited Characters

Familiar examples of sex-limited characters in domestic animals are provided by genetic differences in ability to produce milk, butterfat, and eggs. These are caused by many genes, sometimes called *polygenes*, or *multiple factors* (see Chapter 12), but obviously can be expressed only in females, never in males. Selection for higher productivity is difficult enough because (a) many genes are involved, and (b) the performance (phenotype) is greatly influenced by the environment, without the further complication (c) when it cannot be measured in males. Modern poultry breeders must select not only for more eggs, but also for two-ounce eggs, for the desired shape, for thicker shells, for dense albumen, and for freedom from blood spots. All these objectives are influenced by genes, but they can be measured only in females.

Such restriction to one sex of the expression of some hereditary trait can be said to have resulted primarily from action of the sex chromosomes which determined sex, but debate on that point is unnecessary. Among polygenes affecting production of milk or eggs there are probably some in X-chromosomes and many in autosomes. Some of the examples that follow deal specifically with autosomal traits that can be shown only by one sex.

Cryptorchidism in Dogs

This condition, the failure of one or both testes to descend through the inguinal canal into the scrotum, is an undesirable defect, partly because cryptorchids are temperamental, difficult to handle, and sterile, but also because they are disqualified in the show-ring. Frequently the undescended testis becomes tumorous, and the dog is feminized. Unilateral cryptorchids, often incorrectly called monorchids, are fertile. The defect occurs in swine, horses, and other mammals, but is perhaps most commonly found in dogs.

The defect occurs in many breeds of dogs, and is said by one writer to be more common in those having short skulls. In Germany, Härtl found that 23 percent of 168 male Boxers in 57 litters were cryptorchids, and that the parents were all related. He attributed the high incidence of the defect to inbreeding and resultant concentration of the causative gene or genes. Although some reports suggest that cryptorchidism is a simple recessive autosomal (but sex-limited) trait, the evidence is hardly conclusive to prove that only a single gene (when homozygous) is responsible. One difficulty preventing better understanding of the genetic basis is that genotypes of females can be determined only by progeny tests. Another is that most dog breeders are unwilling to reveal information about hereditary defects in their kennels.

Veterinarians are sometimes asked to overcome this defect by surgery. Such an operation is not *dysgenic* (*i.e.*, bad for the race) if done to correct bilateral cryptorchidism, because the damage done by the higher temperature of the body cavity is irreparable and the dog would be just as sterile after the operation as before. However, correction of a unilateral case *is* dysgenic, because the dog would appear to be normal, and would be fertile like any unilateral cryptorchid.

At one time the British Veterinary Association recommended that dog breeders should not breed from any (unilateral) cryptorchids, their litter-mates, their parents, or any normal males sired by unilateral cryptorchids. This advice (based on the belief that the defect is caused by an autosomal recessive gene) was later withdrawn when the British Kennel Club, unsatisfied that the condition is hereditary, stopped exclusion of cryptorchids (unilateral) from breeding.

While there seems to be more evidence that the defect is hereditary than that it is not, further information from well-conducted genetic tests (selection for and against it) is desirable.

Sex-limited Disorders of Reproduction in Cattle

The best known of these is probably the so-called white-heifer disease which occurs in white Shorthorn heifers with a frequency estimated by different investigators at 10 to 50 per cent. In this condition, the Müllerian ducts of the embryo do not develop and, as a result, the uterus and vagina are abnormal. The genetic basis is not yet known, and the same abnormality can occur in red and roan Shorthorns (and in other breeds), but its frequent association with the white color, which is clearly inherited, shows that white-heifer sterility is genetically determined. White males are unaffected. This peculiar sex-limited trait is considered further in Chapter 22.

Impotency of Friesian bulls resulting from abnormality of the paired *retractor penis* muscles and consequent failure of the penis to straighten out was considered by de Groot and Numans (1946) to be a simple autosomal recessive trait. Their 22 affected registered bulls all traced to one or both of two bulls, but with a single exception all had unaffected sires. Until the genetic nature of this sex-limited defect was recognized in Holland, it had been customary to overcome the abnormality by myectomy of the muscles responsible.

Another sex-limited hereditary abnormality is inguinal hernia, the occurrence of which in swine was discussed in Chapter 5. One must not conclude from these few examples that *all* hereditary abnormalities of the reproductive system must be limited in expression to one sex. The hereditary genital hypoplasia, once so frequent in Swedish Highland cattle, affects both sexes, but is more common in cows than in bulls.

Sterile Friesian Bulls with Knobbed Spermatozoa

One kind of sex-limited sterility results from an abnormality which affects only males. It was first recognized in Friesian cattle in Holland by Teunissen, but later found among descendants of animals imported from that country to England in 1936. In spermatozoa of the sterile bulls, a variable but high proportion shows an eccentric thickening of the *acrosome*, or apical body, of the head, which has led to their being designated as "knobbed." Some spermatozoa of this kind, but fewer than one per cent, are sometimes found in the semen of fertile bulls, but the difference between such animals and the bulls homozygous for this particular genetic defect is so great that accurate identification of the latter type is not difficult. Most of their spermatozoa show the abnormality.

Among 17 such sterile bulls, the pedigrees for which were studied by Donald and Hancock, all but one were sons or grandsons of one bull, A, and most traced on the maternal side to another bull, B. Both A and B were derived from the original importation of 1936. They themselves were fertile, and apparently most of their sons (which would be from crosses of the imported bulls to unrelated stock) were also fertile, but at least five sons of A were proven to have transmitted the defect. Donald and Hancock concluded that this type of male sterility is found in bulls homozygous for an autosomal gene, *kn*, which causes the knobbed acrosomes. As at least one bull with the same defect has been found in Red Danish cattle, the abnormality is not confined to the Friesians.

Another sex-limited character in cattle is the peculiar "streaked hairlessness," to be described in Chapter 9. It exemplifies another reason why a genetic trait may show in only one sex.

The Hen-feathered Sebright Bantams

A sex-limited character which, unlike those mentioned earlier, is not dependent upon sex differences in the anatomy or physiology of the structure affected is demonstrated by Sebright Bantams. In most breeds of chickens, males differ from females in having longer, pointed, fringed feathers in the neck, wing bow, and saddle. They also have long sickle feathers that fall to both sides of the tail (Fig. 8-1). In some breeds the difference between the sexes is heightened further by different colors and patterns.

In the diminutive Golden and Silver Sebrights, and in some strains of Campines, males have plumage identical in structure, color, and pattern with that of females (Fig. 8-2). This henny feathering is caused by a single, autosomal, dominant gene, *Hf*, which, like many

FIG. 8-1 Red Jungle Fowls showing sex dimorphism in plumage that is characteristic of most breeds of the fowl. The cock shows elongated, pointed feathers in the neck, back, and saddle, and very long sickle feathers on each side of the tail.

others, is sometimes incompletely dominant in young heterozygous males. Some of these last may have only a few feathers of the female type in their first adult plumage, but these will show typical hen feathering in the next year, after their moult.

Various experiments, including castration, reciprocal transplantation of gonads between henny males and normal ones, treatment with male hormones, and grafts of skin, which eventually revealed the physiological basis for the peculiar non-conformity of the hen-feathered Sebrights and Campines, are reviewed in detail elsewhere (Hutt, 1949). Contrary to earlier beliefs, the testes of those males and the hormones they secrete do not differ from those in other breeds.

FIG. 8-2 Golden Sebright Bantams, showing hen-feathering in the cock (left) as in the female. Contrast with Fig. 8-1.

Paradoxical as it may seem, hen-feathered males need a normal male hormone to show the female type of plumage. If they are castrated, the feathers grown subsequent to that operation are of the male type, or, more accurately, like those of capons, which are of the same structure as that of normal males, but longer. The henny feathers result from a genetic variation affecting only the feather follicles, and causing them to produce plumage of the female type, so long as the milieu in which the feathers develop includes a male hormone. Danforth found that grafts of skin from hen-feathered Campine males to Leghorn males grew henny feathers.

From all these genetic and physiological studies, the facts emerge that Sebrights (of both sexes) carry an autosomal gene which causes the feather follicles to grow feathers of the female type so long as male or female hormones are present. As hen feathering is normal for females of any breed, the divergence from normal (*i.e.*, the effect of the gene *Hf*) in Sebrights or Campines can be shown only by the males in those breeds. The trait is thus sex-limited.

Sex-influenced Horns in Sheep

Some genetic traits are neither sex-linked nor sex-limited but are manifested differently in the two sexes, or more often in one sex than

in the other. A good example is provided by the type of horns characteristic of the popular Dorset Horn sheep. Both sexes have horns, but
those of the ram are much heavier and more coiled than those of the
ewe (Fig. 8-3).

When Dorset Horns (H^1H^1) are crossed with hornless breeds of
English origin (HH), such as the Suffolk, Shropshire, and Southdown,
the F_1 progeny (H^1H) show horns in males (Fig. 8-4), but not in
females. It is sometimes said that the polled condition is dominant in
females and recessive in males, but, because the horns of those F_1
heterozygous males are much smaller than those of Dorset Horn rams
(Fig. 8-3) and even smaller than in ewes of that breed, it would seem
more correct to say that the polled condition is dominant, but incompletely so in males. While that interpretation is perhaps an over-
simplification, because a few of the heterozygous ewes develop scurs
(vestigial horns), the cross of Dorset Horns with polled breeds provides
in the F_1 generation a good example of a genotype that is manifested
much more prominently in one sex than in the other. Some sheepmen
refer to the small horns of heterozygous males as scurs.

Atresia of the Oviduct in the Fowl

An abnormality studied by Finne and Vike (1949) in Norway is
remarkable because it is lethal to females, and apparently may affect
males slightly, but without lowering their viability. Among daughters
of one male, 50 pullets were normal layers, but 43 were unable to lay.
Dissection of 36 of the latter showed in every case an interruption of

FIG. 8-3 Dorset Horns showing heads and horns of good type in both ewe and
ram. (Courtesy of J. R. Henderson and the Continental Dorset Club, of
Hickory, Pennsylvania.)

FIG. 8-4 Top; Small horns typical of F_1 rams from the cross: Dorset Horn ×
polled breed. Such horns often have to be cut, as in this case, to prevent their
growing into the head. (Courtesy of D. Hogue and Cornell University.) Bottom:
A Polled Dorset ewe. (Courtesy of the Continental Dorset Club, of Hickory,
Pennsylvania.)

the oviduct in the region of the isthmus. As unlaid eggs were returned
to the body cavity the accumulation of such material upset the normal
balance of the body, and the birds assumed an upright posture like
that of penguins. The associated peritonitis caused death.

The sire of these birds were found on autopsy to have an interruption of the right *vas deferens*, but, as the left duct was normal, he was
fully fertile. Among his four sons tested, two had only normal
daughters, a third had two daughters with abnormal oviducts among

26 (not unusual), but the fourth sired 54 daughters among which were 24 non-layers. All but one of these were dissected, and 21 were found to have the typical atresia of the oviduct.

Unfortunately only one male was dissected. The record suggests that he was heterozygous for a dominant gene which, when transmitted to half his daughters, caused in all or most of those receiving that gene a complete atresia of the oviduct. Presumably the gene would also be transmitted to half the sons, although only one of four was found to transmit the defect. Further studies of the effects in males and in homozygotes of both sexes are desirable.

Other Sex-influenced Traits

Hereditary baldness in man is much more frequent in men than in women, and is commonly said to be dominant in males but recessive in females. Asdell and Smith concluded that the beard of goats is caused by a gene which is similarly sex-influenced. Crooked keel (breast bone), a common defect in chickens and turkeys, is caused by both genetic and environmental influences, but is more common in males than in females. The same applies to the undesirable "breast blisters" commonly found in chickens kept beyond three months of age.

The sex difference in the incidence of these defects in poultry is attributable in part to earlier retardation of the rate of growth in females, also to their lighter weight, and, in the case of breast blisters, to their more rapid growth of protective feathers.

Problems

8-1 Investigators are studying genetic variations in the transferrins of the blood serum, and the β-lactoglobulins of the milk in cattle. Which of these two would you consider the more amenable to genetic analyses, and why?

8-2 A mutation causing shells of hens' eggs to be greenish-blue, rather than the normal white or tinted color, is caused by a single dominant gene. Lone mutant fowls laying such eggs have recently occurred in England and Sweden. With one such hen, by what matings and with what evidence could you determine in one generation whether the causative gene is autosomal or sex-linked?

8-3 Koch reported a female spaniel that had in three litters (two by her full brother, the other by the brother of a cryptorchid male) 7 ♀♀ : 7 ♂♂, the latter all cryptorchid. What do these figures suggest about the genotypes of (a) the bitch? (b) the two sires?

8-4 If these sires had shown no sign of the defect, what ratio of normal to cryptorchid would have been expected in their seven sons (a) if the bitch were heterozygous for the defect? (b) if she were homozygous? (c) Which is the more likely?

8-5 Could such a bitch be of any use to a breeder seeking to eliminate cryptorchidism from his kennel? If so, how?

8-6 To identify most easily the genotype of a female suspected of carrying a gene for cryptorchidism, to what kind of dog should she be bred?

8-7 By what kind of analysis can a hereditary defect in cattle, which, like the knobbed spermatozoa, is autosomal, sex-limited, and expressed only in males, be recognized as genetic in origin when it makes the affected bulls sterile?

8-8 How would you eliminate this same defect from a breed? (Remember that it is sex-limited.)

8-10 In purebred Barred Plymouth Rocks, the plumage of males is ordinarily lighter in color than that of females (see Chapter 7). Is this attributable to sex-linked or to sex-influenced heredity? Is the narrower white bar of the female of that breed a sex-limited trait?

8-11 If skin were grafted from a barred female to a purebred Barred Plymouth Rock male, would it grow feathers with narrow white bars like those of the donor, or with wider ones like those of the host?

8-12 Polled Dorset sheep have been bred by sheep breeders who like all characteristics of the Dorset Horns except their horns. As introduction of the poll gene, H, from some other breed must be followed by several generations of backcrossing to pure Dorsets to restore a high proportion of Dorset "blood," a mutation from H^1 to H in pure Dorset Horns could be very useful. One of these occurred in a Dorset Horn ram at the University of North Carolina. How would you recognize such a mutation, and how would you use such a ram to develop a Polled Dorset flock?

8-13 Skin from hen-feathered Campine ♂♂ grafted on White Leghorn ♂♂ by Danforth grew feathers of the henny type. Would you expect feathers of that kind in the reciprocal graft, using males only? Why, or why not?

8-14 What type of feathers should grow on a graft from a Sebright ♀ to a Sebright ♂? To a Leghorn ♂?

Selected References

DE GROOT, T., and S. R. NUMANS. 1946. Over de erfelijkheid der impotentia coeundi bij stieren. *Tijdschrift voor Diergeneeskunde* **71**: 373–79. (In Dutch; original account of sex-limited impotency considered in text.)

DONALD, H. P., and J. L. HANCOCK. 1953. Evidence of gene-controlled sterility in bulls. *J. Agric. Sci.* **43**: 178–81 (About the defective acrosome mentioned in this chapter.)

FINNE, I., and VIKE, N. 1949. En ny subletal faktor hos høner. *Tidsskr. norske Landbr.* **56**: 60–76. For translation into English see *Poultry Sci.* **30**:

455–65, 1951. (A full account of the sex-limited atresia of the oviduct discussed briefly in this chapter.)

HUTT, F. B. 1949. *Genetics of the Fowl.* New York. McGraw-Hill Book Co., Inc. (Chapter 5 reviews genetic and physiological studies of hen-feathering in males.)

STERN, C. 1957. The problem of complete Y-linkage in man. *Amer. J. Human Genet.* **9:** 147–66. (The original address was entitled "On porcupine skin and hairy ears, or, The alleged sins of the Y chromosome.")

CHAPTER

9

Lethal Genes

The Classical Yellow Mouse

A few years after the rediscovery of Mendel's law, when students of heredity were busily fitting their mutations into $3:1$ ratios, the French zoologist, Cuenot, found that when yellow mice were mated *inter se* they never bred true to type but yielded yellow and non-yellow progeny in the ratio of $2:1$. Backcrosses of yellow × non-yellow yielded a ratio of $1:1$. After further experimentation, it was eventually realized that all yellow mice are heterozygous, and that the zygotes homozygous for yellow all die at an early stage of gestation. Litters from such matings are smaller than normal. Cuenot's yellow mice thus provided the first demonstration that a gene could be lethal in the homozygous state.

Assigning the symbol A^y to the gene for yellow coat, and a to its recessive allele (for any color other than yellow), it is clear that the mating of two yellow mice, $A^y a \times A^y a$, should yield the familiar ratio of 1 homozygous yellow ($A^y A^y$) : 2 heterozygous yellow ($A^y a$) : 1 non-yellow (aa). As the first of these classes is not viable, the observed phenotypic ratio is 2 yellow : 1 non-yellow. Some years ago, a summary of results obtained by several investigators yielded the following totals for the two kinds of matings possible with yellow mice:

Parents	Ratio	*Progeny* Yellow	Non-yellow
Yellow × yellow	2 : 1	2,386	1,235
Yellow × non-yellow	1 : 1	2,378	2,398

Better fits of observed ratios to those expected are not likely to be found in such large numbers, nor are we ever likely to have such extensive data for a lethal gene in any mammal bigger than a mouse. Lethal colors in domestic animals are discussed in Chapter 22.

The Abundance of Lethals

With the exception of the "bull-dog" calves produced by Dexter cattle, genetic lethal characters in domestic animals were not recognized as such until after 1920. By 1934 a review of those then known in cattle, horses, sheep, swine, dogs, and the domestic fowl included 30 cases, among which were 11 in cattle and 5 in the fowl. The list given in Table 9-1 (which is not complete) includes lethals for which the genetic basis has been adequately demonstrated, and among these are

TABLE 9-1
Some Monogenic Lethal Characters of Domestic Animals, All Simple
Autosomal Recessives, Unless Marked D (dominant) or S (sex-linked)[a]

Cattle		*Sheep*	
D	Achondroplasia, dominant		Muscle contractures
	Achondroplasia, recessive		Paralysis of hind legs
	Epithelial defects		Cortical cerebellar atrophy
	Hairless		Lethal grey
	Acroteriasis (amputations)		Rigid fetlocks, malformed skull
	Paralysis of hind legs		Amputations
	Muscle contractures		Adactyly (defective hooves)
	Short spine		Agnathia, pharynx occluded
	Congenital dropsy		Myodystrophia
	Short lower jaw			
	Cerebellar hypoplasia	*Domestic Fowl*	*Symbol*
	Missing phalanges	D	Creeper	*Cp*
	Hydrocephaly, internal		Chondrodystrophy	*ch*
	Congenital spasms	D	Cornish lethal, short limbs	*Cl*
	Prolonged gestation		Amaxilla	*mx*
	Adenohypophyseal aplasia		Missing mandible	*md*
DS	Streaked hairlessness		Wingless, lungless syndrome	*wg*
	Ichthyosis congenita		Diplopodia	*dp*
	Keratogenesis imperfecta		Diplodia 2	*dp-2*
	Hypotrichosis and anadontia		Splitfoot	*sf*
D	Fused nostrils, malformed skull		Micromelia	—
	General ankylosis		Short upper beak (semi-lethal)	*sm*
	Atresia ilei	D	Apterylosis (semi-lethal)	*Ap*
	Mannosidosis	S	Naked (semi-lethal)	*n*
	Dermatosparaxie		Dwarfism (thyrogenous)	*td*
			Stickiness	*sy*
Horse			Talpid	*t*
	Atresia coli		Talpid 2	*t-2*
	Flexed forelegs		Congenital loco	*lo*
	Ataxia		Lethal with recessive white	*l*
	Epithelial defects		Crooked-neck dwarf	*cn*
			Liver necrosis	—
			Congenital tremor	—
Swine			Bilateral microphthalmia	*mi*
	Paralysis of hind legs		Short mandible	*sm*
	Hydrocephaly, external	D	Atresia of oviduct	—
	Muscle contractures		Short neck and beak	—
	Thick forelegs		Donald Duck (beaks curled)	*dck*
	Amputations	S	Sex-linked lethal	*xl*
	Epithelial defects	S	Shaker	*sh*
		S	Jittery	*j*
Dog[b]		S	Paroxysm	*px*
D	Hairlessness			
	Bird tongue	*Turkey*	
S	Hemophilia A	S	Partial albinism (semi-lethal)	*al*
S	Hemophilia B		Short spine	—
	Ataxia		Congenital loco	*lo*

[a]Not a complete list
[b]For 13 others in the dog, see Hutt (1979).

no fewer than 25 cases in cattle and 31 in the fowl. Readers should also consult the useful, annotated list compiled by Stormont (1958), which includes several conditions not mentioned in Table 9-1, and also hereditary traits which, though not lethal, are undesirable defects. A similar catalogue of hereditary defects (including lethal ones) compiled by Lauvergne (1968) lists 226 of them in cattle alone.

The conditions listed in Table 9-1 for the various species probably represent only fractions of the total numbers of lethals carried by these animals. Bad genes are brought to light by inbreeding, a practice which is taboo for most animal breeders. In species that have been deliberately inbred to reveal their genetic defects, the frequency of lethal genes is far higher than most people would suppose. In one species of Drosophila, it was found by Dobzhansky that, even among wild flies subject to natural selection, 85 per cent carried at least one chromosome bearing one or more genes that would be lethal to homozygotes. In addition, over half carried at least one recessive gene which, in double dose, would have caused sterility.

We have no proof that the chromosomes of domestic animals are similarly studded with bad genes, but, recalling that animals under domestication are somewhat better protected from natural selection than those in the wild, it seems entirely likely that our cows and chickens carry just as much "genetic junk" as did Dobzhansky's flies. In the sections that follow, examples are cited to show some of the ways in which lethal genes exert their effects, how they are inherited, how they can be detected, and, most important for the breeder of domestic animals, how they may be eliminated.

The Nature of Lethal Genes

If we consider all the genetic variations of form and function that have been discussed in previous chapters, it should not surprise us to find that some genes induce deviations from the normal great enough to reduce viability, or, as in the yellow mice, to cause death.

In the majority of cases such lethal genes are completely recessive, leaving the heterozygous carrier indistinguishable (by inspection) from his fellows that are free from the gene. Some, as with the gene A^y just considered, show their presence in the heterozygote, but do not reduce its viability. Still others are apparently lethal not only to the homozygotes but also, with delayed action, to some of the heterozygotes as well.

Some lethal genes induce gross abnormalities, some interfere with physiological processes, and others kill in ways that have not yet been detected. Some cause death of the zygote in early embryonic life; others may do so (in man) even at ages up to 50 years, or more. *Semi-lethal* genes are fatal only to some of the individuals having the dangerous genotype, but not necessarily to all, and, in such cases, the proportion escaping may depend largely on whether or not the environment is favorable. The term semi-lethal does *not* mean (as more than one student has thought) that the gene leaves the stricken individual half-dead! However, there are genes that lower physiological efficiency either temporarily or permanently, and for traits induced by such genes the term *subvital*, coined by Hadorn, is appropriate.

Time of Lethal Action

Some geneticists classify as *sublethal* any genetic abnormalities not fatal at birth or soon after, but causing death at some time before the age of reproduction. Since (for example) juvenile amaurotic idiocy in man, which causes death in early adolescence, is just as inexorably lethal as the infantile amaurotic idiocy which kills within two years of birth, it would seem preferable, perhaps, to refer to hereditary conditions fatal at later ages as *delayed* lethals.

Lethal genes may kill at various stages during the life span, but there are usually characteristic modal ages for onset and peak of mortality associated with each one. The range in time of lethal action is illustrated by some of the lethals known in the domestic fowl. The gene *Cp*, causing the form of incompletely dominant achondroplasia that is a breed characteristic in Creepers, is lethal to most homozygotes at three to four days of incubation, and the "talpid" mutation causes a peak of mortality five days later. The sex-linked "naked" gene, *n*, is lethal to almost half of the affected chicks during the last two days of incubation. Chicks and poults afflicted with congenital loco hatch all right, but cannot feed or drink, so they die within a few days (Fig. 9-1). Two sex-linked disorders (apparently of the nervous system) are not evident at hatching, may appear at two to six weeks thereafter, and are usually fatal before twelve weeks of age. The hereditary atresia of the oviduct discussed in the previous chapter is fatal after affected hens have begun to ovulate. In contrast, Huntington's chorea in man is not fatal to most of its victims until 30 to 40 years of age, but some die earlier and some even after 50.

It seems probable that in other species, including mammals, there must be a similar wide range in the ages at which lethal genes cause death. Examples already considered include the yellow lethal of mice, effective early in gestation.

FIG. 9-1 Poults showing congenital loco at hatching. Left: at rest; right: after being thrown off balance by an alarm. (From R. K. Cole in *J. Heredity.*)

Autosomal Recessive Lethals

Since any completely dominant lethal mutation would be eliminated at once, most lethal traits must be recessive, and, as autosomes greatly outnumber sex chromosomes, most of them are autosomal. Since recessive traits are brought to light only when animals carrying the same recessive gene are mated together, and because closely related animals are more likely to carry genes in common than are animals unrelated, it follows that lethal characters are commonly brought to light by inbreeding.

Sometimes this happens in somewhat isolated communities where good sires are scarce, and hence are used too long in one herd, or are interchanged among neighbors. Such interchanges are also common in regions that are not isolated. In Iowa, two neighbors each bought a Poland China boar from the same herd and, to avoid close inbreeding, exchanged them in their second year of use. Each boar was thus mated to the daughters of the other. Unknown to their owners, the two were closely related, and each boar carried a recessive gene for amputation of the limbs (Fig. 9-2). The resultant crop of piglets on the two farms in the second year was 207 normal : 25 legless, a close fit to the 7 : 1 ratio expected in such matings.

Another way in which inbreeding brings out lethals is that in which the blood of some famous sire becomes concentrated within a breed, or a section of it, following the all-too-common fondness of livestock breeders for pedigrees showing illustrious names. The 54

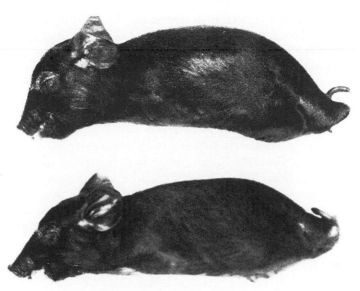

FIG. 9-2   Poland China piglets with hereditary leglessness. (From L. Johnson in *J. Heredity*.)

Holstein-Friesian calves with epithelial defects (raw areas on the legs, in the ears, and in the mouth) found in Wisconsin up to 1928 by Hadley and Cole all traced to one bull, Sarcastic Lad, which had been grand champion at the St. Louis World's Fair in 1904. Other similar cases, in which the glamour of a great name—burnished with additional lustre when the illustrious one was imported from abroad—has resulted in the widespread dissemination of lethal genes, are mentioned later.

The conditions listed in Table 9-1 show that lethal genes may affect the skeleton, the integument, the muscles, the nervous system, metabolism, and reproduction. It would be safe to assume that such genes might cause malformation of any body structure and malfunction of any physiological process. Sometimes the malformation seems to affect the entire skeleton, as in the "bull-dog" calves resulting from achondroplasia. One such lethal affects only the axial skeleton; another amputates the legs, either with associated abnormalities elsewhere (in calves) or without them (in pigs). Another one, found in Sweden by Johansson, eliminates the first two phalanges of the toes, but not the ungual (distal) phalanx, so that the afflicted calves have hooves but cannot stand. This is a good example of a lethal gene with rather localized effect. A lethal in the rat kills in the second week of life, without any visible abnormality whatever, except that the affected animals (which are indistinguishable from normal ones up to nine days of age) stop growing and die about five days later.

The list in Table 9-1 also illustrates the fact that a genetic defect occurring in one species is likely to appear, albeit in modified form, in others. The same applies to breeds. No devotee of one breed need hesitate to report a lethal in his beloved Friesians, because the same defect is equally likely to occur in Ayrshires or Jerseys. This has already happened with the epithelial defects mentioned earlier. In Holstein-Friesians, the affected calves are born alive, but, if not destroyed, die within a few weeks from septicaemia resulting from infection of the raw areas. In Jerseys the defects are so extreme as to suggest that the causative gene may not be the same one as in Holstein-Friesians (Fig. 9-3). The affected calves are born two to three weeks prematurely, sometimes alive, often not. In Ayrshires, however, the areas not covered with skin are small (see Chapter 2, especially Fig. 2-9), and the affected calves live long enough to be sold

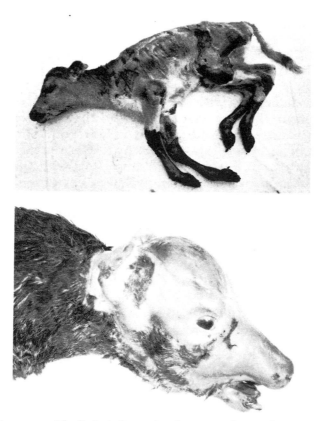

FIG. 9-3 Extreme epithelial defects in Jersey calves. Compare with less severe defects in Ayrshires, in Fig. 2-9. (From W. M. Regan, S. W. Mead, and P. W. Gregory in *J. Heredity.*)

for veal. It is possible that mutations at different loci are responsible for these different degrees of skin defects in the three breeds. Another possibility, equally likely, is that the mutation is the same in all cases, and that other genes, which are responsible for the many differences among these three breeds in size, color, performance, and other traits, are also responsible for the variations in severity of the epithelial defects.

Paralysis in Red Danish Cattle

This autosomal, recessive lethal merits special attention, partly because it has been more thoroughly studied from the genetic point of view than any other lethal in the larger domestic animals, but also because we can learn from this case some ideas that might be helpful in preventing the dissemination of other defects. Affected calves are born at full term and are normal except for inability to stand because of paralysis of the hind legs. These are usually stretched out straight, and, if bent in attempts to get the calf up, are again extended. Efforts to get the calves on their feet are unsuccessful (Fig. 9-4), and sooner or later all of them have to be destroyed.

The first case of this kind was recorded in 1924, and six years

FIG. 9-4   Red Danish calves with hereditary paralysis of the hind legs resist efforts to get them on their feet. (Courtesy of L. J. Cranek, Sr., and N. P. Ralston.)

later it was clear that the defect was hereditary. By 1950 Nielsen was able to show in his extensive report that 63 bulls carrying the causative gene, when mated with daughters of other heterozygous bulls, had sired 1,634 normal calves and 234 with paralysis, an almost perfect fit to the 7 : 1 ratio expected. In the period 1943 to 1947, among 262 bulls registered in the herd books of the Red Danish breed from two provinces of Denmark, the proportion that carried the gene for paralysis was calculated to be 26 per cent. All of the known carriers of the defect traced back to the bull, Tjalfe Kristoffer, born in 1913.

During the course of this study, an interesting record was obtained for 40 proven carrier cows in 7 herds, all of which were mated to carrier bulls, but also (at different times) to bulls free of the gene (Table 9-2).

As expected, the homozygous normal bulls sired no paralyzed calves, but the matings of carrier ♀ × carrier ♂ did not yield the expected ratio of 3 normal : 1 paralyzed. Instead, there were equal numbers of each kind. Why a ratio of 1 : 1? The answer is that it is not really any kind of a Mendelian ratio. The breeders knew all about the hereditary nature of the defect, and, once a cow had produced a paralyzed calf by one sire, she would never again be bred to the same bull, but would be switched to others known to be free of the gene. As a result, matings that would have yielded a ratio of 3 : 1 in larger

TABLE 9-2
Results in Matings of 40 Cows Heterozygous for Paralysis
to Bulls of Two Types

Herd	Carrier ♀♀ (number)	Progeny by ♂♂ Heterozygous for Paralysis		Progeny by ♂♂ Homozygous for Normal
		Normal (number)	Paralyzed (number)	All Normal
A	9	20	11	16
B	7	4	7	15
C	6	8	6	12
D	6	3	4	11
E	6	6	8	10
F	4	1	4	10
G	2	0	2	10
	40	42	42	84

Source: Data of Nielsen, 1950.

numbers were terminated when (by chance) the recessive cropped out in the first or second calf.

After the gravity of this situation was realized, the breeders took the steps necessary to reduce the frequency of the causative gene. With many known heterozygous cows available, young bulls could be mated early to several known carriers, and by 2½ years of age could be proven as free of the gene or not. It became the custom, when bulls were bought by co-operative breeding centres, for the buyer to pay only half the price agreed upon, the remainder being held in escrow and paid only when the animal was proven to be free of the gene. With special attention to pedigrees, families free of the defect were multiplied and known carriers eliminated. Together, all these measures have been successful in reducing the frequency of the gene in Denmark.

Meanwhile, when all this trouble was brewing, but before the frequency of this lethal gene in Red Danish cattle was known, representatives of that highly regarded dairy breed were brought to the United States, and a herd was established by the Department of Agriculture at Beltsville. The stock was multiplied and separate herd books are now maintained for American Red Danish cattle. By 1953, 65 paralyzed calves had been recorded in 27 herds in Michigan, and four years later it was estimated that 25 per cent of the American Red Danish cattle were heterogygous for that defect. Lethal paralysis is not the only lethal that has crossed an ocean.[1]

Hydrocephalic Calves

Another recessive, autosomal lethal is worthy of special note because of an experiment, unusual for large animals, by which its hereditary nature was demonstrated. This is hydrocephaly in calves, more specifically *hydrocephalus internus*, the accumulation of fluid in the ventricles of the brain (Fig. 9-5). Affected calves are usually born alive at full term, show enlargement of the cranium, are unable to stand, and die within two days. In New Mexico, 19 such calves in one herd of Herefords traced on both sides of their pedigrees to one and the same bull.

In Nebraska, Baker *et al.* (1961) rounded up from interested breeders 40 cows and one bull (δW), all known to have produced hydrocephalic calves. Half the cows were mated to δW, and the other half to δU, which came from an inbred line believed free of the defect. In the next year, the bulls were interchanged, so that eventually all 40

[1]See page 188.

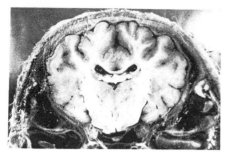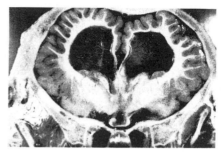

FIG. 9-5 Cross-sections through normal cerebrum of calf (left) and one with ventricles enlarged by hydrocephaly (right) (From M. L. Baker *et al.* in *J. Heredity.*)

cows were bred to both bulls. The results (Table 9-3) were in accord with expectation for a simple recessive character.

Autosomal Dominant Lethals

The short-legged, diminutive Dexter cattle have long provided for texts on genetics a standard example of an incompletely dominant gene that is lethal to the homozygote. They are the bovine counterparts of the achondroplastic dwarfs in man. All Dexters are heterozygotes, which when mated *inter se* yield a ratio of 2 Dexter : 1 long-legged, normal : 1 extremely abnormal, non-viable "bull-dog" calf. The last of these is homozygous for the lethal gene.

Extremely abnormal bull-dog calves (of this type) are aborted, usually before the eighth month. They have very short (phocomelic) legs, so that in some cases the toes seem to project from the body. The cranium is vaulted, and the head shortened so that the tongue protrudes. The body tissues are so swollen that the tail seems to arise from far up the back. (Fig. 9-6).

The mutation that gave rise to the Dexter breed is believed to have originated in the Kerry breed of Ireland, but there is no reason why it could not occur in any other. The specimen illustrated (Fig. 9-6, left) resulted from a mutation (from normal to heterozygous state) in Jerseys, which so pleased the owner that he undertook to breed dwarf Jerseys.

A recessive type of achondroplasia resembles the other, but is less extreme (Fig. 9-6). Affected calves are usually born alive at full term, but are unable to stand, and die within a few days. This is sometimes called the Telemark type, because it was first observed in that

TABLE 9-3

Segregation of Hydrocephalic Calves from Carrier Cows Each Mated to Two Bulls

Year	Mating	Calves Normal	Calves Hydrocephalic
1959	♂ W (carrier) × ♀♀ 1 to 20	12	4
1960	♂ W (carrier) × ♀♀ 21 to 40	14	2
		26	6
1959	♂ U (free) × ♀♀ 21 to 40	19	0
1960	♂ U (free) × ♀♀ 1 to 20	18	0
		37	0

Source: Data of Baker *et al.*

Norwegian breed by Wriedt, but it has since been found in several others.

The Dexters provide one of several cases in which an incompletely dominant lethal gene is the distinguishing characteristic of a breed. The gene *Cp* causes an achondroplasia in the fowl which breeders in three continents have preserved to make Creepers (Fig. 9-7), Scots Dumpies, and Japanese Bantams. All of these are heterozygous for a gene wich is lethal to most of the homozygotes at

FIG. 9-6 Dominant (left) and recessive (right) achondroplasia, both in Jerseys. (From S. W. Mead *et al.* in *J. Heredity* **37:** 183–88, and G. W. Brandt in *J. Heredity* **32:** 183–86.)

FIG. 9-7   A Creeper hen, heterozygous for lethal achondroplasia. (From I. E. Cutler in *J. Heredity.*)

three of four days of incubation. Among hatched chicks and embryos dying late in incubation, Landauer and Dunn counted the following numbers in two kinds of matings:

| | Chicks | |
Parents	Creeper	Normal
Creeper × normal (*Cp cp* × *cp cp*)	1,676	1,661
Creeper × Creeper (*Cp cp* × *Cp cp*)	775	388

In the good old days before huge incubators, when a hen could satisfy her natural yearning for maternity if she wished to do so, Creepers were specially valued by some poultrymen as mothers for brooding chicks because the proximity of their warm and sheltering plumage to the ground provided an ideal haven for chilled or frightened chicks. The figures cited above show that Creeper hens could be good mothers in backcrosses to non-Creepers, but when mated to Creeper males, a quarter of their potential offspring are doomed to early death.

Other completely dominant lethal genes provide the distinguishing feature in Grey Karakul (Shiraz) sheep, Platinum foxes, and also in Silver Sable (Blufrost) and Heggedal minks. All of these breeds (or color varieties) resemble the yellow mouse in showing by their color that they are heterozygous for a lethal gene. We shall consider them again later on.

Sex-linked Recessive Lethals

Hemophilias A and B in dogs and the hypotrichosis with anadontia of calves provide examples in mammals of sex-linked, recessive, lethal traits that kill their victims at some time after birth. Nine sex-linked lethals have been identified as such in the fowl. Other suspects are awaiting further proof. All such conditions, whether in mammals or birds, are easily recognized as being sex-linked, because the abnormality is visible and (except in special matings contrived with defective animals kept alive for that purpose) is confined to the heterogametic sex.

Sex-linked lethal genes effective before birth are more difficult to identify. It is to be expected that there are such genes, and dozens of them are known in Drosophila. Suspected cases have been reported in cattle, the horse, and the fowl. The evidence in such cases rests mostly on a deficiency at birth of males (in mammals) or females (in birds). As a sex-linked lethal gene would be carried only by the homogametic sex, and should eliminate half of the opposite sex, the expected sex ratio would be 2:1. As chance deviations from the normal sex ratio are common enough in small numbers, one must be careful about attributing them to the action of sex-linked genes.

To provide incontrovertible proof of the operation of a sex-linked lethal gene effective before birth, evidence of the following kinds would seem desirable:

1. A sex ratio at birth of 2:1, with significant deficiency of the heterogametic sex.
2. Reproduction in the suspect families only 75 per cent as efficient as in those not carrying such a lethal. Efficiency can be measured by litter size in multiparous mammals, by intervals between births or numbers of services required in others, and by hatchability of fertile eggs in birds.
3. A peak of pre-natal mortality at some modal age.
4. An excess of the heterogametic sex among dead embryos or fetuses.
5. Breeding tests with siblings of the homogametic sex born from the suspect matings, and evidence therefrom that half of them carry the suspected sex-linked lethal gene.

Evidence of these kinds is difficult to get in large domestic animals; it is somewhat more easily found in swine, dogs, and cats, and most readily available in domestic birds. It is easy (by daily candling of incubating eggs) to detect peaks of embryonic mortality, and domestic birds reproduce in the comparatively large numbers necessary to reveal significant deviations from normal sex ratios.

A Sex-linked Dominant Lethal in Cattle

Eldridge and Atkeson studied in Holstein-Friesians a hereditary condition which they called streaked hairlessness. Hair was completely lacking on vertical streaks over the hip joints and sometimes extending up the sides (Fig. 9-8). Similar streaks appeared on the legs of some cows. Variation was extreme, and some cows with streaks had not even been recognized as such by the owner. The extent of the bare areas was best seen after clipping, but affected calves were usually recognizable by their rough coats. Both white and black areas were streaked. Cows thus streaked were more sensitive than others to cold and to scrubbing.

All of the 17 known affected animals were females; all were

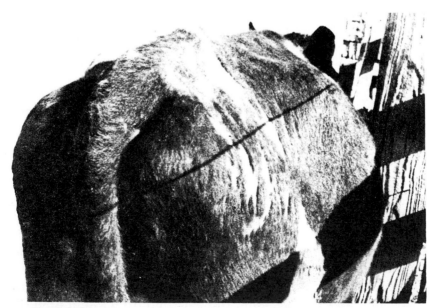

FIG. 9-8 Streaked hairlessness; sex-linked, dominant, lethal in males. (From F. E. Eldridge and F. W. Atkeson in *J. Heredity*.)

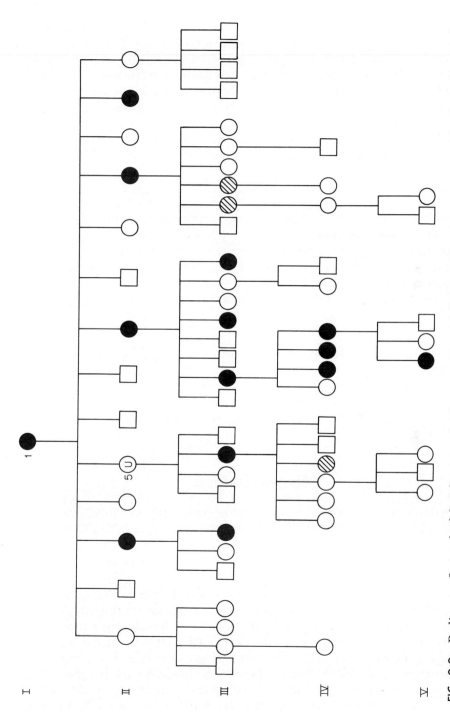

FIG. 9-9 Pedigree of streaked hairlessness in five generations showing dominant sex-linked inheritance and deficiency of males. II-5 was unknown. The three animals shaded were reported as affected, but were not seen by the investigators. (From F. E. Eldridge and F. W. Atkeson in *J. Heredity*.)

descended from one cow. No streaked males were found. Affected ♀♀
× normal ♂♂ yielded a ratio in female offspring of 15 streaked : 13
normal. This indicated that the peculiarity was dominant and caused
by a single gene. The absence of streaked males was at first attributed
to the sale of bull calves for veal before the character could be
recognized, but, when the whole record was unfolded (Fig. 9-9), it
became clear that there was a significant deficiency of males among
calves from streaked cows. Sex ratios in various groups, most of them
shown in Fig. 9-9, were as follows:

	♂♂	♀♀
In all calves descended from I−1	23	44
In calves from normal cows of Generations I	24	34
In calves from cows descended from I−1		
(a) From streaked cows	12	34
(b) From normal cows, excluding the unknown II−5	9	8

The highly significant deficiency of male offspring from streaked
cows, together with the other evidence, led to the conclusion that
streaked hairlessness is caused by a gene that is sex-linked, incompletely dominant, lethal to the hemizygote, and hence sex-limited. It
is to be hoped that further studies of this remarkable lethal trait will be
made if it should appear in other herds.

Subvital Defects

It is difficult to draw any dividing line between lethal characters and
defects that may be lethal in a poor environment, but not in a sheltered
one. Some of these conditions may be lethal only to a few animals, but
can lower the physiological efficiency (*i.e.*, the performance) of the
rest.

A good example is provided by the sex-linked naked mutation in
the fowl (Fig. 9-10). Almost half of the naked chicks die during the last
two or three days of incubation. If those that hatch are brooded at
temperatures adequate for normal chicks (90°−95° F. at the level of
the chick's back), about half of the naked ones die before six weeks of
age. However, if brooder temperatures are raised by 10° F. above the
usual heat, many more of the naked chicks can be brought safely
through the critical first few weeks. By four or five months of age,
most of them have grown a coat of sparse plumage and can withstand

FIG. 9-10 A chick of three weeks showing the sex-linked, semi-lethal mutation, naked. (From F. B. Hutt in *Genetics of the Fowl*, McGraw-Hill Book Co., Inc.)

fairly low temperatures. They do not lay as well as birds with normal plumage, and it is inevitable that a high proportion of their feed must be used to maintain body temperature.

Apart from their susceptibility to chilling, naked fowls have only a few short wing-feathers, and are unable to fly. Obviously, the mutation would be completely lethal in nature, and, for that reason, it should be classified as a lethal character. For the naked hens that the senior author has kept to three years of age or more, it is only subvital.

Another example of such a defect is congenital cataract in cattle. This has been found to be inherited as a simple recessive trait in Holstein-Friesians and in Jerseys. Some of the affected calves are blind at birth (Fig. 9-11), but others apparently do not become completely blind. Gregory and his associates found that the cows with defective vision were somewhat nervous, had difficulty in competing with others in the feeding corral, and produced less milk and butterfat than the other cows. Here is another subvital defect that would be lethal in the wild. The same probably applies to the dwarf calves considered in Chapter 3.

Many hereditary lethals and subvital defects are known in domestic animals. A number of these are discussed at appropriate

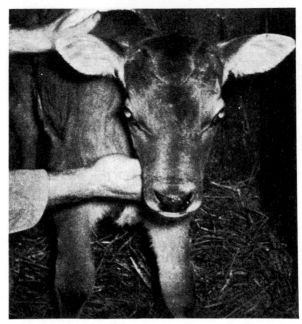

FIG. 9-11   Jersey calf with congenital cataract in both eyes. (From L. Z. Saunders and M. G. Fincher in *Cornell Vet.* **41:** 351–66, 1957.)

places throughout this book (and brought together, by species, in the index), but it is beyond the scope of this text to describe them all.

Dissemination of Lethal Genes

Cattle breeders should recognize that in these days of artificial insemination, when one bull can have up to 50,000 offspring per year, the risk of widespread dissemination of bad genes is much greater than ever before. A sire heterozygous for a lethal gene would pass it to half his progeny, but, unless enough of the cows to which he is bred carry that same gene to yield the homozygous defective calves in significant numbers, its transmission by the carrier bull might not be recognized until several generations later.

Even before artificial insemination was so commonly used, lethal genes managed to migrate from herd to herd, and, as we have seen in the case of the Red Danish cattle, from one country to another. Classical examples include the Percheron stallion, Superb, sent from Ohio to Japan in the 1880's and found (some 30 years later) to have carried a gene for lethal *atresia coli*. Afflicted foals have a closure of the ascending colon in the region of the pelvic flexure, and die within two to four days after birth. Forty years after the importation of Superb, 26 per cent of the stallions of heavy breeds in Hokkaido carried his name in their pedigrees.

The Friesian bull, Prins Adolph, imported into Sweden from Friesland in 1902, was considered an outstanding sire of his breed. It is not surprising that his blood was widespread in Swedish Friesians 26 years later, when it was shown that he had carried a gene for a type of hairlessness that is lethal. Similarly, an Ayrshire bull, Dunlop Talisman, imported into Finland in 1923, brought with him a gene for lethal congenital dropsy. By the end of 1937, he had 84 sons and 189 grandsons registered in the herd books, and Korkman found that 82 of those grandsons came from sires that carried the lethal gene.

These few examples will suffice to show that lethal genes are remarkably adept at getting around the world, and that eternal vigilance is necessary if they are to be circumvented.

Detection of Lethals

There is little difficulty in recognizing as hereditary defects any consistently recurring abnormalities of form at birth, or of function thereafter. One must remember, however, that not all congenital defects are genetic in origin. Many of them result from arrests of development at critical stages, some of which may have resulted from illness of the mother, some from causes unknown, and still others—as has been so unhappily attested by the victims of thalidomide—from the use of drugs. Lethals effective at ages before birth can sometimes be spotted by a deficiency of some expected class (as with the homozygous yellow mouse), by an abnormal sex ratio, or by evidence of unusually high pre-natal mortality.

With multiparous animals like swine, dogs, cats, and rabbits, simple recessive defects lethal at birth or later should appear in about 25 per cent of the progeny from parents that are both heterozygous for the causative gene. When only small litters are considered, the proportion reported to show the abnormality is often higher than 25 per cent because breeders commonly consider only the litters in which at least one defective animal has occurred, and fail to mention litters of 4, 5, 6, or more, all normal, which are to be expected by chance from those same parents. In controlled, experimental matings between parents both known to be heterozygous, better fits to the expected 3:1 ratio will be observed, especially when every offspring is accounted for.

With larger uniparous animals like cattle, reports from breeders are not so likely to deal with 3:1 ratios, because: although the sire is

FIG. 9-12 Short spine. Before this condition was shown to be hereditary, farmers thought such calves were sired by elks. (From O. L. Mohr and C. Wriedt in *J. Genetics.*)

heterozygous, not all of the cows in the herd are likely to be the same. As was explained in Chapter 2, a 7 : 1 ratio is more likely in such cases. It occurs most commonly when a carrier sire is mated to daughters of another carrier, or (less often) to his own. Here, too, an excess of abnormals is sometimes evident in reports from the field because breeders remember them better than the normal progeny. The Oplandske bull, Amor,[2] which sired 11 calves with short spines but normal limbs (Fig. 9-12) among his calves from his 27 daughters, yielded a ratio (44 : 11) that fits 3 : 1 better than 7 : 1, but (as Problems 4-14 to 4-20 were intended to make clear) the former was to be expected only if *all* Amor's 27 daughters carried the gene for short spine. Actually, about half should have done so. Regardless of the fact that 44 : 11 resembled a 3 : 1 ratio, a 7 : 1 ratio was to be expected. The excess of short-spined calves could have resulted from failure to record or recall all the normal ones, from chance segregation, or from both causes together.

A deficiency of mutant types below the numbers expected in Mendelian ratios for segregating lethals and defects may sometimes result from the condition's being lethal to some of the zygotes at early stages of development.

---

[2]A name of which (if you remember your Latin) any bull could once have been justifiably proud, but one now rendered somewhat obsolete, alas, by the impersonal relationships pervading much of modern bovine reproduction.

A Digression into Probabilities

In Chapter 2, we dealt briefly with a simple probability necessary for our understanding of backcrosses and test-crosses, *i.e.*, that of some one event occurring or not. To understand (a) Mendelian ratios in litters of different sizes, which were mentioned in the previous section, and (b) progeny tests of sires, to be considered in the next one, we should have a better understanding of the frequencies with which two events are likely to be combined.

In litters of pigs or puppies from parents both heterozygous for a simple autosomal recessive gene, the chance that any one individual will show the defect is 1 in 4, or ¼, or 0.25. The chance of not doing so, *i.e.*, of being normal, is 3 in 4, or 0.75. We can compute the chance that in a litter of four pups all will be normal by multiplying together the separate probabilities for each one: $(¾)^4 = 0.3164$. This means that, in 31.6 litters among 100 litters of four, we would expect all four pups to show the dominant character. Conversely, if the probability of any one pup's showing the recessive allele is ¼, the chance that all four will do so becomes $(¼)^4$ or 0.0039. Only about one in 256 such litters might be expected to have all four showing the recessive trait.

In between these extremes of all one kind, or all the other, are the possible combinations—3 : 1, 2 : 2, and 1 : 3. We might like to know how often they should occur. This we can find at once, and also all possible combinations in larger litters, from the terms in the expansion of the binomial $(a + b)^n$ in which

a = probability of one event

b = probability of the other event

n = number of individuals to which either event might apply

For the litter of four just considered, the expanded binomial is:

$$a^4 + 4\,a^3b + 6\,a^2b^2 + 4ab^3 + b^4$$

Substituting for a the value ¾, and for b that of ¼, we can not only compute, as before, the expected frequency of litters all showing the dominant character, or all the recessive one, but also those for any combination of the two. For example, the expected frequency of two dominant and two recessives is

$$6a^2b^2 = 6(¾)^2(¼)^2 = {}^{54}/_{256} \text{ or } 0.2109,$$
$$\text{or 21 times in 100 such litters}$$

If we were concerned, not with a 3:1 segregation, but with a ratio of 1:1, as would be the case if we were calculating expected frequencies of males and females in a litter of four (*if* the sex ratio were 1:1, which it isn't, exactly, in dogs), then the probabilities of the two events become ½ and ½, and the expected frequency of equal numbers of males and females in a litter of four is:

$$6a^2b^2 = 6(½)^2(½)^2 = {}^6/_{16}, \text{ or } 0.3750,$$
$$\text{or } 37.5 \text{ times in } 100 \text{ such litters}$$

When the number of individuals concerned is six, the number of possible combinations is increased. Expansion of $(a + b)^6$ gives:

$$a^6 + 6\,a^5b + 15\,a^4b^2 + 20\,a^3b^3 + 15\,a^2b^4 + 6\,ab^5 + b^6$$

The coefficients of any term in such expansions can be determined by multiplying the coefficient for the previous one by the first exponent in that previous term and dividing the product by the total number of previous terms. For example, the coefficient for the third term of the expansion of $(a + b)^6$ is $6 \times 5 \div 2 = 15$.

Alternatively, all the coefficients for any number of terms (or pups in the litter) can be easily taken from Pascal's triangle:

n	Coefficients
1	1 1
2	1 2 1
3	1 3 3 1
4	1 4 6 4 1
5	1 5 10 10 5 1
6	1 6 15 20 15 6 1
7	1 7 21 35 35 21 7 1
etc.	etc.

In this arrangement, each coefficient is the sum of the two nearest ones above it. While the student should know how expectations for various combinations are derived, it is fortunate that he need not work out every case, for with calculations for more than ten individuals the figures begin to resemble the national debt. The expected frequencies for three different Mendelian ratios, with eight individuals concerned, as given in Table 9-4, are taken from Warwick's useful tables.

An important point to remember is that the two probabilities in the binomial that is to be expanded must together equal one, *e.g.*, ¾ + ¼, ½ + ½, ⅞ + ⅛, etc. Similarly, when all the terms of the expanded

TABLE 9-4

Probabilities (per cent) of Getting Various Combinations when Mendelian Ratios Are Expected in Families of Eight Individuals

Combination Dominant : Recessive	Expected Ratio		
	1 : 1	3 : 1	7 : 1
8 : 0	0.39	10.01	34.36
7 : 1	3.13	26.70	39.27
6 : 2	10.94	31.15	19.63
5 : 3	21.87	20.76	5.61
4 : 4	27.34	8.65	1.00
3 : 5	21.87	2.31	0.11
2 : 6	10.94	0.38	0.01
1 : 7	3.13	0.04	—
0 : 8	0.39	—	—
	100.00	100.00	99.99

Source: From Warwick, 1932.

binomial are worked out, the frequencies of the different combinations must add up, as in Table 9-4, to 100 per cent (or, if expressed in decimals, to 1).

Elimination of Lethals

Contrary to some beliefs, lethal genes do not eliminate themselves. As with some of the examples cited in this chapter, and the dwarfism of cattle considered in Chapter 3, bad genes can even accumulate within a breed to a level at which concerted action must be taken by the breeders to reduce their frequency.

The big problem in doing so is to identify the animals, particularly sires, that carry the unwanted gene. When that gene is completely recessive, the heterozygotes can usually be identified only by progeny tests. Unfortunately these are not as easy, when lethal genes are concerned, as with traits like the unwanted red color in Holstein-Friesians, because the simplest kind of test-cross—a backcross to the homozygous recessive type—is impossible. The next best thing is to mate the suspect sire to known heterozygous females.

When this is done, if as many as eight offspring are all normal, such an event would be expected, as we see in Table 9-4, in about 10 per cent of similar trials, even if the suspect sire were heterozygous. By calculation of the probabilities for larger numbers of offspring, it is found that, if a male thus tested sires none of the recessive type among 11 offspring, the chance that he could still be heterozygous is reduced to <.05, and with 16 such offspring it becomes <.01. The only other kind of breeding test possible, that of the suspect sire to his own daughters, is scarcely practicable with large animals. For one thing, it necessitates delay until the daughters are old enough to breed. For another, the expected ratio of 7:1 would require no fewer than 23 normal daughters to reduce to <.05 the probability that the sire under test does not carry the unwanted gene, and 35 would be needed to bring that chance down to <.01.

It is possible that the current widespread use of artificial insemination in animal breeding may lessen somewhat the risk that some lethal gene (or any other that is equally undesirable) may be spread far and wide, as in some of the examples given earlier. Johansson (1961) calculated that if such a gene were carried by 5 per cent of the females served, any sire carrying that same gene would be shown up as a carrier by the time he had sired 200 successive progeny, and that his chance of not being so revealed would be less than 0.01. Any sire found to be a carrier should presumably be culled at once, but that is not always done.

With frequencies of carriers at higher levels in the females served, fewer progeny would be needed to reveal any carrier sire. Culling of carrier sires would not eliminate a bad gene but might help to keep it at a low frequency in a large population (thousands of animals), so long as enough females carry that gene to serve as testers. It would not stop a new mutation in the sire from spreading afar in a population that provides no female carriers to serve as testers.

Genes causing lethal or otherwise undesirable defects may be eliminated in some cases by use of laboratory tests rather than by the much slower breeding tests. This applies particularly to alleles in which the two partners are codominant, with heterozygotes intermediate between the two homozygotes. This is the case in various physiological functions, and seems to be the rule with defective enzymes that are responsible for disorders of metabolism. The number of such diseases currently known in domestic animals is far fewer than the 170 metabolic disorders now known in man, but three of them in which the carriers can be identified by quantitative tests for activity of the defective enzyme are described in Chapter 20. Two are in dogs (hemolytic anemia and ceroid lipofuscinosis) and one is in cattle. Two are lethal and one is subvital. (See also the section on codominance in Chapter 3.)

Lethal Characters Affecting Both Progeny and Dam

In a class by themselves are several lethal conditions in which the abnormality of the offspring frequently prevents normal birth and, hence, is sometimes fatal to the mother. This is the case with prolonged gestation, now known as a simple recessive, autosomal trait in at least three breeds of cattle. Although it is not established that all three breeds have the same mutation, conditions common to all three include the prolongation, gigantism of the calf, dystocia, and death of the calf either before, during, or soon after birth. Nine such calves in Swedish Red and White cattle were carried for periods varying from 332 to 510 days, and six of them were delivered only after slaughter of their dams. The average period overdue for nine Ayrshire calves was 80 days, and the mother of one died from dystocia. The other eight cows recovered and later produced normal calves by other bulls. The genotype responsible for the prolongation is clearly that of the homozygous calf, and not that of its heterozygous dam. In Holstein-Friesians the calves were not carried so long, but there was a record of dystocia, which proved fatal to the cow in at least one case, and other cows were slaughtered.

A type of hereditary prolongation of gestation (called adenohypophyseal aplasia) has also been found in Guernseys and Jerseys. In those breeds the fetuses show several abnormalities,

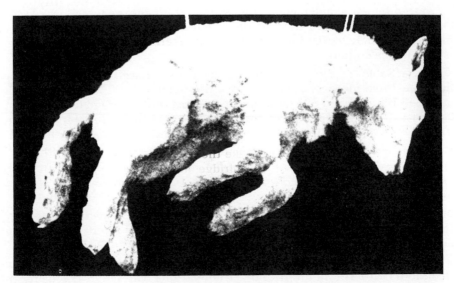

FIG. 9-13 Lamb showing bent fore limbs and wry neck caused by hereditary muscle contractures. (From J.A.F. Roberts in *J. Genetics.*)

including various degrees of hypotrichosis, and are smaller than normal. It is attributed to failure of the hypophysis to develop. All these types of prolonged gestation are considered and compared by Wilson and Young (1958).

Other genetic disorders sometimes causing death of the dam include muscle contractures, a simple recessive lethal in sheep, cattle, and swine (Fig. 9-13). In the first two of these species, the fore limbs are often bent, with joints rigid. This and the accompanying wry neck make delivery difficult. Three of five such calves in one herd had to be dismembered, and one of the cows was fatally injured in the process. In sheep, affected lambs are rarely born alive, but none is born prematurely, and some of the ewes die as well.

Problems

9-1 A Telemark bull carrying the type of recessive achondroplasia found in that breed was imported from Norway to England by T. H. Riches and there mated to Dexter cows in order to test any possible relationship between the two types of achondroplasia. Among the 24 offspring there were no bull-dog calves, but some had long legs, and others were of the Dexter type. When matings were made between these Dexter-like crossbreds, the progeny consisted of 5 Dexter type (viable), 2 long-legged, 3 Dexter-type bull-dog calves, and one calf with the Telemark type of achondroplasia.

What do these results reveal about the genetic bases for these two kinds of achondroplasia?

9-2 Assigning for convenience the symbol D to the gene for the Dexter achondroplasia, and t to that of the Telemark, assign genotypes to all the animals that Riches used for breeding, and determine how well the observed results fit expectations.

9-3 Silver Sable (or Blufrost) minks have silvery guard hairs and light blue under-fur. From matings of Blufrost $\male$ × normal dark $\female \female$, Moore and Keeler found 345 Blufrost and 325 dark in the progeny, with average litter size 5.11 kits. When Blufrosts were mated together, the offspring were 19 Blufrost and 10 dark, with average litter 3.65 kits. How do you interpret these findings?

9-4 By the chi-square test, determine the levels of statistical significance at which the sex ratios found by Eldridge and Atkeson deviated from the expected $1:1$ ratio in their family of Holstein-Friesians showing streaked hairlessness (see page 186).

9-5 By expansion of the binomial $(a + b)^n$, determine the expected frequencies of all the different combinations of $\male \male$ and $\female \female$ pups in litters of five, taking the sex ratio as $1:1$.

9-6 Similarly, what are the expected frequencies of an autosomal, recessive defect, lethal at birth, in litters of five from parents both heterozygous for the causative gene?

9-7 Any recessive mutation carried by an animal is more likely to be found also in near relatives than in unrelated animals. Under what conditions would the mating of a known carrier to unrelated animals provide more conclusive proof of the genetic basis for a defect than would mating to relatives?

9-8 Goodwin and his associates studied a condition in which, at various ages after 23 days, Leghorn pullets (previously in good health) were found in a semi-comatose state in the mornings and usually died within a few hours. It is called the *xl* lethal. When the sire of these chicks was mated with 14 unrelated hens, among 32 ♂ ♂ and 42 ♀ ♀ offspring, four or five of each sex died from various causes, but 21 showed the *xl* lethal, all ♀ ♀. What genetic basis was indicated?

9-9 How many of Goodwin's surviving chicks would you expect to carry the gene *xl*?

9-10 How could they be identified as possible carriers?

9-11 By what test could the identity of carriers be definitely established?

9-12 From a pen of Leghorns headed by a male heterozygous for a sex-linked gene that is lethal during the last three days of incubation, 200 fertile eggs are still alive (in one setting) at 18 days of incubation.

If the mortality thereafter from causes other than the lethal is 12 per cent, how many chicks would you expect to find in each of the following classes:

Dead-in-shell: ♂ ♂, ♀ ♀?
Hatched chicks: ♂ ♂, ♀ ♀?

9-13 By what criteria, including test-matings of various kinds, would you determine the manner of inheritance of abnormalities that eventually prove to be:

(a) A sex-linked character in swine that is lethal during the first week of gestation?

(b) A sex-linked character in the fowl that is lethal at four days of incubation?

(c) An autosomal, recessive character in cattle, lethal within three days of birth, but not previously recognized as being hereditary?

(d) A non-genetic, congenital abnormality in a puppy?

9-14 From matings *inter se* of turkeys heterozygous for the recessive gene, *lo*, Cole hatched in two years 788 poults, among which 199 showed typical congenital loco (Fig. 9-1). To what extent, if any, would you consider that *lo* might be lethal to homozygotes during incubation, and why?

Selected References

BAKER, M. L., L. C. PAYNE, and G. N. BAKER. 1961. The inheritance of hydrocephalus in cattle. *J. Heredity* **52:** 134–38. (With data supplementing the experiment cited in the text.)

COLE, R. K. 1957. Congenital loco in turkeys. *J. Heredity* **48:** 173–75. (A thorough genetic study of the abnormality illustrated in Fig. 9-1.)

DOBZHANSKY, T. 1957. Genetic loads in natural populations. *Science* **126:** 191–94. (Lethals in fruit flies and man.)

ELDRIDGE, F. E., AND F. W. ATKESON. 1953. Streaked hairlessness in Holstein-Friesian cattle. *J. Heredity* **44:** 265–71. (Details of the sex-linked dominant lethal discussed in text.)

HADLEY, F. B., AND L. J. COLE. 1928. Inherited epithelial defects in cattle. *Wisconsin Agric. Exper. Sta. Bull.* 86 (Good example of extensive investigation of a lethal.)

HADORN, E. 1961. *Developmental Genetics and Lethal Factors*. New York: John Wiley & Sons. London: Methuen & Co. Ltd. (Indispensable reference book on effects of lethal genes on embryonic development, particularly in invertebrates.)

HUTT, F. B. 1961. Identification and elimination of defects in animals. In *Germ Plasm Resources*, pp. 355–69. Washington. Amer. Assoc. Adv. Sci.

HUTT, F. B. 1979. *Genetics for Dog Breeders*. San Francisco. W. H. Freeman & Co. xiv + 245 pp. (Appendix 1 lists 18 lethal characters.)

JOHANSSON, I. 1961. *Genetic Aspects of Dairy Cattle Breeding*. Urbana, Ill. Univ. of Illinois Press. xii + 259 pp. (Progeny tests to reduce undesirable genes; 225–28.)

JOHNSON, L. 1940. "Streamlined" pigs. *J. Heredity* **31:** 239–42. (About the legless ones described in the text.)

LAUVERGNE, J. J. 1968. Catalogue des anomalies hereditaires des bovins (*Bos taurus L.*). Inst. Nat. Recherche Agronomique Paris, Bull. Tech. Departement Genetique Animale No. 1. (A useful catalogue.)

NIELSEN, J. 1950. *Arvelig lamhed hos kalve*. Copenhagen. Andelsbogtrykkeriet i Odense og det Danske Forlag. 173 pp. (Lethal paralysis in Red Danish cattle; English summary.)

STORMONT, C. 1958. Genetics and disease. *Advances in Vet. Sci.* **4:** 137–62. (Annotated list of lethals and defects known in domestic animals, with citation of literature thereon.)

WARWICK, B. L. 1932. Probability tables for Mendelian ratios with small numbers. *Texas Agric. Exper. Sta. Bull.* 463.

WILSON, A. L., and G. B. YOUNG. 1958. Prolonged gestation in an Ayrshire herd. *Vet. Rec.* **70:** 73–76. (With comparisons of the abnormality in other breeds, and references.)

10

Linkage, Crossing-over, and Chromosome Maps

The Inevitability of Linkage

In Chapter 4 we learned that, according to Mendel's second law, genes in any one pair of alleles segregate independently of those in other pairs of alleles. We were warned, however, of exceptions to that rule to be considered later, and here they come.

Independent assortment occurs when the two or more pairs of alleles concerned are on separate chromosomes. (It can also appear to occur even if the genes are in one and the same chromosome, but a long way apart, but this is a special situation to be considered later in this chapter.) Can there be a separate chromosome for each gene? Obviously not. In *Drosophila melanogaster,* well over 500 genes are known, but there are only four pairs of chromosomes and one of these is only a comparatively tiny dot. In maize, at least 112 genes have been located in its ten pairs of chromosomes, and others have been identified, though not yet assigned to specific chromosomes. Each chromosome must carry many genes.

An inevitable corollary is that, when homologous chromosomes disjoin in the meiotic divisions of gametogenesis, genes in one chromosome will tend to go together. As a result, the characters they induce will be associated, or *linked.* We have already considered genes located in the sex-determining chromosomes, and hence sex-linked. In its broader aspects, linkage is an extension of that same concept to all chromosomes. Since genes cannot be assigned to specific autosomes as easily as to the X- and Y-chromosomes, we classify genes linked in autosomes as belonging to specific *linkage groups,* and, with favorable material, can sometimes eventually identify some of those linkage groups with specific chromosomes.

Linkage, Complete and Partial

In males of Drosophila and of some other flies, and in females of Bombyx, the silkworm, linkage is complete. In other words, if genes *A* and *B* (both in the same chromosome) enter a cross from one parent, and their recessive alleles *a* and *b* from the other, the resulting dihybrid male fly or female silkworm moth produces only two kinds of gametes: *A B* and *a b*. There is none with the combination *A b* or *a B*. In most species, however, linkage is not complete, and, in a case like the foregoing one, the dihybrid would produce four kinds of gametes, but more of the *A B* and *a b* types than of the new combinations, *A b* and *a B*. These last are made possible by interchanges between chromatids of maternal and paternal origins, the mechanism for which will be considered later.

Linkage of *F* and *I* in the Fowl; Coupling Phase

It will be easier, perhaps, to see how behavior of linked genes differs from independent segregation if we consider first a typical case of the latter. Among several crosses made by the senior author in quest of linkage of the gene, *F*, causing frizzled plumage in the fowl (see Chapter 3), was one in which *F* was tested for linkage with the gene *R* (rose comb). To that end, dihybrid females were backcrossed to a double recessive male, with the following results:

	Parents			*Progeny*	
♀ ♀ *Ff Rr*	×	♂ *ff rr*	13	Frizzled, rose comb	*Ff Rr*
			9	frizzled, single comb	*Ff rr*
(frizzled,		(normal,	11	normal, rose comb	*ff Rr*
rose comb)		single comb)	13	normal, single comb	*ff rr*
			46		

Since the number to be expected in each class with independent segregation was $^{46}/_4$, or 11.5, it is clear that the numbers obtained fit well the hypothesis that *F* and *R* (or their recessive alleles) segregate independently. They show no sign of linkage.

Now let us consider in another way what is behind that independent segregation in the backcross. The gametes of the double recessive male carried no dominant genes—they were all *f r*. Accordingly, those male gametes could not mask or obscure in any way the combinations of dominant and recessive genes in the gametes produced by the females. The backcross was therefore really a test-cross that revealed, not only what *kinds* of gametes were formed by the females, but also the *proportions* in which the four different kinds—*F R, F r, f R,* and *f r* —were produced. Therefore, to say that the characters concerned segregated independently is equivalent to saying that these four kinds of gametes were formed in approximately equal numbers. The cross to be considered next was one in which four kinds of gametes were also produced, *but not in equal proportions.*

In tests for possible linkage of *F* and *I*, the gene causing dominant white plumage, crosses were made as follows:

P₁: *FF II* (frizzled, white) × *ff ii* (normal, colored)
$$\downarrow$$

F₁: *Ff Ii*

Backcross: ♀ ♀ *Ff Ii* × ♂ *ff ii*
$$\downarrow$$

Progeny: $\left\{\begin{array}{ll} \text{15 frizzled, white} & Ff\,Ii \\ \text{2 frizzled, colored} & Ff\,ii \\ \text{4 normal, white} & ff\,Ii \\ \underline{\text{12}}\text{ normal, colored} & ff\,ii \\ \text{33} \end{array}\right.$

Here is a different story. Although the numbers are few, it is clear that they do not fit the 8.25 in each class that would be expected if *F* and *I* had segregated independently. Surely the ratios of frizzled to normal (17:16) and of white to colored (19:14) fit well enough the 1:1 ratio expected for each pair of alleles in the backcross, but there is a marked excess of the two *parental combinations* (*F* with *I*, and *f* with *i*) over the *new combinations* (*F* with *i* and *f* with *I*). In other words, *F* and *I* are linked. The strength of that linkage is measured by the proportion of exceptions to it, in this case six new combinations among 33 individuals, or 18.2 per cent.

If we now look back at these two tests for linkage of *F*, the one with *R* and the other with *I*, the conditions under which one can readily determine whether two pairs of genes are independent or linked are seen to be as follows:

1. We must have an animal that is heterozygous for each pair of genes. (These convenient creatures don't often occur by chance; they usually have to be made to order.)

2. We then arrange a mating that will reveal what kinds of gametes are formed by the dihybrid, and in what proportions.

3. To that end, the dihybrid is backcrossed to the double recessive type.

In this backcross of dihybrid females to the *ff ii* ♂ , the gametes of the double recessive type are all alike, and (most important) they carry no genes that can obscure the combinations of genes in gametes of the dihybrid. Accordingly, the counts for the 33 backcross progeny given above constitute the classification of a sample of 33 gametes produced by their dihybrid mothers. These showed the ratio of parental combinations to new combinations to be 27:6. When linkage has been demonstrated, the new combinations are commonly referred to as *cross-overs*.

Coupling and Repulsion

In the test for linkage of F and I considered above, both dominant genes entered the cross in a chromosome from one parent, and both recessive alleles came in from the other parent. In genetic parlance, these genes were linked in the *coupling phase* of linkage. We owe that term to Bateson and Punnett who, in an age when most genetic characters were still dutifully segregating independently in nice Mendelian ratios for neo-Mendelians, found pairs of alleles in their sweet peas in which the dominant genes stayed *coupled* in crosses more often than not. When their crosses were made in such a way that one dominant gene came in from each parent, then the dominant genes tended to *repel* each other, *i.e.*, to stay apart in crosses. In such a case, we now say that the linked genes entered the cross in the *repulsion phase*[1] of linkage.

It is helpful in most studies of linkage, and essential in some, to know beforehand in which phase of linkage the genes under test entered the cross. When that information is known, it can be indicated by using a simple shorthand device for that purpose. In the case just considered, the parents of the dihybrid can be designated either as *FF II* and *ff ii*, or as

$$\frac{FI}{FI} \text{ and } \frac{fi}{fi}$$

These last two arrangements of genes help us to visualize the fact that there are two homologous chromosomes, that one parent has F and I in both, and that the other parent has the two recessive alleles in both. Then, when the F_1 hybrid is considered, its genotype must be

$$\frac{F\ I}{f\ i}$$

a designation which tells not only that it is doubly heterozygous, but also that it carries the two pairs of genes in the coupling phase.

---

[1] Not, as some bright lads write on their examination papers, the "repulsive" phase.

Conversely, when F and I enter a cross from different parents, as in the linkage test to be reviewed in the next section, then the original parents are

$$\frac{Fi}{Fi} \text{ and } \frac{fI}{fI}$$

and the hybrid must be

$$\frac{F\ i}{fI}$$

That formula for its genes shows that it is a dihybrid, with the two pairs of alleles in the repulsion phase of linkage. After getting accustomed to this way of writing genotypes, most geneticists eliminate one of the horizontal lines (chromosomes) and show the formula as

$$\frac{F\ i}{fI}$$

Crossing-over

The process by which new combinations of linked genes are formed in the gametes is called *crossing-over,* hence the designation of such gametes (and of individuals arising from them in test-crosses) as *cross-overs.* It occurs during the first meiotic division, at the time when pairs of homologous chromosomes lie close together in synapsis. As we learned in Chapter 6, that stage begins after the leptotene (or thin-thread) stage early in meiosis. By the time the diplotene stage is reached, each member of a homologous pair has formed a duplicate of itself, or, as is commonly said, it has split into two strands. A pair of such chromosomes is thus comprised of four strands, or chromatids. From critical tests in favorable species, it is now believed that crossing-over occurs at this stage and that it results from an interchange of segments between two strands. That interchange, in turn, is believed to result from (a) one strand's lying across another, (b) a break in both at the point of junction, or chiasma, and (c) subsequent fusion of the broken end of one chromatid with the broken end of the other.

If the two chromatids that interchange parts of themselves are both of maternal origin, or both of paternal origin, the newly formed, fused chromatid will have exactly the same genes as were in the parental chromosomes. However, when the broken end of a maternal chromatid joins up with a segment of a paternal chromatid, a new combination of genes can result. Moreover, since bits and pieces of the broken chromatids do not ordinarily wander off by themselves, each break results in an interchange affecting *two* chromatids, not just one. There are altogether four broken ends. These fuse to re-form two chromatids, each of which carries a segment of maternal origin and one of paternal origin. This is shown diagrammatically in Fig. 10-1.

Utilization of Crossing-over

An important point to be made clear at this stage is that mathematical determination of the proportion of all chromatids carrying specific linked alleles which undergo exchanges of parts can tell us something about the locations of the linked genes in the chromosomes concerned. Just as there is more chance of getting a kink in a 30-foot garden hose than in one only 10 feet long, similarly, cross-overs between chromatids should occur more frequently in long distances than in shorter ones. It follows that, by measuring the amount of crossing-over between two linked genes, we can get some idea of the distance between their loci in the chromosome to which they both belong. As additional linkage relationships in that same chromosome become known, the relative positions can be determined for several or many genes, and these can be combined to make a map showing the loci of all genes proven to lie in that chromosome. The knowledge afforded by such a map increases the predictability of results in matings that involve linked genes for which the loci are known.

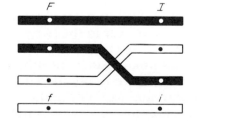 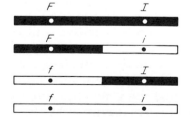

FIG. 10-1   Diagram showing how crossing-over, the interchange of matching segments between two chromatids, results in two new combinations of genes, both different from the original parental combinations.

Linkage of *F* and *I*; Repulsion Phase

To review and consolidate the many new points already brought out in this chapter, let us consider another test for linkage with the genes *F* and *I*. Dihybrid females carrying those two genes in the repulsion phase were backcrossed to a double recessive male, and 157 chicks were classified. Behavior of the chromosomes and genes in this linkage test is shown in Fig. 10-2.

As is shown at the bottom of the figure, crossing-over between *F* and *I* was 19.7 per cent in the 157 gametes measured in this test. That figure is very close to the 18.2 per cent found for the same relationship in the test cited earlier, in which these same genes were in the coupling phase.

The fact that the two parental classes both had the same number of chickens (63) was a rather unusual coincidence, as identical numbers in both parental classes or both cross-over classes are not often found. It is true that each *F I* cross-over should have been matched by one of the type *f i*, but, although the two kinds are produced in equal numbers, not all gametes form zygotes (*i.e.*, some eggs are not fertilized), and not all zygotes survive to an age at which they can be classified (*i.e.*, some embryos die during incubation). It will be noted that the ratios of 81 frizzled : 76 normal, and 81 white : 76 colored, conform well to the 1 : 1 ratio expected within each pair of alleles. Such an agreement assures us that there was no differential mortality reducing the numbers in any one of the four classes. In other words, the chickens that were classified did provide a fair sample of the kinds of gametes produced by their dihybrid dams.

Considering now both tests for linkage of *F* and *I*, one in the coupling phase and the other in repulsion, let there be no confusion because the parental combinations in the former (*F I* and *f i*) are the cross-over combinations in the latter (Fig. 10-2). The important point is that, whether the linked genes enter the cross in one way or in the other, they tend to stay together in the same combinations as those in which they entered the dihybrid. Furthermore, as is indicated by the determinations of 18.2 and 19.9 per cent crossing-over in the two tests just considered, the exceptions to linkage (*i.e.*, the cross-overs) occur with about the same frequency in one phase of linkage as in the other.

Another point to keep straight is that what are really linked are not just the genes that have here been designated as *F* or *f* and *I* or *i*, but the loci occupied by these genes in the chromosome. Accordingly, we could refer to the linkage of *f* and *i* just as well as to that of *F* and *I*. In the repulsion phase the genes actually linked are *F* with *i*, and *f* with *I*, but one could still designate the linked loci as *F* and *I*. This is permissible because it is customary to label a locus according to the

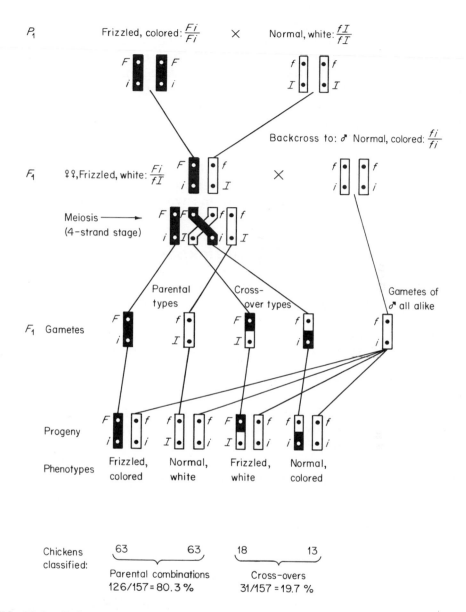

FIG. 10-2 Behavior of chromosomes and genes in a typical test for linkage, here illustrated by *F* and *I* carried in the repulsion phase. Numbers given at the bottom are those actually found in one such test.

nature of the mutant gene thereat. It happens that both F and I are dominant mutations (*i.e.*, dominant to the normal, standard, or *wild type*, as it is often called), hence the loci that they occupy are designated by capital letters. When the linked genes are both recessive, as are px and n in the fowl's sex chromosome, we designate their loci with small letters, and refer to the linkage of px with n. In that same chromosome there is about 7 per cent crossing-over between dw and K.

Another reason for thinking in terms of linked loci rather than of specific linked genes is that, as we shall see later, a single locus may be occupied by any one of several different allelic genes. All of these should show the same linkage relations with other loci in the same chromosome.

Testing for Linkage

The simplest and most direct tests for linkage are backcrosses of dihybrids to double recessives, like the two considered thus far in this chapter. Complications sometimes arise when interactions of genes prevent recognition separately of the two parental and two cross-over classes, but such apparent difficulties are sometimes not as serious as they may at first seem. A case in mice will serve as an example.

Albinotic mice (cc) have pink eyes, but there is another recessive mutation, called pink-eye (pp), which causes extreme dilution of pigment and pink eyes, so that homozygotes are difficult to distinguish from albinos, especially in young mice. In a test for linkage of these two genes, Dunn made the following matings:

$$P_1: \quad \text{Albino: } cc\ PP \times \text{Pink-eye: } CC\ pp$$

$$F_1: \qquad \frac{cP}{Cp} \times \text{double recessive: } \frac{cp}{cp}$$

Progeny: $\left\{ \begin{array}{l} \text{1,262 Pink-eyed, including albinos} \\ \text{107 Dark-eyed, colored} \end{array} \right.$

If c and p are linked, the progeny to be expected from the test-cross would be classified as follows:

$$\left. \begin{array}{l} c\ P\ /\ c\ p : \text{Albino} \\ C\ p\ /\ c\ p : \text{Pink-eye} \end{array} \right\} \text{Parental combinations}$$

$$\left. \begin{array}{l} C\ P\ /\ c\ p : \text{Dark-eyed, colored} \\ c\ p\ /\ c\ p : \text{Albino and pink-eye} \end{array} \right\} \text{Cross-overs}$$

It is now evident that three of the four expected phenotypic classes would be indistinguishable, but all is not lost. Since the numbers in one cross-over class should be about the same as in the other, and since we can identify the 107 dark-eyed mice as comprising one cross-over class, it can be assumed that the total number of cross-overs is about 214 among 1,369 gametes tested, or 15.6 per cent.

Other examples of interactions that complicate slightly the measurement of crossing-over will be found in the problems at the end of this chapter.

Plant breeders often find it easier to measure crossing-over in an F_2 generation than in a backcross. This is particularly so with species in which self-fertilization is normal, when backcrosses could be made only by artificial pollination of many flowers that would first have to be emasculated. With such species, an F_2 population has the advantage of easily providing many seeds (or plants) without any special effort except to protect the self-fertilizing flowers from pollen from other plants. The drawback to determination of crossing-over in F_2 is that, since both F_1 parents produce parental and cross-over gametes, the proportions of these cannot be seen directly in their F_2 progeny, but must be calculated from the degree to which observed numbers deviate from those expected in the normal phenotypic ratio of $9:3:3:1$. There are statistical procedures for such determinations, but they are somewhat complicated. Rather than to attempt them, animal breeders should be grateful that vertebrates are not self-fertilizing, and that they (the animal breeders) can usually stick to backcrosses for measurement of crossing-over.

At the same time, one must remember that the double (or triple, etc.) recessives necessary for such test-crosses are sometimes not available. One cannot test for linkage of an autosomal lethal gene in a backcross, except in cases when a few animals homozygous for that lethal manage to survive (perhaps by special protection, such as an extremely favorable environment) to breeding age. Otherwise, to test for linkage of an autosomal recessive lethal gene, one would have to do so, willy-nilly, in an F_2 population. In such a case, one whole class would be eliminated by the lethal.

Special techniques for measuring crossing-over are necessary with species in which extensive breeding experiments are impossible (man), or impracticable (cattle and horses). In such cases, linkage can sometimes be detected, and estimates of crossing-over can even be made, by determining the degree to which two different traits are associated in pedigrees. Techniques suitable for such studies are given by Stern (cited in Chapter 7), and other methods applicable in special cases, including F_2 populations, are explained by Mather (1951).

Testing for Linkage of Sex-linked Lethals

In contrast to the difficulties confronting tests for linkage of autosomal lethal genes, crossing-over between sex-linked lethals and other genes in the same chromosome can be measured fairly easily. The procedure can be illustrated with an example from the domestic fowl, because it has already been proven practicable with sex-linked, lethal genes in that species.

One must first identify individuals of the homogametic sex (males) that are heterozygous for both the genes to be tested. The males to be used can have received the lethal gene only from their sires, and only half of them are likely to carry the lethal. Accordingly, breeding tests will be necessary to identify the heterozygotes. If the other gene in the test is not lethal, one may know, from the parentage, or by the appearance of the males, which are heterozygous for that other gene; otherwise a breeding test may be necessary. Having identified the dihybrid males, these are mated with almost any females available, except those carrying autosomal genes that might mask the expression of sex-linked genes (*e.g.,* dominant white from White Leghorns that would be epistatic to barring and silver). From these test matings, all the male chicks are discarded, because they must all have received from their dams a sex chromosome not carrying the lethal gene. Crossing-over is measured *only in the female progeny.* As these are hemizygous, and get their one sex chromosome from their sires, they (unlike their discarded brothers) have no protection against the lethal, and will therefore show, so far as the sex chromosome is concerned, the kinds of gametes formed by their dihybrid sires.

Using this procedure, Cole (1961) tested for crossing-over between sex-linked albinism, *al,* and paroxysm, *px.* The latter is lethal; the former is not. When about two weeks old, or later, chicks afflicted with paroxysm react to unusual noise, sudden bright light, or other alarms, by starting to run and then falling as in a tetanic seizure, with legs rigidly extended, head thrown back, wings beating violently, and the body in a state of tremor (Fig. 10-3). After about 10 seconds, the birds relax, lie quietly, and eventually stagger off. Most of them die before 15 weeks of age, and none can be used for breeding.

An *Al Al* male known to carry *px* was mated to an albinotic female, *al Px—,* and, by breeding tests, three sons of that mating were proven to carry *px.* Their genotypes had to be *Al px/al Px.* When these were mated to normal females, *Al Px—,* the results (Table 10-1) showed 28 cross-overs in 268 measurable gametes, or 10.5 per cent.

In mammals, very few sex-linked lethals are known, but more

FIG. 10-3 Typical paroxysm (sex-linked) in a chick of seven weeks, showing legs and toes rigidly extended and head thrown back in opisthotonic posture. (From R. K. Cole in *J. Heredity*.)

TABLE 10-1
Progeny from Cole's Test for Linkage of Albinism and Paroxysm in the Fowl

Males	Females		Paternal Gamete Indicated
Phenotype and Possible Genotypes	*Phenotype and Number*		*Paternal Gamete Indicated*
None albinotic, no paroxysms			
Al px/Al Px	Not albinotic, paroxysm:	119	*Al px*
al Px/Al Px	Albinotic, no paroxysm:	121	*al Px*
Al Px/Al Px	Not albinotic, no paroxysm:	17	*Al Px*
al px/Al Px	Albinotic, paroxysm:	11	*al px*
All discarded		268	

must eventually be recognized. Tests for linkage of these could be made in the same way as that illustrated above for the fowl, except that the gametes to be tested would have to come from females, and crossing-over could be measured only in the male progeny.

Three Linked Genes, and a Map

At about the time when the genes F and I of the fowl were found to be linked, other investigators showed that F is linked with Cr, the dominant gene that causes the crest, or top-knot, of feathers on the head in Houdans, Polish, and other breeds (Fig. 4-5). Now, if the theory expounded in previous chapters is correct—that genes are arranged in linear order in the chromosome—then, if F and I are linked, and F and Cr are linked, so also should I and Cr be linked. Two questions arise: (a) In what order are the three genes arranged? (b) What distances separate them? These questions can be answered by tests to determine the linkage relationship still unknown—that of I and Cr.

Crossing-over between F and Cr is about 27 per cent. When many measures of crossing-over between F and I were put together, it was found that, while determinations for different sires varied from 7.5 to 26 per cent, a total of 1,105 gametes tested showed 17.1 per cent cross-overs. For arrangement of the three genes, there are two possibilities, and two only. If the order is F-I-Cr, then the locations of the genes are approximately as follows:

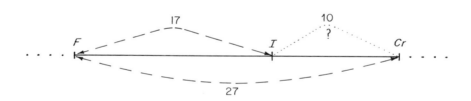

In this little *map*, the amounts of crossing-over already known are shown, not as percentages, but as measurable distances between the linked genes—17 and 27; the probable distance between I and Cr is indicated as the difference between those figures. Geneticists usually measure distances between genes in cross-over units, each unit being equal to one per cent crossing over. If one cross-over unit is considered as the distance on the chromosome within which a single cross-over can occur, then, when 27 such units of distance have been measured between F and Cr, and 17 between F and I, it would be logical to expect the distance from I to Cr to be 10 cross-over units, as is shown (tentatively) above.

Conversely, if the order of the genes is Cr-F-I, then their locations must be approximately like this:

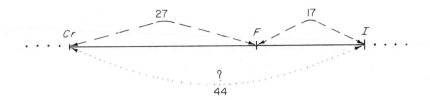

With this arrangement, the cross-over distance between *I* and *Cr* should be approximately the *sum* of the other two. When crossing-over between *I* and *Cr* was measured by Warren and Hutt (1936), it was found to be 12.5 per cent. This is close enough to the 10 per cent postulated for the order *F-I-Cr* to show that it is the correct one, and that the alternative possibility is eliminated.

The result of all these tests is that determinations of three different relationships among three linked genes have given us a *map* of that portion of the chromosome in which those genes lie. It is not necessary to draw a map when only a few genes are concerned; the order of the genes and the distances between them can be written simply as:

F 17 *I* 12.5 *Cr*

This simple little exercise with the three genes has illustrated three principles which are important enough to warrant the following enumeration:

1. When three genes, *A*, *B*, and *C* are linked, and the amounts of crossing-over between *A* and *B*, and between *B* and *C*, are known, then crossing-over between *A* and *C* should be either the sum, or the difference, of the other two. (The inevitable exceptions follow in the next section!)

2. Such a relationship is possible only if the genes lie in linear order in the chromosome; *ergo*, evidence that the relationship postulated is actually realized constitutes proof of that linear order.

3. By studies of linkage and crossing-over, some genes can be assigned to a string of genes that all belong in one chromosome, and to their loci in that chromosome in relation to other genes.

Theoretically, all genes might eventually be thus mapped, but Tennyson's reminder (in "Locksley Hall") that "Science moves but slowly, slowly, creeping on from point to point" will be endorsed by every geneticist who has ever quested after linkage in any animal bigger than a shrew. Most of us will be content, when the game is finished, if we have been lucky enough to have found in a lifetime one case of linkage for every decade of our working years.

Double Crossing-over

Complications arise when linked genes are far apart. In the fowl, duplex (bifurcated) comb, D, was found to show 28 per cent crossing-over with M, a dominant gene causing multiple spurs, which is a breed characteristic in Black Sumatra Games. M, in turn, proved to be loosely linked with Po, another dominant gene causing polydactyly, or duplication in the short inner toe, or hallux. Polydactyly is a breed characteristic in Dorkings and Houdans (Fig. 4-5). Crossing-over between M and Po was 33 per cent.

To find the relative positions of these three genes, a test for linkage of D and Po was made. Fortunately D, M, and Po are all expressed in both sexes and can be classified at hatching, or even in late-dead embryos, and 517 gametes were tested in the backcross. According to the results and principles explained in the previous section, the amount of crossing-over between D and po should have been either $33 - 28 = 5$ per cent, or $33 + 28 = 61$ per cent. The former would have shown the order of the genes to be Po-D-M, while the larger figure would indicate a D-M-Po arrangement. When the test was made, the amount of crossing-over between D and Po proved to be 42 per cent. This completely ruled out the Po-D-M order, but was far short of the map distance to be expected with the other. Why? Because of *double crossing-over* between D and Po.

Just as the number of cross-overs between pairs of linked loci increases with the map distance between those pairs, when that distance becomes long enough, there may be *two* cross-overs, at different locations, between two genes far apart. The effect of these is illustrated diagrammatically in Fig. 10-4. To simplify matters, no single cross-overs between D and Po are shown, but only a double cross-over between two chromatids, one carrying D, M, and Po in the coupling phase of linkage, and the other d, m, and po.

Comparison in Fig. 10-4 of the two new gametic combinations resulting from the double cross-over with the two parental combinations reveals that the only difference is in the gene in the middle—M or m. The two terminal, or outside, genes are identical with those in the original, parental chromosomes. In fact, if M and m were not considered, the gametes resulting from double crossing-over would be indistinguishable from those of the parental type.

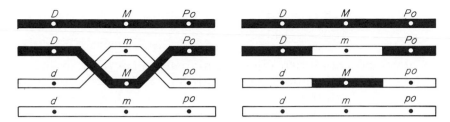

FIG. 10-4 Double cross-overs between two strands of a tetrad leave the genes at the ends in the parental combinations (upper and lower chromatids), but interchange the genes in the middle.

This is exactly the case with the data given above for linkage of *D* and *Po*. Nothing was said about *M* and *m*. The *visible* cross-overs between *D* and *po* were only 42 percent, but these were all single cross-overs: *D po* and *d Po*. The total amount of crossing-over was considerably greater, but was not evident because the double cross-overs (when *M* and *m* were not considered) were indistinguishable from the parental types. Another way of saying the same thing is that the total amount of crossing-over, cross-over units, or map distance between *D* and *Po*, though estimated from its two component segments at 61 units, was reduced by double cross-overs to an apparent distance of only 42 units.

Actually, in this test for linkage of *D* and *Po*, the other gene, *M*, was not ignored, and classifications for multiple spurs or normal ones were carefully made with every bird. Unlike the situation shown in Fig. 10-4, the trihybrid females that were backcrossed to a triple-recessive male carried *D* and *Po* in the coupling phase, but *M* in the repulsion phase. Their genotypes were

$$\frac{D\ m\ Po}{d\ M\ po}$$

Double cross-overs would therefore show either *D M Po* or *d m po*. The number of these in the 517 birds classified was 48, or 9.28 per cent, a figure almost identical with the number expected. This last can be calculated easily, provided that the order of the genes in the chromosome is known.

Recalling that the probability of two events occurring together is the product of the separate probabilities for each, if the frequency of cross-overs in the region *D* to *M* is 28 per cent, and that in the region *M* to *Po* is 33 per cent, then the probability of two cross-overs occurring in one tetrad, one in each region, becomes

$$\frac{28}{100} \times \frac{33}{100} \times 100, \text{ or } 9.24 \text{ per cent}$$

Conversely, when two adjoining map distances are known (as for *D* to *M* with 28 units, and *M* to *Po* with 33), and when double cross-overs involving both regions can be identified, and have been counted (as 9.28 per cent), then the amount of visible (*i.e.*, single) crossing-over between the two genes farthest apart (*D* and *Po*) can be predicted fairly accurately. It should be:

Sum of the two distances = 28 + 33   61.0 units
Less 2 × observed frequency of double c.-o. = 2 × 9.28 = 18.5 units

Expected: 42.5 units
Actually observed: 42.0 units

In making this calculation, students are often mystified by the subtraction of *twice* the observed frequency of double cross-overs. One need only recall that every double cross-over means two interchanges between chromatids, equivalent to two single cross-overs, between the two genes at the extremes.

Some General Considerations of Crossing-over and Mapping

The example considered above has shown that, because of double crossing-over, determinations of crossing-over between genes far apart do not give an accurate measure of the actual distance between those genes. Obviously the true distance in such cases is best measured by adding up the separate determinations for several component segments of the whole, each so short as to be little affected, or not at all, by double crossing-over. In Drosophila, it has been found that double cross-overs do not occur within distances of 10 units, and in some regions not even within 15 units.

When the visible cross-overs reach 50 per cent, it is impossible to say whether the genes concerned are linked, or are segregating independently. At that figure, the four kinds of gametes produced by a dihybrid would show up in backcrosses in the ratio of 1:1:1:1 expected with independence. Another way of saying the same thing is that crossing-over measured between any two loci should not exceed 50 per cent. It can be calculated, by considering all the possibilities for interchanges between any two strands of a tetrad, that the number of parental combinations plus double cross-overs cannot exceed the number of single cross-overs within one tetrad, but we shall not go into that here.

It follows that two loci in one chromosome may be so far apart that they show no linkage whatever, and the genes occupying those loci would therefore appear to be independent of each other. Crossing-over between B and S in the fowl is about 48 per cent, so close to independence that hundreds of gametes have to be measured to show a significant deviation from 50 per cent. Their linkage might never have been known, were it not that B and S had earlier both been shown to belong in the sex chromosome. In the same species, the mutation, jittery, is also known to lie in the sex chromosome, but the causative gene is so remote from other sex-linked genes with which it has been tested that no linkage has been found.

A corollary to all this is that, while many hundreds of gametes have to be measured to prove statistically that genes far apart are really linked and not independent, close linkages are evident in comparatively small numbers of test animals. With only 35 hens, among which were two cross-overs and the rest of the parental types, it was shown that P (pea comb) and O (blue egg) are so close together in one chromosome that crossing-over between them is less than 6 per cent. With distances shorter than that figure, it would again be necessary to test larger numbers to find whether or not any cross-overs occur at all. Landauer's determination of 0.39 per cent crossing-over between Cp (creeper) and R (rose comb) was derived from 7,408 chicks and embryos. He could have classified 300 or 400 without finding a single cross-over or perhaps only one!

Although crossing-over between genes far apart cannot exceed 50 per cent, the total map distance can easily do so, as it did with D and Po in the previous section. Similarly, crossing-over between speck (body) and star (eyes), at opposite ends of the second chromosome of Drosophila, is about 50 per cent, but the map distance between those genes, as established from separate determinations of many short segments within it, is about 105 units.

When testing for linkage, it is desirable to use some gene or genes already known to belong to some specific linkage group, and also one or more "unknowns." The former serve as *markers* for one group (or chromosome). If the unknown gene (*i.e.*, not yet assigned to any linkage group) proves to be independent of markers for the known linkage groups, then it might well be tested with other unknowns in the hope of finding a new linkage group. One should remember that there is a special advantage in testing in one cross for possible linkages of three genes, or more. If such a test should reveal a *three-point linkage*, as with *D*, *M*, and *Po*, then, if any double cross-overs can be identified, comparison of these with the parental combinations will reveal the order of the genes. The gene in the double cross-overs that differs from the parental types must be between the other two, as we saw in Fig. 10-4.

Crossing-over has been shown to occur occasionally in somatic cells. It is revealed by "twin spots" showing side-by-side two genetically different kinds of cells, both of these differing from the other cells of the individual affected. Thus a plant or animal of the genotype *A B/a b* would show (in a twin spot) tissues recognizable as *A b* and *a B*. Such cases are rare.

Interference; Variations in Crossing-over

Just as a twist of two strands of rope, the one around the other, physically prevents another similar twist from being made close to the first, so does a chiasma at one point prevent another from being formed close by. In other words, a cross-over at one locus prevents another from occurring right next to the first. The degree of such *interference* with double crossing-over varies with the distance in cross-over units. Thus, interference is complete in Drosophila within 10 map units, and decreases in longer distances up to 50 per cent crossing-over, when none is found. The degree of interference is measured by *coincidence*, which is the ratio of observed double cross-overs to those expected from calculations.

In our example with *D*, *M*, and *Po*, in a map distance of 61 units there was apparently no interference whatever, as the double cross-overs agreed with the proportion expected. However, in a three-point linkage of *F*, *I*, and *Cr* (17.1 + 12.5 units), Warren and Hutt calculated the theoretically possible, double cross-overs at 2.13 per cent (17.1 × 12.5), but found none at all among 284 gametes tested. Presumably interference was responsible for some of the discrepancy, and, perhaps, deviation from expectation by chance in smallish numbers for the rest.

In mice and rabbits, crossing-over seems to run slightly higher in females than in males, but this does not apply to the fowl. In Drosophila it is influenced by age, by temperature, and also by mutations that prevent crossing-over in an entire chromosome, or in a region of one. There seems also to be considerable variation among individuals; hence it is desirable to sample the gametes produced by several animals before considering any determination of crossing-over to be a reliable measure of the distance between the genes concerned.

A Three-Point Linkage Test in a Mammal

Enough of fowl genetics! To consolidate our position again, let us see how what we have learned about linkage in examples from the fowl applies to a three-point test for linkage in a *real* animal. It would be very satisfying to be able to use for this purpose a case in the horse or in cattle, in sheep or in swine, but (alas!) only a few cases of linkage, mostly in genes for red-cell antigens, or enzymes, have been established in any of those otherwise commendable species. The best we can produce is a rabbit, which, though not as well explored genetically as the house mouse, has some advantage over the mouse in size, and, moreover, enjoys a reputation as a respectable domestic animal in that great outer world beyond the ivy-mantled towers of genetical laboratories.

The late W. E. Castle, whose valuable contributions to our knowledge of animal genetics covered a span of six decades, once tested, by backcrossing trihybrids to triple recessives, for linkage relations of the following genes in the rabbit:

Colored:	C	Black:	B	White fat:	Y
Himalayan:	c^h	Brown:	b	Yellow fat:	y

The 908 rabbits from that test-cross (Table 10-2) showed clearly that these three pairs of alleles were not segregating independently. If they had done so, there should have been 113.5 in each class, and although two of the eight phenotypes appeared in numbers close to that figure, none of the others did so. Obviously, some of the genes, at least, were linked. We are not told in what phase or phases of linkage these three pairs of alleles entered the cross (*i.e.*, the parental combinations), nor is their arrangement in the chromosome known. The problem is to see how much can be learned from those 908 rabbits about the relationships of the three pairs of alleles involved.

Assuming as a working hypothesis that all three genes are linked, the first objective is to determine, if possible, in what order they lie in the chromosome. Although they are shown in Table 10-2 in the order $c^h \, b \, y$, that arrangement may not be the correct one. If we recall that, in a test for three-point linkage, the order of the linked genes can be learned from comparisons of any double cross-overs with parental combinations, and that such comparisons show an interchange in the pair of alleles that lies between the other two, the first step is to identify, if possible, which gametes are most likely to be of the parental types, and which are most likely to be double cross-overs. As

TABLE 10-2

Classifications of Rabbits in Backcross of Castle's Three-Point Test for Linkage

Phenotypes	Corresponding Gamete from Trihybrid Parent	Number of Rabbits
Himalayan, black, white fat	$c^h B Y$	276
" , " yellow fat	$c^h B y$	7
" , brown, white fat	$c^h b Y$	125
" , " yellow fat	$c^h b y$	46
Colored, black, white fat	$C B Y$	55
" , " , yellow fat	$C B y$	108
" , brown, white fat	$C b Y$	16
" , " , yellow fat	$C b y$	275
		Total: 908

the former must outnumber any one kind of cross-over gamete, the parental combinations are easily recognized as $c^h B Y$ and $C b y$.

Similarly, if there are any double cross-overs, these must be far fewer in number than any single cross-overs, and are therefore quite likely to be represented by the combinations $c^h B y$ and $C b Y$, which appeared in only 7 and 16 rabbits, respectively. Together these made 2.53 per cent of the total number. These rabbits were like the parental combinations with respect to two pairs of alleles, but differed in having Y and y transposed. The correct order of the genes is, therefore, not the one shown in Table 10-2, but $c^h Y B$ in one parental combination, and $C y b$ in the other.

It is now clear that y and b are carried in the coupling phase, and the other pair of alleles in the repulsion phase. To work out the distances separating the loci of the three alleles, it may help us to visualize their relationships at this stage if we draw a tentative map showing the arrangement of the genes in the two parental combinations, even though distances between them cannot be given. It will look like this:

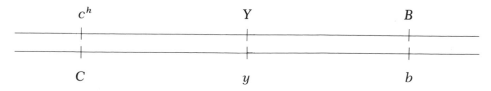

If we were to follow a convention that is common usage with

geneticists, we could designate the wild-type (normal) genes by + signs, and write the parental combinations thus:

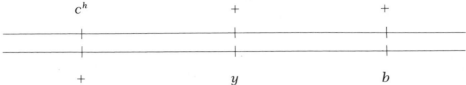

When we are learning, we may do so more easily by writing out all three symbols, as above. For most of us, subsequent calculations will also be facilitated if, in our working papers, we cover up or rub out the genotypes shown in Table 10-2, and rewrite them with the genes in their correct order, *i.e.*, with Y and *y* in the middle.

We can now calculate the number of single cross-overs in the two regions shown in our tentative map, the one to the left of the *y* locus, and the other to the right of it. Single cross-overs in the former would yield gametes carrying $C\,Y\,B$ and $c^h\,y\,b$, and these two classes are seen to contain 55 and 46 rabbits, respectively, or 101 cross-overs in the 908 gametes sampled.

Similarly, cross-overs in the other region (the only two classes not previously considered) are 125 + 108, a total of 233. Before converting these numbers for the two regions to percentages, and thus to map distances, we must remember that there were also 23 double cross-overs, each involving both regions. Hence, to determine the total number of cross-overs in either region, we must include not only the visible, single cross-overs, but one more for each double cross-over.

The amounts of crossing-over in the two regions thus prove to be as follows:

$$\text{Between } c^h \text{ and } y: \ \frac{101 + 23}{908} = 13.65 \text{ per cent}$$

$$\text{Between } y \text{ and } b: \ \frac{233 + 23}{908} = 28.19 \text{ per cent}$$

Total distance: 41.84 per cent

The completed map for the segment of the chromosome containing these three genes can now be drawn as follows:

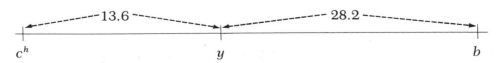

The total distance between c^h and b could have been determined without first making separate measures of the two component segments. The total number of single cross-overs (*i.e.*, in four classes only, excluding the two kinds of double cross-overs) was 334, or 36.78 per cent of all gametes tested. Adding to that figure twice the observed proportion of double cross-overs (*i.e.*, 2 × 2.53 per cent) or 5.06, one gets 41.8, which agrees with the sum of the two component parts, as calculated earlier.

The proportion of double cross-overs to be expected in this test if there were no interference was

$$\frac{13.6}{100} \times \frac{28.2}{100} \times 100, \text{ or } 3.83 \text{ per cent}$$

The observed proportion was 2.53 per cent. Calculation shows coincidence—the ratio: 2.53/3.83 × 100—to be 66 per cent, which indicates that some double crossing-over was prevented by interference.

Maps of Chromosomes

Evidence that two genes are linked provides some information—a tiny advance, perhaps—in one frontier of biology. Still more is gained when three genes can be assigned to one linkage group, with the distances between their loci established, as with the three just considered in the rabbit. When still more genes are added to known linkage groups, the groups are extended, gaps are filled in, and several or many genes really justify our designation of the linkage group as a map for some chromosome. Theoretically, it should eventually be possible to spot most of the genes that occur in a chromosome, but, in practice, progress is very slow, especially with any animal bigger than a fruit fly, and there is no way of knowing at any time how many more genes remain to be recognized.

Thanks to the extensive studies of many investigators, the chromosomes of *Drosophila melanogaster* (the fruit fly) have been more thoroughly mapped than those of any other organism. One factor facilitating that study is the fact that all the genes of that species have to belong in one or another of only four chromosomes, one of which, a mere dot under the microscope, contains only five mapped genes. Two of the others run to a little more than 100 map units, and the sex chromosome has 66 such units. An important confirmation of all genetic theory is provided by the fact that the lengths of the four chromosomes in Drosophila, as determined from studies of crossing-over, and hence measured in map units, correspond very closely with the relative lengths of those same chromosomes as seen under the microscope and measured in microns.

The number of linkage groups that can be found in any one species should be the same as the haploid number of chromosomes for that species. In *Drosophila virilis*, which has six pairs of chromosomes, there are six linkage groups.

The *Biology Data Book* (1972) lists 1,115 mutations in Drosophila, which are distributed as follows:

In the sex chromosome	547	In Chromosome 3	230
In Chromosome 2	296	In Chromosome 4	42

Loci in the fly's map are given for all of these except 37 genes (mostly lethals) assigned to Chromosome 4, but not mapped.

A Chromosome Map for the Mouse

Among mammals, the most extensive chromosome map available is that for the mouse, *Mus musculus*, which has 20 pairs of chromosomes. It also has 20 linkage groups. Thanks to the fact that the individual chromosomes can now be recognized by the banding techniques, and the fact that *translocations* can also be identified, it has been possible to assign each linkage group to its proper chromosome. Translocation is the term used by geneticists for accidents in the cell by which a part (or all) of one chromosome becomes attached to some other chromosome. This subject is covered in a later chapter. By determining from cytological examination which chromosomes are involved, and observing which linkage groups (formerly separate) have been combined in the mice carrying the translocations, it has been possible, by combining the results from many dedicated mouse geneticists, to assign each of the 20 linkage groups to its proper chromosome. The resultant map, compiled by T. H. Roderick and Muriel T. Miller (of the Jackson Laboratory, Bar Harbor, Maine), shows the loci for no fewer than 470 genes (Fig. 10-5a). Figure 10-5b shows a single chromosome and names of mutations assigned to it.

Other Maps

After the mouse, the most extensive chromosome map for any domestic animal is that of the silkworm moth, *Bombyx mori*. In that species,

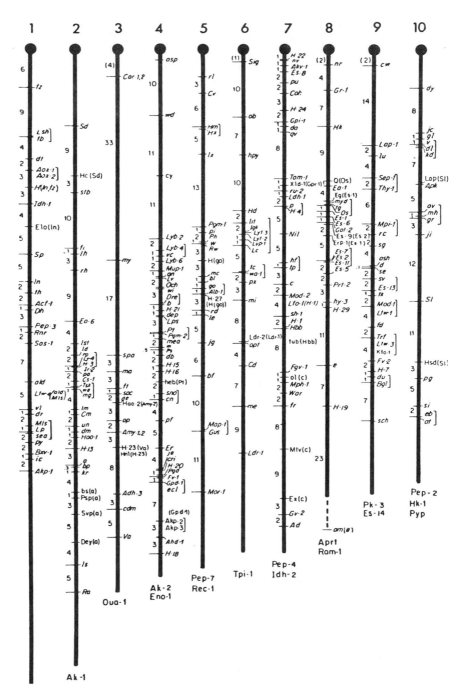

FIG. 10-5 (*a*) A chromosome map for the mouse. The black balls represent the centromeres. Chromosme 12 lacks one because it is not yet known at which

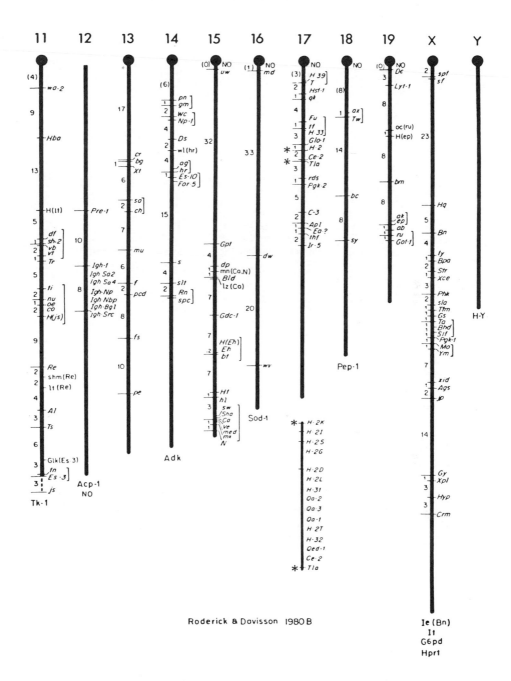

Roderick & Davisson 1980 B

end the centromere is located. (Courtesy of Muriel T. Davisson and Thomas H. Roderick, The Jackson Laboratory, Bar Harbor, Maine, USA, from whom a key for the symbols may be obtained.)

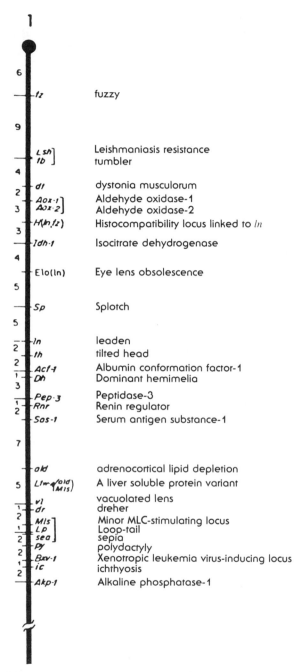

FIG. 10-5 (*b*) The genes known in Chromosome 1 of the mouse. The diversity of the characters that they influence should be noted. (Courtesy of Muriel T. Davisson and Thomas H. Roderick, The Jackson Laboratory, Bar Harbor, Maine, USA.)

Japanese investigators have assigned 168 mutations to 27 linkage groups (*Biology Data Book*, pp. 50–58). As the silkworm has 28 chromosomes (haploid number), this leaves only one more linkage group to be identified in that insect. It has 109 known mutant genes that have not yet been assigned to any group.

In the rabbit, six small linkage groups are known, and the rat has five.

A Chromosome Map for the Fowl

Since the first case of linkage in the fowl was recognized in 1917, progress in spotting genes among that bird's 39 chromosomes has been relatively slow. That is not surprising, because it is more difficult to find linkage groups among 39 chromosomes than in 20 (mouse) or four (Drosophila). However, as many of the fowl's chromosomes are little more than tiny dots, it is probable that most of its genes are in the seven longest ones.

Newer maps for the fowl's chromosomes have extended considerably the one shown in the first edition of this book. In that of Etches and Hawes (1973) two linkage groups once thought to be independent were combined to make one long group with a total length of 221 cross-over units. It is now known to belong to Chromosome 1, the fowl's longest (Zartman, 1973). Some of the loose linkages (over 40 per cent crossing-over) in the map of that chromosome have yet to be confirmed, so revisions of map distances are to be expected.

The latest map, compiled by Somes (1981), is shown in Fig. 10-6. In the sex chromosome (fifth in length), 16 loci are known, and at five of these there are multiple alleles (see Chapter 11). About 12 other genes known to be sex-linked have yet to be mapped. Chromosomes VI, VII, and VIII each have single genes assigned to them (by special techniques), but linkage of any of these genes with others in the same chromosome has not yet been found. Similarly, an enzyme is assigned to a microchromosome not yet identified, but so far it has that chromosome all to itself. Group X belongs in some chromosome ranking from 15th to 18th in size. Although the exact chromosome is not yet identified, it has been shown by special techniques to carry both the locus for the B blood groups (*Ea-B*) and a region that organizes the nucleolus.

This latest map for the fowl is small in comparison with that for the mouse, but it is the most extensive map for any domestic animal (except man) larger than a mouse. It represents contributions from many investigators during some 60 years. Anyone hoping to expand it might save some time and energy by first looking up the tests for linkage in earlier years that have yielded only negative results.

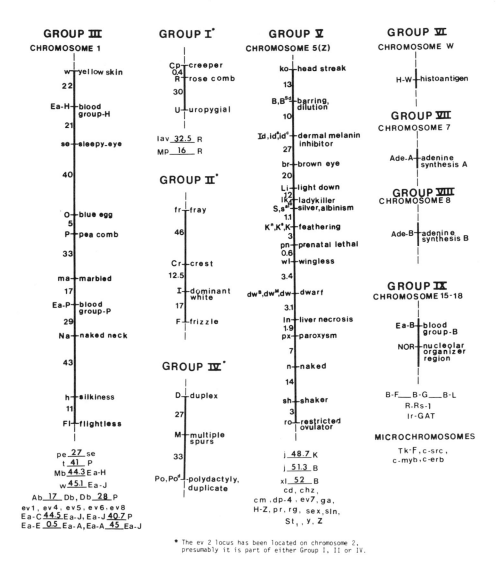

FIG. 10-6 Chromosome map for fowl. (From Somes, R. 1981 International Registry of Poultry Genetic Stocks. *Bull. 460*, Storrs Agric. Exper. Sta. University of Connecticut.)

Summaries of those tests compiled by Warren (1949), and by Etches and Hawes (1973) for trials since 1950, will be useful. It would also help to consult the alphabetical list of the fowl's genes compiled by Somes (1980).

Chromosome Maps and Predictability

Students sometimes ask, "What good is a chromosome map? What can you do with it?" As these same questions might equally well be asked about Mendel's laws, we might pause a bit to see how far we have come.

Everything learned thus far about heredity has either increased slightly our control over that process, or has revealed limitations beyond which we cannot go in attempting such control. Mendel's laws taught us the proportions in which various hereditary traits could be expected to reappear in different kinds of matings. When a sire is proven to carry a recessive defect or lethal, we can predict what proportion of his offspring will carry the gene, and even that in one type of mating it will reappear in a 3:1 ratio, and in another with a frequency of only 1 in 8. When hemophilia is shown to be sex-linked in the dog, we can predict not only the proportions in which it will appear in litters from heterozygous females, but also which of her offspring might transmit it, and which can not. Such predictability is obviously highest for monogenic traits, and for those clearly expressed. For characters subject to suppression in whole or in part by environmental influences, or dependent on many genes for their expression, predictability is considerably less.

Linkage relationships tell us that certain traits are more likely to be associated than to be separated. When one of two such linked traits is undesirable, the breeder looks for individuals showing the trait that he wants, but not the other. In such cases he is really looking for a cross-over (if single genes are concerned), but many a breeder has sought his desired combination without ever hearing of a cross-over. Knowledge of the linkage relationships of two genes, and, hence, of the kinds of matings and numbers of progeny necessary to produce new combinations, can accelerate attainment of the objective desired.

As an illustration of this fact, consider the poultry breeder who might want to transfer the gene O, for blue eggs, from the pea-combed Araucanas, in which it is most readily available, to his White Leghorns, which are single-combed. He wants to eliminate the pea comb, but P is so closely linked with O that crossing-over between them is only about 5 per cent. What crosses does he make, and how many chickens must be raised to get the combination that is desired: $pp\ OO$?

The cross of Araucana × White Leghorn ($PP\ OO \times pp\ oo$) should yield (if the Araucanas are homozygous) dihybrids: $P\ O\ /\ p\ o$. Comparatively few females (perhaps 40) are needed in the F_1 generation. Among them the breeder eventually identifies those that lay blue eggs. If the original Araucana parents were OO, all the F_1 females would lay blue eggs, but some of the Araucanas might have been *Oo*; hence specific identification of their blue-egg daughters is desirable. Such daughters can then be backcrossed to Leghorn males (*P O / p o* × *pp oo*). The resultant progeny will be distributed approximately as follows:

Parental Combinations		*Cross-overs*	
P O p o (pea, blue)	47.5%	*P o p o* (pea, white)	2.5%
p o p o (single, white)	47.5%	*p O p o* (single, blue)	2.5%

Clearly, to get a dozen females that are single-combed and lay blue eggs, the breeder would have to raise over 400 pullets. In this case, it would pay him to discard approximately half of the pullets—those with pea-combs—and to trapnest only the single-combed females. Among these about 5 per cent should lay blue eggs. To be sure of raising 200 single-combed females to laying age, after allowing for some mortality during the growing period, the breeder would have to hatch nearly 900 chicks! Even after finding a dozen blue-egg hens from that backcross, he would still have to make his stock homozygous for *O*, but that would be easier than finding those cross-overs.

Here is a case where, knowing the linkage relationships of the genes involved, the breeder can predict results, can determine the simplest way to attain his objectives, and can see in advance the scope of the operations necessary to that end. Many of the genes in chromosome maps are economically insignificant, but, in the fowl at least, several of the genes spotted in Fig. 10-6 qualify as genes important to the poultry industry. These include *B*, *S*, and *k* in the sex chromosome, also *I* and *P*. Some of those that were once considered only playthings for geneticists became very important almost overnight, when white or silver plumage was made indispensable for broilers by dictates of the market-place. The special values of *k* for ensuring good feathering in broilers, and of *B* for sex-linked crosses, were discussed in Chapter 7.

Apart entirely from such utilitarian applications of the information available in chromosome maps, there is surely some value in ascertaining, as an advancement of knowledge, the loci occupied by specific genes. Such studies are to some extent like ultra-microscopic explorations within chromosomes, and like any other study of cells and their contents, they have an important role in the advancement of science.

Problems

10-1 In rabbits, recessive white spotting of the "Dutch" type (d) is linked with the recessive gene that causes the long hair of the Angora rabbit (l). Crossing-over between d and l is 14 per cent.

Assuming that spotted rabbits homozygous for short hair are crossed with Angoras homozygous for the wild type (non-spotted), write out the genotypes for the P_1 and F_1 animals, showing whether the linked genes are carried in the coupling phase or in repulsion.

10-2 When those F_1 rabbits are backcrossed to Dutch Angoras, what phenotypes are to be expected, and in what proportions, among their 86 progeny?

10-3 From backcrosses of yellow (A^y), non-pallid (Pa) mice to agouti (A^+), pallid mice (pa), Roberts and Quisenberry obtained the following:

Yellow, non-pallid:	188	Agouti, non-pallid:	52
Yellow, pallid:	41	Agouti, pallid:	174

Linkage is evident, but what are the parental combinations and how much was crossing-over between the linked genes?

10-4 A backcross of yellow mice heterozygous for albinism (c) to albinos homozygous for agouti, A^+A^+cc, yielded 47 yellow, 49 yellow agouti, and 97 albinos. Is any linkage indicated here? Why are there only three classes?

10-5 Considering together the results from Problems 10-3 and 10-4, would you expect to find linkage between pallid and albino?

10-6 In tests for linkage of silver (s) and chocolate (b) in the rat, Castle backcrossed dihybrids carrying these two genes in the coupling phase to chocolate silvers, and got the following:

Black:	$bb\ Ss$	195	Chocolate:	$Bb\ Ss$	18
Black silver:	$bb\ ss$	9	Chocolate silver:	$Bb\ ss$	181

How close are b and s in their chromosome?

10-7 In other tests Castle found s to show 43.5 per cent crossing-over with curly, Cu, and Cu to show 45.2 per cent with b. Draw a map showing the most likely positions of s, Cu, and b in their chromosome.

10-8 When barred (B) silver (S) fowls were crossed with non-barred (b) gold (s) ones, and the $F_1\ \delta\ \delta$ were backcrossed to non-barred, gold females, the following chicks were obtained:

Barred, silver	282	Non-barred, silver:	226
Barred, gold:	206	Non-barred, gold:	266

What numbers would be expected in each class if the genes concerned were independent?

10-9 In Problem 10-8, what amount of crossing-over between B and S was shown?

10-10 Applying the chi-square test to a four-fold contingency table, determine whether or not the linkage determined in Problem 10-9 was statistically significant from independent segregation, and, if so, at what level of significance.

10-11 If a third mutation, K, shows in one test 5 per cent crossing-over with S, and you have only the data of Problems 10-8 to 10-10 as a guide, would you expect K to show crossing-over with B? If so, how much? If not, why not?

10-12 When six male fowls heterozygous for sex-linked dwarfism (dw) and silver (S) were mated by the senior author to normal hens, all the male chicks were discarded, and the daughters, when classified at five months, showed the following peculiar distributions:

	Dw S	dw s	Dw s	dw S
From sires J, H, and K	153	127	13	11
From sires M 54, M 55, and U	12	12	184	174

How do you account for these differing results?

10-13 What do they tell to the nearest decimal about the relationship of dw and S?

10-14 A test for linkage of dw with B (barring) yielded 160 parental combinations and 199 new combinations. What do these numbers suggest, when you consider also the findings in Problems 10-8 and 10-12, about the relationship of dw, B, and S?

10-15 Castle crossed two kinds of rex rabbits, which, although phenotypically indistinguishable, were separate recessive mutations, r-1 and r-2. The F_1 rabbits were all normal. When these were backcrossed to double recessives, the resultant population of 384 young contained only 33 normal; the rest were all rexes.

What numbers of normal and rex were to be expected if r-1 and r-2 were independent genes?

10-16 How do you account (to the nearest decimal) for the deviation of the numbers observed from those expected with independence?

10-17 In the fowl, px and al are sex-linked recessive genes causing paroxysm and albinism, respectively. R. K. Cole mated a dihybrid male

$$\frac{px\ Al}{Px\ al}$$

with normal females, not albinotic, and hatched 74 chicks among which were 16 albinesses. Assuming that half the chicks were females, that none died early, and that crossing-over between px and al was 10 per cent, by four weeks of age, when chicks subject to paroxysm should have shown it, what *numbers* would you expect to have done so among (a) the albinotic chicks? (b) those not albinotic?

10-18 As px is lethal, it can be carried from one generation to the next only by males, but the male chicks produced in Problem 10-17 would show neither px

nor *al* because of the *Px Al* genes received from their dams. What proportion of those male chicks should have carried *px*? Carried *al*?

10-19 Tobiano coat spotting is inherited as a simple, autosomal dominant trait in horses; tobiano horses are *To To* or *To to* and solid colored horses are *to to*. Ann Trommershausen-Smith found that test-crosses of a stallion heterozygous for tobiano and for type of albumin (*To to Alb^A Alb^B*) bred to doubly homozygous mares (*to to Alb^B Alb^B*) produced 10 tobiano foals with albumin BB and 5 solid-colored foals with albumin AB. Why, do you think, were there no tobiano foals with albumin AB, and no solid colored ones with albumin BB?

Selected References

ALTMAN, P. L., and D. S. DITTMAR, eds. 1972. *Biology Data Book*. Bethesda, Md. Federation of American Soc. Exper. Biol. Vol. 1, XVI + 606 pp. (Linkage groups for vertebrate and invertebrate animals, with brief descriptions of mutants, pp. 15–58.)

CASTLE, W. E. 1936. Further data on linkage in rabbits. *Proc. Nat. Acad. Sci.* (Washington) **22:** 222–25. (About the 908 rabbits described in this chapter to illustrate three-point linkage.)

COLE, R. K. 1961. Paroxysm—a sex-linked lethal of the fowl. *J. Heredity* **52:** 46–52. (A good account of the peculiar lethal considered in this chapter.)

ETCHES, R. J., and R. O. HAWES. 1973. A summary of linkage relationships and a revised linkage map of the chicken. *Can. J. Genet. Cytol.* **15:** 553–70. (Chromosome maps and useful data on linkage tests.)

HUTT, F. B. 1949. *Genetics of the Fowl*. New York, McGraw-Hill Book Co., Inc. (Chapter 14 shows a pictorial linkage map and reviews literature on linkage, with references thereto, including work cited in this chapter.)

HUTT, F. B. 1960. New loci in the sex chromosome of the fowl. *Heredity* **15:** 97–110. (Reports studies of linkage in sex chromosome; adds five new loci to map of 1949.)

HUTT, F. B., and C. D. MUELLER. 1943. The linkage of polydactyly with multiple spurs and duplex comb in the fowl. *Amer. Naturalist* **77:** 70–78. (Original data on linkage of *D M Po* discussed in this chapter.)

MATHER, K. 1951. *The Measurement of Linkage in Heredity*. 2nd ed. London: Methuen & Co. Ltd. New York: John Wiley & Sons. (Covers *inter alia* techniques for special situations, as when alleles do not appear in expected proportions.)

ROBINSON, R. 1958. Genetic studies of the rabbit. *Bibliographia Genetica* **17:** 229–558. The Hague. Martinus Nijhoff. (A thorough review of genetic variation in this useful domestic animal.)

ROBINSON, R. 1971. *Gene Mapping in Laboratory Animals*. London and New York. Plenum Press, Part A: Methods, Part B: Linkage in several rodents and in cat, dog, mink, 479 pp.

RODERICK, T. H., and M. T. DAVISSON. 1979. The linkage map. In *Genetic Variants and Strains of the Laboratory Mouse*. M. C. Green, ed. Stuttgart. Gustav Fischer Verlag.

SOMES, R. G., JR. 1980. Alphabetical list of the genes of domestic fowl. *J. Heredity* **71:** 168–17.

SOMES, R. G., JR. 1981. In *International Registry of Poultry Genetic Stocks*. Bulletin 460, Storrs Agric. Exper. Station, Univ. Connecticut. (Source of Fig. 10-6).

TROMMERSHAUSEN-SMITH, A. 1978. Linkage of tobiano coat spotting and albumin markers in a pony family. *J Heredity* **69:** 214–16. (Source of data in Problem 10-19.)

WARREN, D. C. 1949. Linkage relations of autosomal factors in the fowl. *Genetics* **34:** 333–50. (Includes a valuable summary of tests that have *not* shown linkage, knowledge of which saves future investigators from repeating work already done.)

YOKOYAMA, T. 1959. *Silkworm Genetics Illustrated*. Tokyo, Japan Society for the Promotion of Science. (With detailed map and list of all mutants known in the silkworm; many are beautifully illustrated.)

ZARTMAN, D. L. 1973. Location of the pea-comb gene. *Poultry Sci.* **52:** 1455–62. (Procedures by which *P* was assigned to the longest autosome.)

Multiple Alleles

In this chapter we come to another situation which, at first glance, appears to be an exception to what was learned earlier, but which is really just a simple extension of the facts presented in preceding chapters. Heretofore we have considered genes in pairs of alleles, with only two alternate forms possible at any given locus in a chromosome. Now we come to series of multiple alleles, some of which include three alleles, others six or a dozen, and some even more than a hundred.

However, just as with the alleles that are so far known to occur only in pairs, no matter how many genes belong to a series of multiple alleles, only one of them at a time can occupy the special locus in a particular chromosome to which they all have a claim. One individual may, therefore, be homozygous for any one allele of the series, or heterozygous with respect to any two, but no normal diploid animal can carry more than two of them at a time. In the herd, flock, or breed to which that animal belongs, different individuals can have different combinations (in pairs) of the genes that comprise a series. When these result in different phenotypes, the group or population in which these differing effects of multiple alleles at one locus are evident is said to show *polymorphism* (many forms) with respect to the characters influenced by genes at that locus.

From what we learned in Chapter 6 about the complex structure of the gene, it should not surprise us to find that changes in a gene (*i.e.*, mutations) can occur in various ways. If three different kinds of mutations occur in one gene, the result is a series of four alleles. One of the four is the original normal condition, often called the wild type; the others arose from it by mutation. All four occupy the same locus.

The Albino Series of Alleles in the Rabbit

A well-known series of four genes in the rabbit will serve to show how multiple alleles behave in crosses. These genes have been assigned the symbols, C, c^{ch}, c^h, and c in accordance with standard practice among geneticists of assigning superscripts to symbols for multiple alleles. As before, the capital letter, C, indicates a dominant gene, and the small letter c a recessive one, but here C is dominant to all those below it in the series, and c is recessive to all those above it. The colors and patterns that these four genes induce when homozygous (Fig. 11-1) are as follows:

$C:$ Full color, which may be solid black, or black as changed to agouti, brown, lilac, or any other shade or pattern, by genes not in this series. Wild rabbits have full color. Their somewhat grizzled appearance results (like that of other wild

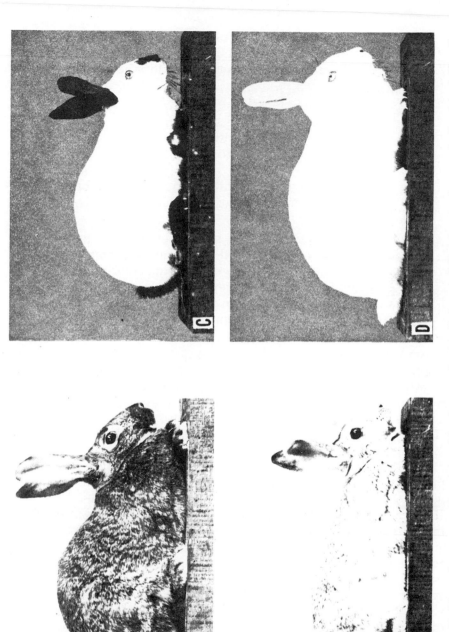

FIG. 11-1 Rabbits showing colors caused by four multiple alleles. A full color (and agouti); B, chinchilla; C, Himalayan; D. albino. (From C. E. Keeler and V. Cobb in *J. Heredity.*)

rodents) from the subterminal band of brownish-yellow pigment induced in each hair by a gene which causes that agouti coloration in rabbits that would otherwise be black.

c^{ch}: Chinchilla; eliminates all yellow pigment from the fur. Its action is best seen in rabbits that would otherwise show the agouti coloration. With the subterminal band in each hair changed from yellow to white, the pelage has a silvery-grey appearance that is very attractive. There are really three different alleles, $c^{ch\ d}$, $c^{ch\ m}$, and $c^{ch\ l}$, which induce dark, medium, and light chinchillas, respectively, but for our purposes, it is simpler to consider only one of them at this time.

c^{h}: Himalayan; black pigment is restricted to the extremities: the ears, feet, nose, and tail. Himalayan rabbits are small, but Californians, which have the same color pattern, are as big as any other breed, except the "giants." A very similar restriction of black pigment to the extremities is seen in Siamese cats and in the plumage of Light Brahmas, Light Sussex, and other fowls that show the Columbian pattern.

c: Albino; pure white with pink eyes.

It will be noted that the last three of these four genes have one thing in common: they all cause successive reductions in pigmentation of the fur.

The cross $CC \times cc$ yields fully colored rabbits in the F_1 generation and the familiar ratio of 3 colored to 1 albino in the F_2. Similarly, chinchilla × albino yields in the F_1 only chinchilla rabbits, and in the F_2 approximately 3 chinchillas to 1 albino. Clearly, both C and c^{ch} behave as simple dominant alleles to c.

This is a situation not previously encountered in this text. Hitherto we have considered only pairs of alleles at any one locus, with one member dominant to the other, or partially so, or as in the M and N blood antigens of man, with neither dominant over the other, and each inducing its characteristic effect. In these rabbits we have two distinctly different phenotypes, each behaving as a simple dominant allele to a third. This situation could arise if the chinchilla color were not an allele of albinism, but epistatic to it, and induced by a gene at some other locus. In that event, two separate loci would be involved.

Obviously, the case can be cleared up by making a third cross, that of colored × chinchilla. If two pairs of genes are operating to cause the phenotypes considered thus far, one would expect in the F_2 generation a ratio of $9:3:3:1$. Actually the cross $CC \times c^{ch}c^{ch}$ yields in the F_1 colored rabbits and in F_2 a ratio of 3 colored:1 chinchilla. It is clear, therefore, that full color, chinchilla, and albino constitute a set of three alleles, with the first two both dominant to albino and full color dominant to chinchilla. Appropriate tests have shown that the Himalayan pattern ($c^h c^h$) also belongs in this series, that it, too, is dominant to albino but that it is recessive both to full color and to chinchilla.

Order of Dominance in Multiple Alleles

In general, multiple alleles can be arranged in a series in which any one member is dominant to all those below it, but recessive to all above it. Dominance is often incomplete. There is no difficulty in distinguishing the four phenotypes of rabbits homozygous for C, c^{ch}, c^h, or c, and, as the crosses just reviewed have shown, these four alleles are here given in descending order of dominance. Specialists in rabbit genetics can easily distinguish dark, medium, and light chinchillas, each type caused by a separate allele, and have shown that the order of dominance for all six alleles of this series is C, $c^{ch\,d}$, $c^{ch\,m}$, $c^{ch\,l}$, c^h, and c. Even a chicken expert might be able (with practice) to distinguish the three kinds of chinchillas in homozygotes, especially if he had enough of each kind to make three distinct classes, but it is doubtful that he could do so with heterozygotes, in which dominance is incomplete, with resultant overlapping of intermediate types with homozygotes.

A series of at least 11 sex-linked alleles is responsible for different colors of the eye in Drosophila. These range from the normal red of wild flies through shades designated as wine, blood, coral, cherry, eosin, apricot, buff, ivory, and honey, down to white. In general, each of these shades is dominant to those named after it, but different authorities give these shades in different orders. Few of us who are not addicted to Drosophila would attempt to distinguish even phenotypes of homozygotes in this series, and most of the heterozygotes are intermediates.

In other series of alleles, particularly in those determining the production of blood antigens and polymorphism in proteins, it is common to find in heterozygotes that each allele produces its specific antigen regardless of its partner. Such heterozygotes show codominance, about which we learned in Chapter 3. Obviously, in a series of codominant alleles there is no order of dominance.

Numbers of Genotypes and Phenotypes

If we assume that each allele in a series induces, when homozygous, a specific phenotype, the number of phenotypes possible for any series must be at least the same as the number of genes it contains. That figure in turn will be raised by the number of heterozygous genotypes that can be recognized as having a specific phenotype of their own. Our rabbits, if we count all three shades of chinchilla, must have at least six phenotypes in the albino series. A specialist would probably recognize more.

The number of genotypes possible is much greater. For a small series, one can easily write them all down. By so doing we find that the six genes of the albino series in rabbits yield 21 possible genotypes (Table 11-1). Seven such genes in a series would permit 28 genotypes, and the number possible from 11 sex-linked genes for eye-color in Drosophila is 66. Writing them down would be quite a chore for the 300+ alleles that Stormont had found by 1961 to determine blood antigens at the B locus in cattle. Fortunately we can determine the number of genotypes theoretically possible with a series of n alleles by a very simple general formula which may have become evident from the other examples. It is

$$\tfrac{1}{2}n(n + 1)$$

or, in this case,

$$\tfrac{1}{2}\,(300 \times 301) = 45{,}150$$

Some of the genotypes theoretically possible from different combinations of multiple alleles will never be found if that particular combination is lethal. The gene causing yellow coat (A^y) is one of a series of multiple alleles in the mouse, but, because it is lethal to homozygotes at an early stage of development, the homozygote is never seen. A similar case in foxes will be considered later.

TABLE 11-1
Genotypes Possible in the Albino Series of Alleles in the Rabbit

Phenotype	Possible Genotypes		Total
	Homozygote	Heterozygotes	
Full color	CC	$Cc^{ch\,d}$, $Cc^{ch\,m}$, $Cc^{ch\,l}$, Cc^h, Cc	6
Dark chinchilla	$c^{ch\,d}c^{ch\,d}$	$c^{ch\,d}c^{ch\,m}$, $c^{ch\,d}c^{ch\,l}$, $c^{ch\,d}c^h$, $c^{ch\,d}c$	5
Medium chinchilla	$c^{ch\,m}c^{ch\,m}$	$c^{ch\,m}c^{ch\,l}$, $c^{ch\,m}c^h$, $c^{ch\,m}c$	4
Light chinchilla	$c^{ch\,l}c^{ch\,l}$	$c^{ch\,l}c^h$, $c^{ch\,l}c$	3
Himalayan	c^hc^h	c^hc	2
Albino	cc	none	1
			$\overline{21}$

General Characteristics of Multiple Alleles

Before proceeding to consider a few special cases, it may be helpful to recapitulate some of the characteristics of multiple alleles.

1. They usually affect only one character, causing different degrees of expression, or, in the case of codominant alleles, variations in effects of different genotypes.

2. In many series, the genes can be arranged in order of dominance, so that any one is dominant to those below it, and recessive to those above.

3. Any one, normal individual can carry only two genes of a series, but may be homozygous or heterozygous at the locus concerned.

4. A flock, breed, or population may carry several different genes of one series of alleles.

5. Since multiple alleles all occupy the same locus, all genes of a series should show the same linkage relations with other genes in the chromosome.

6. The wild-type allele is usually dominant to all others, but not always.

Platinum Foxes and Others

The standard black fox (commonly called silver) has been a fully accredited domestic animal for many years. It differs from the wild-type, red fox in having the guard hairs of the fur black, instead of red, and the ventral surface black, where that of the red fox is tawny. After its domestication, selection was practised to reduce the degree of black pigmentation and thus to increase the silvering. The latter appearance results from a mixture of light-banded guard hairs among the black ones.

Two mutations—platinum and white-faced silver—have been discovered in the silver fox since its domestication. Both are dominant to silver, and these, together with the standard silver, comprise a series of three alleles. Six genotypes are theoretically possible, but, because the genes causing platinum and white-faced silver are both lethal to homozygotes, also in combination with each other, only three genotypes are viable. Designating silver as W^+, platinum as W^p, and white-faced as W, the genotypes and corresponding phenotypes are as follows:

W^+W^+	: Silver (wild type)	WW	: dies
W^pW^+	: Platinum	W^pW^p	: dies
WW^+	: White-faced silver	W^pW	: dies

In platinum foxes, the pigmented area is reduced, the black seems to be diluted, and white spotting is increased to show a white snout and a white blaze running up the face to join a white collar. The breast and a broad strip on the belly are white (Fig. 11-2).

White-faced silvers are intermediate between the standards and platinums. They have the white blaze and white collar, but otherwise are almost as dark as standard silvers. All three types show variations in the intensity of pigmentation. Platinums intergrade with light silvers, and some white-faces are difficult to distinguish from dark platinums.

The potential value of the platinum mutation was first recognized in Norway by Hans Kjaer of Rossfjord in Troms. (For details see Mohr and Tuff, cited at the end of this chapter.) In 1934, Kjaer mated Mons, a light-colored mutant from standard silver parents, to an unrelated silver vixen and obtained, in the resultant litter of seven,

FIG. 11-2 Phenotypes of foxes determined by genes at the W locus. A: platinum (W^pw); B: white-faced (Ww); C: Georgian white (W^gw); D: embryos from white-faced x white-faced at 30 days of gestation; note two dead embryos. (From Belyaev *et al.* in *J. Heredity* as cited. Copyright © 1975 by the Amer. Genetic Association.)

four cubs that were light like their sire. By 1936, pelts showing the new mutation, to which Kjaer assigned the name Platinum, were sold at good prices in Oslo. In the following year, two light platinum pelts brought about $1,000 or 40 times the price for standard silvers, and the breeders of platinum foxes were on their way to riches. Male platinums were sold for breeding at prices as high as $5,000 and pairs were quoted at $7,500. In 1940, a single skin (from Norway) was reported to have sold for $11,000 in New York.

As the stock of platinums increased, prices became more realistic. Similar foxes were found and multiplied on fur farms in the United States and in Canada. There can be little doubt that they had been seen there earlier, but had been considered as "freaks" or "sports," and, in accordance with common practice among knowledgeable breeders of fur, flesh, or fowl, had been ushered into the next world before the arrival of the certifying government inspector. At any rate, when the wheel of feminine fashion brought short furs to the fore and put the long furs into eclipse, the mink breeders were not caught napping, and every last mutation in color was carefully preserved, multiplied, and tested in crosses with the hope of finding a new winner in the market.

From segregations of silver, platinum, and white-faced silver as recorded on fur farms in Wisconsin, Cole and Shackelford concluded that these types are caused by three alleles, with W and W^p both dominant to silver (W^+) and lethal when homozygous or in combination. This was confirmed by Gunn in Prince Edward Island and by Johansson's studies of extensive data in Sweden (Table 11-2).

Further evidence that W and W^p are lethal to homozygotes, or in combination, was provided by the litter sizes in different matings. Whereas the average litter from matings of silver × silver, or silver × mutant, was 4.48, that from matings of mutant × mutant in any combination was reduced to 3.56.

TABLE 11-2
Segregation of Color Alleles in the Fox

Mating	Expected Lethal (per cent)	Silver (Number)	Platinum (Number)	White-face (Number)
Silver × platinum	0	4,157	3,842	—
Silver × white-face	0	3,038	—	2,986
Platinum × platinum	25	58	127	—
White-face × white-face	25	267	—	483
Platinum × white-face	25	167	182	188

Source: Data of Johansson, 1947.

An interesting point brought out in Johansson's studies is that W^p, and, to a lesser degree, W, are lethal not only to the homozygotes, but also to some of the heterozygotes carrying either mutant along with W^+. Mortality of pups (per cent) during the first three months after birth was: for silvers, 6.91; for white-face, 9.24; but for platinums, 13.13. This higher mortality among the platinums accounts in large measure for the deficiency of that type below expectation as shown in Table 11-2 for the mating of silver × platinum. Such a deficiency was not found in litters from the other matings involving the platinum allele, but these were made by more-experienced fox farmers, who not only recorded colors earlier, but also lost fewer pups.

Apart from their lethal effects, the W and W^p alleles were found to reduce somewhat the breeding efficiency of the vixens carrying them. Among these, sterility and aborted litters were more frequent (in Johansson's data) than in standard silvers. As frequently happens, the breeders became aware of that fact before the geneticists did, and therefore they used mutant males for breeding more frequently than mutant vixens.

Uncertainty about the time of lethal action of the genes W and W^p in homozygotes was ended when Belyaev *et al.* (1975), working at Novosibirsk in Russia, found that embryos of the genotype WW and W^pW^p die at about 30 days of embryonic life. They found that a fourth allele, W^g, induces in heterozygotes a beautiful phenotype called Georgian white, with the coat predominantly white except for black spots on the feet, face, and body (Fig. 11-2). Like other dominant alleles at this locus, the gene W^g is lethal to homozygotes. Most of them die before implantation, but embryos that survive beyond that critical stage develop normally. At birth, these homozygous pups are pure white. After 30 days of age, they lose weight, become comatose and die.

Oddly enough, the gene W^g when in combination with either W or W^p exerts a sparing action on such heterozygotes. Even though they carry two genes, each of which is lethal to embryos homozygous for it, they do not die *in utero*, but are born alive as pure white pups. Their double dose of lethal genes is inescapable, however, and most of them die within two months after birth. Of 43 such foxes, only three lived to attain sexual maturity.

The Russian investigators found that from matings of Georgian whites *inter se*, homozygous (white) pups were more frequent in large litters (over five) than in smaller ones. There were also more of them in litters from vixens mated in early March than from those mated a month earlier, when both groups were maintained under natural daylight. From sample counts of *corpora lutea* in relation to sizes of litters, they concluded that, as the day lengthened, secretion of hormones from the *corpora lutea* was increased and that this enabled more of the $W^g W^g$ embryos to survive the critical period before implantation. That view was supported by the fact that, when the short days of early February were supplemented with artificial light, the numbers of white pups subsequently born were even greater than from the March matings.

This remarkable improvement of embryonic viability by an environmental influence (light) was proven to be effective only in embryos homozygous for Georgian white and to have no such effect on those homozygous for the white-face or platinum alleles. The difference is that the lethal period for the WW and $W^p W^p$ genotypes is at 30 days of embryonic life, whereas the lethal period for the $W^g W^g$ embryos is before implantation.

Altogether these mutations at the W^+ locus in the fox provide examples of

1. Multiple alleles in a domestic animal.
2. Genes completely lethal to homozygotes and to some of the heterozygotes.
3. Genes reducing viability and reproductive efficiency of heterozygotes.
4. An association of reduced physiological efficiency with a visible external indicator—in this case, the color pattern of the fur.
5. Interactions between some lethal alleles that prolong life beyond the usual time of lethal action.
6. Similar sparing action on viability of embryos by an environmental agency—increasing light.
7. Phenotypes demanded by the market, and hence preferred by the breeder, in spite of the fact that their distinguishing characteristics are caused by lethal genes.

Blood Groups in Man and Animals

Thanks to the curiosity of Karl Landsteiner, who, working in his laboratory in Vienna, in 1900 discovered the ABO blood groups, it is now known that virtually all human beings belong to one or another of four blood groups, 0, A, B, and AB. The differences among us in these four groups are caused by a series of multiple alleles, I^A, I^B, and i. The genes I^A and I^B are codominant, so that the relationships between genotypes and phenotypes (blood groups) are:

Genotype	Blood Group
$I^A I^A$, $I^A i$	A
$I^B I^B$, $I^B i$	B
$I^A I^B$	AB
$i\ i$	O

A and AB are often further subdivided into A_1 and A_2 and A_1B and A_2B, and even more subdivisions may be made by blood-grouping experts; these subtypes are determined by additional alleles at the ABO locus.

The Rh locus in man has a large number of multiple alleles, and there are many examples of multiple alleles for blood groups in animals. The B blood-group locus in cattle has been shown to have at least 500 different alleles. The blood factors by which these alleles are identified are, like A and B and Rh factors in man, codominant. Blood groups in domestic animals will be discussed in more detail in Chapter 21.

Other Multiple Alleles in Domestic Animals

In the rabbit, in addition to the albino series of alleles considered earlier, there is a series of five alleles which cause differing degrees of extension of black pigment throughout the coat. Resultant phenotypes grade from dominant black, through steel, normal extension, and Japanese brindling, down to non-extension. C.C. Little used three series of multiple alleles, in addition to several pairs of genes, to account for various coat colors in dogs. A fourth series of four genes determines different degrees of white spotting. Several series of multiple alleles, most of them influencing color, are known in the house mouse.

One must not think, however, that multiple alleles are more likely to affect coat color than anything else. In domestic animals and man, many series of multiple alleles are known to determine differences in blood antigens. One series in the fowl induces different degrees of polydactyly. Another causes tardy, retarded, or normal feathering.

In sheep, the presence or absence of horns in different breeds may depend upon a series of three alleles. Warwick and Dunkle came to that conclusion from crosses involving the Dorset, Rambouillet, and Merino breeds. In the last two of these three breeds, the rams have horns, but the ewes normally have only knobs that do not come through the skin, or small scurs. Dolling, who studied in Australian Merinos a dominant gene that caused the polled condition in rams, and another that induced horns in ewes, found that his results could also be explained by three alleles at one locus, but he did not rule out the possibility that two pairs of closely linked genes could have had the same effect. His P^1 gene that made the Merino ewes grow horns was considered the same as that in Dorsets.

By using the earlier symbols of Warwick and Dunkle for these three alleles, their distribution and effects in various breeds and combinations are as follows:

HH or Hh: Both sexes polled, as in Suffolks, Southdown, Shropshires, and in Dolling's polled Merinos; depressions at the site of horns

H^1H^1: Dorset type; both sexes horned

H^1H: Small horns in rams; ewes polled

hh: Rams horned; ewes polled, but with knobs or scurs, as in Merinos and Rambouillets

H^1h: Horns in both sexes

In cattle, one series of alleles determines the production of differing forms of transferrins, proteins found in blood serum that bind iron. Another series causes variations in the lactoglobulins of cows' milk. We shall consider them further in Chapter 21.

Problems

11-1 For the cross in rabbits: full color × Himalayan, write out the genotypes and phenotypes expected in the F_1 and F_2 generations.

11-2 Do the same for the cross: Chinchilla × Himalayan.

11-3. If the F_1 rabbits of Problems 11-1 and 11-2 were mated together, what phenotypic ratio would be expected in their progeny?

11-4 If a doe kindles a litter of five kits that includes three chinchillas, one Himalayan, and one albino, what must have been the genotypes and phenotypes of the buck and the doe?

11-5 Which of the matings considered above provide evidence that multiple alleles are involved, and why?

11-6 Would it be more profitable for a fox breeder wishing to get as many platinum foxes as possible to mate platinum *inter se*, or to backcross platinum × silver, and why?

11-7 If he chose to backcross, should he mate silver ♂ × platinum ♀, or *vice versa*, and why?

11-8 What is the probability that the deviation from expectation recorded in the first line of Table 11-2 could have occurred by chance?

11-9 Among litters numerous enough to yield accurate figures, Johansson found the average size of litter to be 4.48 when at least one parent fox was silver, but 3.56 when both parents were mutants. How much (per cent) was litter size reduced by lethal action of the mutant alleles, and why was that reduction not 25 per cent? What conditions would have to prevail to make it that figure?

11-10 Apart from the platinum, white-faced, and Georgian foxes, in what other breeds of domestic animals is the distinguishing characteristic caused by a lethal gene? By a heterozygous state?

A series of autosomal multiple alleles in mice induces coat colors as follows:

A^y: yellow (lethal to the homozygote)
A^w: white-bellied agouti
A^+: grey-bellied agouti (wild type)
a^t: black and tan
a: non-agouti (black)
A^e: extreme non-agouti (intense black)

As listed above, each gene in the series is dominant to those below it. The series is shown in Fig. 10-5a in the fifth linkage group, but without the wild type. The lethal yellow was discussed in Chapter 9.

11-11 What ratios of viable phenotypes are to be expected from the following matings:

1. yellow $(A^y A^+)$ × yellow $(A^y a^t)$
2. yellow $(A^y a^t)$ × yellow $(A^y a)$
3. yellow $(A^y a)$ × grey-bellied agouti $(A^+ a)$

11-12 What would the genotypes of parent mice have to be to yield litters containing: (a) wild type, black and tan, and black, (b) yellow, wild type, and intense black?

11-13 How many different genotypes are possible within this series of alleles in the mouse, and how many of them are viable?

Selected References

BELYAEV, D. K., L. N. TRUT, and A. O. RUVINSKY. 1975. Genetics of the W locus in foxes and expression of its lethal effects. *J. Heredity.* **66:** 331–38.

COLE, L. J., and R. M. SHACKELFORD. 1943. White spotting in the fox. *Amer. Naturalist* **57:** 289–321. (Data on these mutations as bred on fur farms in Wisconsin.)

DOLLING, C. H. S. 1961. Hornedness and polledness in sheep. IV. Triple alleles affecting horn growth in the Merino. *Austral. J. Agric. Res.* **12:** 353–61. (All three studied within the one breed.)

GUNN, C. K. 1945. Genetics of some new type foxes. *Dom. of Canada, Dept. of Agric., Publ. No.* 768, Tech. Bull. No. 53. (Evidence from Prince Edward Island on inheritance of the platinum series of alleles.)

JOHANSSON, I. 1947. The inheritance of the platinum and white face characters in the fox. *Hereditas* **33:** 152–74. (Summarizes extensive data from Swedish fur farms on the inheritance of the platinum alleles, and their relation to reproduction; source of data cited in this chapter.)

MOHR, O. L., and P. TUFF. 1939. The Norwegian Platinum fox. *J. Heredity* **30:** 226–34. (An interesting account of the first utilization of this mutation and of its value in earlier years.)

STERN, C. 1973. *Principles of Human Genetics*. 3rd ed. San Francisco. W. H. Freeman & Co. x+891 pp. (See this book for further information about human blood groups.)

WARWICK, B. L., and P. B. DUNKLE. 1939. Inheritance of horns in sheep. *J. Heredity* **30:** 325–29. (With good illustrations of the different types.)

CHAPTER 2

Quantitative Characters

Thus far we have been dealing mostly with inherited characters influenced by single genes, or pairs of genes, or by simple interactions between two pairs of alleles. Cattle had horns or didn't, and were black or red. Combs of chickens were rose, pea, walnut, or single, and, where lethal genes were concerned, homozygotes had either one allele and lived, or the other allele and perished. With incomplete dominance, a single pair of genes could induce three phenotypes, but each was different from the other two.

In all such cases, the variation among the animals concerned was *discontinuous*. With few exceptions, those animals could be classified by inspection as falling into one of two classes, or, including incomplete dominance, into one of three classes. Even where multiple alleles are operating to induce gradations in one trait, as with all the shades of eye color between red and white in Drosophila, a specialist familiar with those colors could differentiate them, and, knowing the genotypes of the parents, could sort the progeny into one or another of the two or three classes expected in different matings of such parents. Similarly, the experienced breeder could (with few exceptions) classify his rabbits as dark, medium, or light chinchilla, even though to an inexperienced eye they might appear to show a continuous gradation from light to dark.

Continuous Variation

In this chapter we deal with characters which, unlike those considered earlier, show only very slight gradations, often imperceptible, among individuals within the range from least to most. Stature in man is a good example. Within a group of students there may be a difference of 17 inches or more between the tallest and the shortest, but, apart from the extremes, differences between two members may be so slight that they can be discovered only by measuring. Classes can be established within such a population, but that can be done only arbitrarily. For example, one might put all those shorter than 5' 6" in one class, and all those above that height in another. If the students were sorted out to the nearest inch in height there could be 17 different classes. To arrange them thus, each would have to be measured separately.

Hereditary characters like stature in man, body weight in man or in domestic animals, and others to be considered later, are said to show *continuous* variation because the graduation from one individual to the next is imperceptible (in large groups) except by measurement or weighing. Such traits are designated as *quantitative* characters, because individuals differ in the amount or degree to which they manifest the trait.

That designation permits a distinction between quantitative and *qualitative* hereditary characters. The latter term is restricted to those in which variation depends on the presence or absence of a certain quality or property, not on its amount. For example, the difference between black and red cattle depends on qualitative characters, but, within either color, variations in size, rate of growth, milk production, and fat content of the milk all represent quantitative traits.

In general, quantitative variations influence, size, conformation, yield, strength, and viability. Some people think of them, therefore, as fundamental characters, and regard as somewhat superficial the qualitative traits—the variations in color and minor variations of structure that are so often caused by single genes. Such a distinction is quite unwarranted. There is nothing superficial to the victim of hemophilia about his unpleasant affliction, and many a single gene is lethal in the homozygous state.

A good idea of the kinds of things dependent on quantitative inheritance is provided by a sample of such traits in the domestic fowl. These include rate of growth, age at first egg, ability to lay eggs, size of eggs, thickness of egg shell, color of egg shell, proportions of thick and thin albumen in the eggs, body weight, resistance to disease, and other characters.

When members of any flock, herd, or other population are measured individually with respect to some quantitative trait, and then assigned to classes delineated by equal intervals between the midpoints of those classes, the resulting distribution, which shows the frequency in each class, conforms more or less closely to what is called a *normal* curve of *distribution*. When the population is not large enough, the distribution may be somewhat irregular. When it contains a number of aberrant individuals at one end of the range from least to most, a *skew curve* results. However, with large numbers, and with normal variation, one gets an approximately normal curve, *i.e.*, with equal numbers above and below the mean, and no excessive deviations in any one class.

Inheritance of Quantitative Characters

Whereas discontinuous variations (when genetic in origin) tend to segregate in nice, recognizable ratios in succeeding variations after a cross, quantitative traits do not conform to that pattern. Crossing big × little, early × late, high yield × low yield, etc., usually results in an F_1 population that is comparatively uniform, and intermediate between the parental types, but not necessarily exactly at the mid-point between the two parents. Its range may overlap that of either parental stock, or both. Furthermore, such a blending of two different parental types is to be expected only when the individuals crossed are typical representatives of breeds or stocks that are genetically quite different with respect to the character under consideration. It is not to be expected in every case when single individuals are crossed, because one of these, or both, may be atypical.

Following such a cross, if members of the F_1 generation are mated *inter se,* the resulting F_2 generation usually has a mean, or average value, much the same as in the F_1 but with a greater range of variation than in either the P_1 or F_1 generations. It may even include some individuals that exceed the range in the original parental stocks.

During the first decade of this century, when biologists were busily fitting variations of animals and plants into satisfying 3:1 Mendelian ratios, the non-conformity of characters like size and yield was difficult to explain. The lack of segregation displayed by such traits (which were not then designated as quantitative characters) led for a while to the view that *blending inheritance,* as it was called, did not conform to Mendel's laws and, hence, that some other mechanism still unknown might be involved. This situation was corrected about 1910, when the Swedish botanist, Nilsson-Ehle, showed that blending inheritance did, after all, result from the same kind of hereditary transmission as that responsible for Mendelian ratios.

A Note on Terminology

Good students, who will consult other texts to see whether or not the authors of this one have their facts straight, should be told what to look for. While some geneticists write about quantitative characters and quantitative inheritance, others discuss *polygenic* characters and polygenic inheritance. Statisticians deal in *metric* traits influenced by *additive* genes, and some biologists write about *polymery* and *polymeric* genes. Once upon a time, when hereditary determiners were still referred to as *factors,* blending inheritance was shown to depend upon the cumulative action of *multiple factors,* each adding its mite of effect, and all together pushing the *multifactorial* character either in one direction or the other. Its expression was for many years known to depend on *multifactorial* inheritance.

The student who encounters in other texts any of the terms italicized in the previous paragraph should stop, look, and remember that, although the words are different, the music there should be not only somewhat the same as in this chapter, but also, perhaps, better played. While some of the newer terms may have resulted (in part, at least) from reasoning similar to that which converts some undertakers into "morticians," we do know more about genes than we did 50 years ago, and a good, descriptive adjective like *polygenic* is to be preferred, perhaps, over any other. Geneticists interested more in final products than in initial causes may find it simpler to think, as do these authors, in terms of hereditary, quantitative differences in ability to produce meat, milk, eggs, or wool.

Mendelian Basis for Quantitative Characters

In studying different crosses between red and white wheats, Nilsson-Ehle found that in most cases the F_1 generations were intermediate. In the F_2, however, the ratio of red or reddish kernels to white ones was from some crosses $3:1$, for others about $15:1$, and in some cases approximately $63:1$. This led to the discovery that the red color was caused by one, two, or three independent pairs of genes, with the dominant alleles of each pair inducing color and their effects approximately equal and cumulative. Recessive alleles of each pair had no effect on color; hence simple, double, and triple recessives were white. Dominance within any pair of alleles was incomplete; thus homozygotes for a dominant allele were darker red than heterozygotes.

The effects of such genes on color of the wheat kernel are best understood by following through to an F_2 generation a cross of dark red × white parental types. Let us assume, for simplicity, that only two pairs of genes are involved. These can be designated as *A-a* and *B-b*. Results of the cross are shown in Table 12-1.

In this dihybrid cross, as in those considered in Chapter 4, all possible combinations of the two pairs of genes are found in the F_2 population. The difference is that, whereas in the cases considered earlier, independent expression of the two pairs of alleles resulted in a phenotypic ratio of $9:3:3:1$, the genes inducing red color are not independent, but, rather, they have cumulative effects. Accordingly, not only is the homozygote for either pair darker than the heterozygote, but the individual that gets from both pairs of alleles together as many as three or four that induce red color is proportionately darker. As a result, the F_2 generation contains five different classes, four of which show differing intensities of red, while the fifth is white.

Instead of writing down the numbers of individuals in each of those four classes, as is done at the bottom of Table 12-1, their frequencies can be shown graphically in a histogram (Fig. 12-1). When the mid-points of the tops of the five columns are joined, the resultant figure is a normal curve of distribution, a better-than-average sample of the kind of curve that is characteristic of the distribution of quantitative characters in unselected populations. Actually, the "curve" in Fig. 12-1 looks more like a triangle than a curve,

TABLE 12-1

Effects of Two Pairs of Cumulative Genes on Color of the Wheat Kernel

Parents:		Dark red	×	White
Their genotypes:		AA BB		aa bb
Genes inducing red:		4		0
F$_1$:			Aa Bb	
Genes inducing red:			2	
Phenotype:			Medium red	

Genotypes in F$_2$	Individuals (number)	Genes inducing red (number)	Phenotype
AA BB	1	4	Dark red
AA Bb	2	3	Medium dark red
Aa BB	2	3	Medium dark red
Aa Bb	4	2	Medium red
AA bb	1	2	Medium red
Aa bb	2	1	Pale red
aa BB	1	2	Medium red
aa Bb	2	1	Pale red
aa bb	1	0	White

Shades and numbers in F$_2$:

Dark red	Medium dark red	Medium red	Pale red	White
1	4	6	4	1

but that is inevitable when the population is divided into only five classes. That situation, in turn, resulted here from our desire to explain the cumulative action of polygenes with the simplest example possible.

With these five classes of color, one might have little or no difficulty in sorting all the red or reddish kernels into one or another of the four classes other than white. Accordingly, one could say that here again we are dealing with discontinuous variation. However, had we used as an illustration three pairs of quantitative genes rather than two, that F$_2$ generation would have contained seven different phenotypic classes, with six of them showing different intensities of red color. Any attempt to sort the kernels into those classes would have been more difficult. Eventually, with increasing numbers of genes

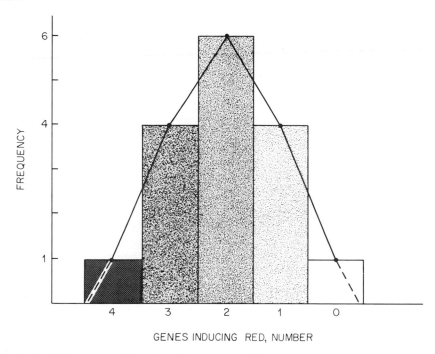

FIG. 12-1 Histogram showing the expected frequencies of individuals carry-ing 0, 1, 2, 3, or 4 genes for a quantitative character (in this case, red color in wheat) in the F_2 generation from a cross between the extreme types, *AA BB* × *aa bb*.

influencing some quantitative trait, accurate visual distinctions among phenotypes nearly alike but resulting from differing genotypes become practically impossible. Variation is then continuous, and differences among individuals within any one part of the long range from one extreme to the other can be detected only by careful mea-surements of one kind or another.

Estimating Numbers of Genes Determining Quantitative Characters

In the example worked out above, with results in the F_2 generation as shown in Table 12-1 and Fig. 12-1, the number of pairs of genes exerting equal cumulative effects was two. When only one pair of genes is concerned, and dominance is incomplete, the number of phenotypic classes in F_2 is three, and the minimum number of indi-viduals needed to permit all possible combinations is four.

When three pairs of genes with equal cumulative effects influence a quantitative character, the situation becomes more complicated, but we can still write out all the combinations possible in an F_2 generation from trihybrid parents. As each such parent could produce eight kinds of gametes, the number of possible combinations is 64. When we write them all out, and then add up the numbers that have all six dominant alleles, or all six recessive ones, or any combination of the two kinds between those extremes, the number of different phenotypic classes is found to be seven, and the frequencies in those classes are

$$1:6:15:20:15:6:1$$

These, we recognize, are the coefficients of the terms in the expansion of $(a + b)^6$, or:

$$a^6 + 6\,a^5b + 15\,a^4b^2 + 20\,a^3b^3 + 15\,a^2b^4 + 6ab^5 + b^6$$

The number of individuals in the F_2 generation having all the six cumulative genes that induce the character is thus identical with the number that lack all of them, which is 1 in 64. Another way of saying the same thing is that a population of at least 64 individuals would be necessary to permit all possible combinations and to manifest the two extremes of the range for the particular quantitative character under consideration. Expectations for getting other degrees of that character (*i.e.*, other combinations of dominant and recessive genes) are indicated by the coefficients of the other terms in the expansion.

Having come thus far by slow and easy stages, we can proceed to Table 12-2, in which are given simple, general formulae (derived from the previous studies with 1, 2, or 3 pairs of alleles) by which we can quickly determine, for any number (n) of pairs of genes exerting equal cumulative effects on some quantitative character, just how much variation, and how great a range between extremes, one might expect in an F_2 population.

From such figures some people have attempted to estimate how many pairs of genes might be influencing different quantitative traits. If, from a cross between two extremes with respect to some such character, we could regularly recover those same extremes in an F_2 population with a frequency of 1 in 16, it would not be unreasonable to assume that only two pairs of genes are concerned. The same line of reasoning might also be valid when only three or four pairs of genes are concerned. However, as is evident in Table 12-2, when the number of pairs exceeds four, any F_2 population that could yield the extremes would have to be so large that no one who breeds animals bigger than Drosophila would venture to produce it.

TABLE 12-2

Theoretical Expectations in an F_2 Population When Pairs of Alleles Exert
Equal, Cumulative Effects on a Quantitative Character

Pairs of Alleles (number)	Size of F_2 Necessary for all Combinations	Phenotypic Classes (number)	Chance of Getting Either Extreme of the Character
1	4	3	1 in 4
2	16	5	1 in 16
3	64	7	1 in 64
4	256	9	1 in 256
.	.	.	.
.	.	.	.
.	.	.	.
n	4^n	$2n + 1$	1 in 4^n
.	.	.	.
.	.	.	.
.	.	.	.
8	4^8	17	1 in 65,536

None need bother to do so. There is no assurance that all quantitative characters conform to Nilsson-Ehle's wheat in having their causative genes produce exactly equal and cumulative effects. Some may have more influence than others. Epistasis might interfere in some cases. Some of the genes might be linked with others, in which case certain combinations could produce disproportionate effects. More important, so far as practical breeding for quantitative traits is concerned, it matters little whether the number of loci concerned is 8 or 100. In either case, the breeder will have to select, by the best methods he can use, for his desired objective.

Transgressive Inheritance

Sometimes, following crosses between parents that differ in some polygenic trait, one finds in the F_2 generation a few exceptional individuals that exceed the range of variation found in both parental types together. To illustrate, if the parents differed in size, the exceptional descendants might be even smaller than the small parent, or larger than the big one. When such cases are clearly not caused by environmental influences that affect one generation more than others, they are said to illustrate *transgressive inheritance*.

Sometimes extreme deviations of this sort are not so exceptional as they may first seem. Thus, if one parent were considered to represent a large race, that animal could still not be the best representative. His actual size might be considerably below the potential for his genotype. Such a parent could easily contribute to the cross more genes for size than are indicated by his phenotype, and it would not be surprising if some of his F_2 descendants should exceed him.

Otherwise, transgressive inheritance can result from one parent's contributing genes for some trait that are different from those of the parent showing more of that trait. Thus, if, in addition to genes *A-a* and *B-b* inducing red kernels in wheat, there were a third pair of alleles *C-c* with similar effect, then a cross between dark red and light red wheats could yield in the F_2 generation some individuals still darker than the dark red parent:

P₁:

	Dark red		Light red
	AA BB cc	×	*aa bb CC*
	(4 *genes for red*)		(2 genes for red)

F_1:

Aa Bb Cc
(3 genes for red, hence
lighter than the darker parent)

F_2:

27 different genotypes, including:

AA BB CC } both darker than the
AA BB Cc } dark red parent

Similarly, a small rabbit could contribute to a cross genes for size different from those of its larger partner, and from such a cross one could get in the second generation some rabbits bigger than the larger parent.

Relation of Quantitative Inheritance to Selection

Because we have dwelt at some length on the range of variation in quantitative traits that can be found in F_2 generations from crosses between widely divergent types of parents, it seems desirable to emphasize the fact that similar variation can usually be found in populations not thus derived from matings specially designed to produce it.

For example, if we consider as one population all the dairy cows in a large herd, or all the Holstein-Friesians in the State of New York, there is a wide range between the poorest milk producer and the best.

The breeder of dairy cattle makes no matings to produce in some F_2 generation poor producers as well as good ones. He selects as breeding stock only those animals which he hopes will produce offspring capable of high yields. The amount of variation existing in his herd, or in other herds of the same breed within his territory, is adequate to facilitate that selection. A curve showing distribution of individual yields around the mean for a large population approaches very closely a normal, symmetrical curve.

Similarly, among the senior author's 2,150 White Leghorns hatched in 1961, some laid over 290 eggs by the time they were 500 days old, but others produced fewer than 50. Those birds were not products of any crosses between good and poor layers. They came from matings of the best breeding stock available, designed to the best of our ability to produce only good layers—no poor ones.

We leave for a later chapter any consideration of different methods of selection. It is pertinent to our understanding of quantitative inheritance to point out that selection for improvement in yield, performance, or any other quantitative character is fundamentally an attempt on the part of the breeder to accumulate in the progeny more effective genes, or better ones, than those of the parents. To that end, outstanding parents are used, from the upper range of the distribution. Presumably they have the kind of genes needed for good results, and have them in greater number than do animals closer to the average, or below it. Most of their progeny should be similarly endowed, and, "wiv' a little bit of luck," some may get—from the genes of both parents together—an even better combination than that of either parent alone. Such a result would be somewhat similar to the transgressive inheritance considered earlier.

With respect to most quantitative traits of domestic animals, or all of them, the limits to be attained by such selection have not yet been reached. They are not even known, with certainty. These facts, in turn, suggest that the numbers of genes influencing those traits are far beyond any simple three, four, or five pairs of alleles. If only a few genes determined genetic differences in milk production, butterfat, egg production, and other quantitative traits, homozygosity for those genes would have been reached long ago, and further productivity could be attained only by improving the environment. There is every reason to suppose that many quantitative characters of domestic animals are influenced by dozens of genes at least, and more likely by hundreds.

Selection of this sort is complicated by the fact that, in general, quantitative characters are somewhat more subject to environmental modification than are qualitative characters. As a result, genotypes with respect to quantitative traits are more difficult to identify by inspection, or even by measuring, then are genotypes for color, or for variations in hoof, hide, or hair. A chick hatched with a single comb, pure dominant-white plumage, and four toes is immediately recognized as having the genotype *pp rr II popo,* and, so far as those four pairs of alleles are concerned, one knows right then exactly what genes it could eventually transmit to its offspring. On the other hand, the questions whether or not that same chick is genetically resistant to leukosis, can begin laying at about 160 days, and is able to lay 250 eggs, can be answered only if environmental conditions permit any answers. To answer all three, the environment must provide exposure to the disease, suitable conditions of daylight (or artificial light) when the pullet approaches five months of age, and adequate diet, housing, management, and freedom from disease thereafter. Even if all three questions are answered in the affirmative, there remains still a fourth. To what extent can the bird pass on its genes for good performance? That question can be answered only by her progeny.

Curves Askew or Bimodal

Frequency curves for the distribution of quantitative characters do not always conform to the symmetrical normal curve. The most common reason for failure to do so is probably an inadequate number of individuals, with resultant chance fluctuations in some classes. Skew curves result when the population measured contains a number of individuals whose performance deviates considerably from the average. Such deviations are sometimes brought out by one environment, but not by another.

An example of a skew distribution thus induced is provided by distributions of ages at first egg for White Leghorns hatched at Cornell University in 1956 (Fig. 12-2). Age at first egg is a typical quantitative character, and several experiments have shown that by appropriate selection the breeder can advance or retard the average onset of laying in different strains. It is also subject to environmental modification, and is influenced considerably by the amount of light to which the birds are exposed as they become mature, or even earlier.

For the 369 pullets hatched on March 9, the distribution is fairly symmetrical, and the *modal* class (with the greatest number) is close to the average. As the curve shows, there was a disproportionately high frequency in the earliest class, the average for which was 21 weeks. Actually, ten of that group laid at 144 days, which was their first day in the laying house. It is clear that a number of the birds had begun to lay while still on the rearing range, and that true ages at first age were not obtained for the earliest-maturing pullets of that flock.

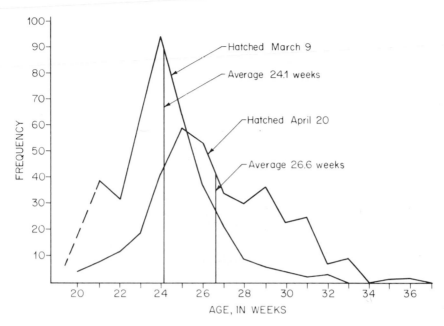

FIG. 12-2 Skewed distribution of age at first egg for White Leghorns of the second of two hatches six weeks apart. Environmental conditions distorting the symmetry of both curves are considered in the text.

The average of 24.1 weeks for the whole group is therefore higher than it would have been if the birds had been housed two weeks earlier.

For the 354 pullets hatched six weeks later, average age at first egg was 2.5 weeks above that for the flock hatched earlier. Such a difference was to be expected because of the shorter days and reduced light to which these birds were exposed as they matured. The shape of the curve indicates that some of the flock were retarded more than others, and that the stragglers were numerous enough to pull the distribution farther to the right of the modal class and the average than in the curve for the birds hatched earlier. The first layers began five weeks earlier than the modal class (25 weeks), but the last straggler did not start until eleven weeks beyond that mode. Similar, somewhat skew distributions are not uncommon for other quantitative characters.

A *bimodal* curve indicates that what might have been considered as one group showing continuous variation with respect to some trait may consist of two separate but overlapping populations. In our example, the overlapping is so slight, and the differences between means (averages) for the two groups so great, that there is never any question about there being two distinct classes (Fig. 12-3). The birds

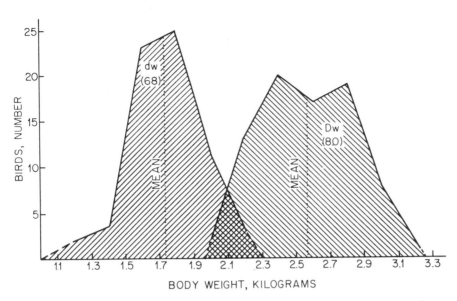

FIG. 12-3 Distribution of adult weights of daughters of one sire showing extreme bimodal frequencies resulting from segregation of a single gene affecting size. (From F. B. Hutt in *J. Heredity*.)

for which adult weights are shown were all daughters of one sire. Body size is a typical quantitative character. Ordinarily their weights should have fitted a fairly symmetrical unimodal curve. In this case the sire was heterozygous for the sex-linked gene, *dw*, which is completely recessive but reduces adult size in hemizygous females by 26 to 32 per cent, and in homozygous males by about 43 per cent (Fig. 12-4). His daughters that got *dw* were therefore so much smaller than their big sisters (and half-sisters) which carried the normal allele, *Dw*, that there was very little overlapping of the two classes. Mothers were the same for both groups.

Apart from the major effects of the *Dw-dw* alleles, the distribution within each of the two groups showed normal, continuous variation from smallest to largest. The gene *dw* provides a remarkable example of a gene that exerts a major effect on a quantitative character ordinarily influenced chiefly by the cumulative effects of many genes. Its location in the fowl's sex chromosome is shown in Fig. 10-6.

Bimodal distributions for other quantitative characters are likely to show much more overlapping of the two component groups than that in Fig. 12-3. Such a curve is to be expected for distribution of

FIG. 12-4 Hereditary dwarfism (left) induced by a gene with major effects on what is ordinarily a quantitative character. These full brothers (one gold, the other silver) were hatched on the same day, but at 30 weeks the dwarfed male (*dw dw*) weighed 1,870 grams and his big brother (*Dw dw*) 3,200 grams. (From F. B. Hutt in *J. Heredity.*)

heights in a large class of students that contains both men and women. Heights of men would be distributed around one mean and heights of women around another a few inches shorter, but the two classes would overlap to make one bimodal distribution. Sometimes one cannot tell by inspection whether or not any apparent bimodality should be attributed solely to chance fluctuations in relatively small numbers. Genes with known major effects on quantitative characters are not common, and any indication that one might be operating deserves further investigation. A bimodal distribution is likely to be the first such indication.

Inheritance of Body Size

A fairly typical example of the behavior of a quantitative character in crosses between parents of different types is provided by a cross made by Punnett and Bailey between a Golden-pencilled Hamburg cock and Silver Sebright hens. Sebrights are the perky little bantams in which the males defy avian conventions and appear in henny plumage (Chapter 8). Females of that breed weigh about 600 grams and males around 750. Hamburg hens are about 500 grams heavier than Sebright females, and the one cock used weighed 1,350 grams. As these sizes are characteristic of the two breeds, which breed true to size (with normal variation therein) as for other breed characteristics, the difference in size between the breeds obviously depends upon their differing genotypes.

To avoid complications because normal sex dimorphism in the fowl makes the males about 25 per cent heavier than females, weights of the latter in the parental and F_1 generations were multiplied by 1.25 to convert them to equivalent weights for males. For the F_2 generation weights of males only were given.

Progeny of this cross were intermediate in weight between the two parents, but closer to the larger one (Fig. 12-5). As only a few F_1 birds were raised, it is not surprising that their distribution scarcely conforms to a normal curve. Among the 112 males of the F_2 generation there was a wide range in size, with several larger than the Hamburg male, and others smaller than the Sebrights. These birds at the two extremes of that range illustrate transgressive inheritance, but one might wonder whether or not the lone Hamburg cock was typical in size for his breed, or the two Sebright hens for theirs.

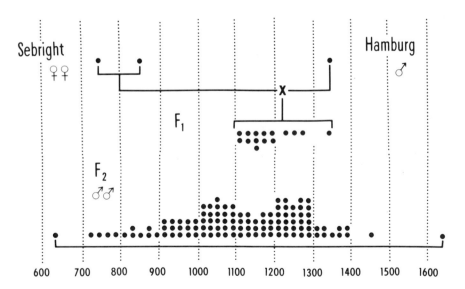

FIG. 12-5 Inheritance of body weight following a cross of Hamburg ♂ × Sebright ♀ ♀. Weights of P_1 and F_1 females are converted to equivalent weights for males. Each dot represents one bird; weights are in grams. (From R. C. Punnet in *Heredity in Poultry*, courtesy of Macmillan & Co., Ltd.)

Polygenes Affecting Levels of Butterfat in Milk

Genetic differences in milk production by cattle, and in the level of butterfat in the milk, are good examples of important quantitative traits in domestic animals. As such, they have been incorporated in varying degrees in different breeds according to the preferences of breeders in different parts of the world. The breeds originating in the Channel Islands of Jersey, Guernsey, and Alderney are rightly famous for the richness of their milk, but other breeds are equally acclaimed for the amount of milk they produce, rather than for its level of butterfat.

We are indebted to Christian Wriedt for summarizing in clear form some results of a cross between Red Danish and Jersey cattle that was begun in 1906 by Count Ahlefeldt-Laurvigen on the Danish island of Langeland. As an illustration of the transmission of a quantitative trait in large domestic animals, the experiment, which was conducted on a scale unusually extensive for such animals, was almost perfect.

The average level of butterfat for 108 Red Danish cows was 3.4 per cent, while that for 66 Jerseys was 5.57 per cent. The range for one breed did not overlap the range within the other (Fig. 12-6). From many crosses between the two breeds, 108 cows were produced in the F_1 generation, and for these, average butterfat was 4.39 per cent, a figure slightly below the mid-point between the two parental averages.

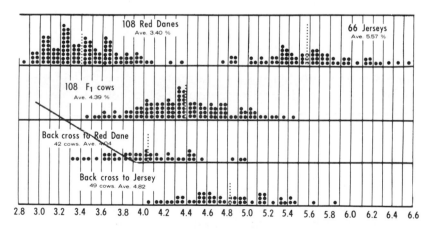

FIG. 12-6 Inheritance of fat content of the milk in F_1 and both backcrosses from the original cross of Red Danish × Jersey cattle. Each dot represents one cow. (From C. Wriedt in *Heredity in Live Stock*, courtesy of Macmillan & Co., Ltd.)

Apparently, no F_2 generation was produced, but backcrosses of F_1 animals were made to both parental breeds, thus yielding cows that were from one cross three-quarters Red Danish, and from the other three-quarters Jersey. It is clear (Fig. 12-6) that the infusion of additional genes from the Jerseys raised the average level of butterfat considerably above that for the F_1 generation, while the other backcross had the opposite effect. Such results are to be expected not only for levels of butterfat, but also for any other quantitative character in which the parent breeds are significantly different.

Polygenes Affecting Size of Hens' Eggs

Transmission of another quantitative trait in a different species was found by Roberts *et al.* (1952) to follow a pattern very much like that just considered. They crossed highly inbred lines of Leghorns that differed greatly in egg size, and had done so consistently for several generations prior to the cross. Average size of egg in the small-egg line was 14.5 grams below that of their big-egg partners in the cross, a difference of a little more than 25 per cent, and the range in either parental stock did not overlap that for the other.

At 52.8 grams, the F_1 were almost exactly at the mid-point between the parental races, and from backcrosses of F_1 to both kinds of parents two different populations were obtained (Fig. 12-7). Although the ranges within these overlapped considerably, their average weights differed by 6.6 grams. The difference between them is indicated even better by the fact that the modal class for one backcross is the same as the largest class of the other.

As before, such results fit well the expectations from matings of these kinds when the hereditary character under consideration is influenced by some undetermined number of genes with cumulative effects.

Exceptions and Complications

Because the typical examples of quantitative inheritance illustrated in Figs. 12-5 to 12-7 provide almost perfect fits of results realized to results expected, it is desirable to point out that exceptions do occur and that under some circumstances those exceptions may be what the breeder is most anxious to get.

Whenever breeding experiments or selection programs extend over several years, there is always the risk that some change in the environment in later years may affect expression of the character

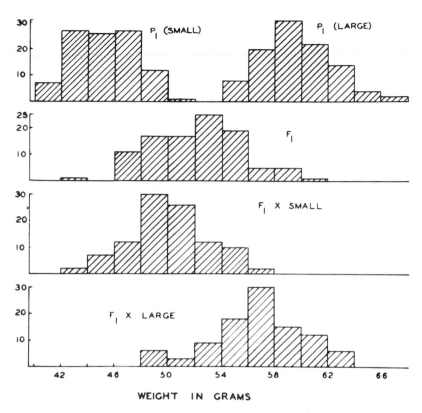

FIG. 12-7 Inheritance of egg size following a cross between parental strains of White Leghorns that differed by 25 per cent in that quantitative character. (From E. Roberts *et al.* in *Poultry Sci.*)

under study, and thus make results in one generation not comparable with those for another. Deviations from typical behavior of quantitative traits may also result in some cases from the action of genes with major effects which might prevent the expression of many genes with lesser ones. A good example of such a complication is provided by the sex-linked gene, *dw*, which reduces size so markedly in the fowl (Fig. 12-4). Ordinarily, body size in that species, as in other animals, depends on an unknown number of genes. When *dw* is not present, size of body conforms in crosses to expectation for any quantitative character. Fortunately, the effect of *dw* is so great that its presence is recognized at once, and, when its effects are known, expectations can be revised accordingly. A similar dwarfing, caused by a deficiency in a pituitary hormone, is known in the mouse, but those affected are sterile. Let us suppose, however, that in some cases other genes exert disproportionate effects not so great as those of *dw*. One of them might reduce size, or milk production, or anything else by only 5 or 10 per cent. In such cases, results could be somewhat confusing until presence of the trouble-maker is detected, and its effects determined.

A common reason for deviations in F_1 generations from a yield or performance about half-way between those for two widely differing parents is hybrid vigor. We shall discuss that subject in a later chapter, but it is pertinent to point out here that, when a considerable degree of hybrid vigor results from a cross, its effect is usually to push the F_1 generation further in the direction of the superior (more desirable) parent. In such situations, distribution of the F_1 generation might not be exactly intermediate—as in Figs. 12-6 and 12-7—but only slightly below that of the bigger or more productive parents. For example, a cross in fowls between breeds weighing 2,000 and 1,200 grams could yield F_1 birds closer to 1,800 grams than to the 1,600 mid-point. As that condition might arise from dominance of a single gene with a major effect, from effects of a few genes or of many genes, all of which might contribute to hybrid vigor, further analysis would be necessary to interpret the results.

Finally, since most traits of economic importance in the larger domestic animals and poultry are quantitative characters, knowledge of the factors affecting the expression and variability of such characters is indispensable for the breeder seeking to make present high yields still higher. To a considerable extent, the problem in pure breeds, or strains, is one of reducing the range between high and low producers, but with the reduction all coming off the lower end of that range! The increasing use of hybrid vigor to induce maximum yields suggests that if we are to make efficient use of that helpful booster of productivity, we must try to learn more about its still mysterious influence on quantitative characters.

Some Statistics Applicable to Quantitative Characters

When dealing with things showing a normal distribution, *i.e.*, a fairly symmetrical frequency curve with approximately equal numbers of individuals on each side of the average for the whole lot, certain special measures are very useful for describing those distributions concisely. The things thus measured need not be hereditary traits. They could be grades of students on an examination, weights of 100 apples or acorns picked at random, or anything else showing continuous variation from least to most. The special measures used are called collectively *statistics*. The group of things to which we apply them is generally known (to statisticians) as a *population*, even though some purists would restrict that term to people, and would never concede it to cows or chickens. Some of these *statistics* we have already considered, and one of them, the average, was an old acquaintance long before we wandered into genetics.

Statisticians deal in *means* rather than averages, but the two terms mean the same thing, if we may say so. Either can be computed by adding up the separate weights, heights, or grades, or whatever they are, for all the members of the population, and then dividing the sum by the number of those individuals, usually indicated as n. To save time and space, the thing measured (weight, height, grade, etc.) is usually designated as x, and the mean as $\bar{x}$ (read as x-bar). Also, instead of writing out "the sum of all the weights," etc., statisticians use the Greek letter sigma (Σ). Accordingly, to give us an easy entrée into the determination of statistics, along with (perhaps) some appreciation for statistical shorthand, we could condense almost everything presented in this paragraph into the simple formula:

$$\bar{x} = \frac{\Sigma x}{n}$$

When the number of individuals measured is large (perhaps anything above 50), it becomes quite a chore to add up all their separate scores, so a short-cut is utilized to find the mean. The determinations are sorted into classes that differ by some constant interval and the mid-value or class centre for each such group is multiplied by the number of individuals in that class; the resulting products are then added up (Σ again!) and their sum divided by n. To put it in simpler terms, we determine a weighted average. An actual example follows shortly.

While the means tell us about how big are our rabbits, how tall our soldiers, or how clever our girls, there are other measures that also give some idea of the population under review. One of these is the mode (mentioned earlier in this chapter), which is the class containing more individuals than any other. As with modes in other matters, it sets the fashion for its group. Another measure is the median. It is not a class, but an individual—the one exactly at the half-way mark in the distribution. Poultry breeders seeking to compare different sire-families with respect to age at first egg of the pullets therein find the median a more satisfactory measure than the mean. To determine the latter, one must wait until the last timorous straggler decides to lay. For the median, which is an equally good measure, one just takes the age at first egg for the 36th pullet to lay in a family of 71 (for example) and lets the 35 remaining birds get down to business when they will.

In a perfectly symmetrical distribution, the mean, mode, and median will coincide. In biological variations this seldom happens. Even when measuring the same thing in two different populations, the mean may be much closer to the mode in one of them than in the other (Fig. 12-2).

The Standard Deviation

While the three statistics just considered tell us something about any population, they do not distinguish between the one in which all members are fairly homogeneous and another in which they are extremely variable. Two populations of equal numbers may have identical means but be radically different (Fig. 12-8). Earlier in this chapter we mentioned the *range* from least to greatest as some indication of the variation in populations. It is not the best measure for that purpose, because it gives no hint whether the extremes (or other

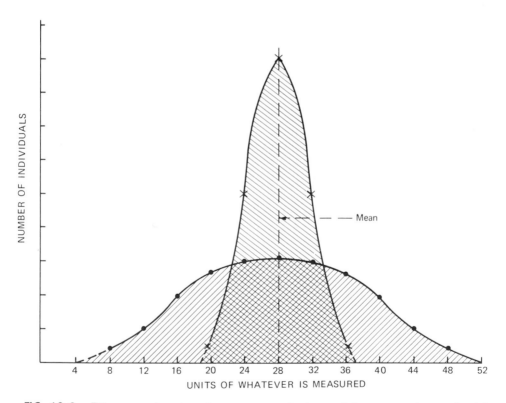

FIG. 12-8 Diagram showing that two populations of the same size and with the same mean may be radically different because one is more variable than the other.

classes remote from the mean) contain several individuals or just one lone exception.

A better measure might be the *average* deviation, which is determined by multiplying the deviation of each class from the mean by the frequency in that class, summing up the products and dividing by *n*. (See Table 12-3). In practice, the average deviation is seldom used because biologists (and especially *biometricians*, who measure living things) prefer to use the *standard* deviation, which is represented in statistical shorthand by the Greek sigma in lower case: σ. It is determined by squaring the deviation of each class from the mean, multiplying by the frequency in that class, summing the products again, dividing by *n* before (or *n* — 1, as statisticians prefer) to find the average squared deviation, and then extracting the square root of that figure. If the frequencies in the various classes are designated as *f*, and deviations of those classes from the mean as *d*, the formula for the standard deviation (S.D.) becomes:

$$\text{S.D.} = \sigma = \sqrt{\frac{\Sigma f d^2}{n-1}}$$

The reason for dividing by *n* — 1 rather than by the actual number of observations was conceded by one statistician to depend on "rather abstruse mathematical theory," but, for our present purposes, it may suffice to say that the degrees-of-freedom concept, which we use in determinations of χ^2, is involved. In any case, this concession to statistical dogma costs us nothing, for it matters little whether we divide by *n* or by *n* — 1; hence we might as well adopt the latter with all the grace we can muster.

One reason for determining the standard deviation as a measure of variation is that it is necessary for calculation of the *standard error*, which is another useful statistic.

Procedures for determining the mean and standard deviation are followed through in Table 12-3, where they are worked out for the age at first egg in the 354 White Leghorns shown earlier in the lower, wider curve of Fig. 12-2. As before the mean is 26.6 weeks, and the standard deviation proves to be 2.86 weeks.

Mathematicians have determined that when a population is distributed symmetrically in a normal curve, about two-thirds of the individuals will fall between one standard deviation above the mean and one below it. About 95 per cent will come within two standard deviations on both sides of the mean and 99 per cent within three (Fig. 12-9). Apparently, this applies equally well to distributions slightly askew, like the one for age at first egg in these Leghorns hatched on April 20. Accordingly, our standard deviation of 2.86 tells us that

TABLE 12-3

Determination of Mean and Standard Deviation for Age at First Egg in 354 White Leghorns

Class Range (weeks)	Midclass Value (V)	Frequency (f)	f × V	Deviation of Class from Mean (d)	d^2	f × d^2
20–21	20.5	10	205.0	−6.1	37.21	372.10
22–23	22.5	29	652.5	−4.1	16.81	487.49
24–25	24.5	100	2,450.0	−2.1	4.41	441.00
26–27	26.5	87	2,305.5	−0.1	.01	0.87
28–29	28.5	66	1,881.0	+1.9	3.61	238.26
30–31	30.5	46	1,403.0	+3.9	15.21	699.66
32–33	32.5	14	455.0	+5.9	34.81	487.34
34–35	34.5	1	34.5	+7.9	62.41	62.41
36–37	36.5	1	36.5	+9.9	98.01	98.01
		$n = 354$	$\Sigma fV = 9,453.0$			$\Sigma fd^2 = 2,887.14$

$$\text{Mean} = \bar{x} = \frac{\Sigma fV}{n} = \frac{9,453}{354} = 26.6 \text{ weeks}$$

$$\text{S.D.} = \sqrt{\frac{\Sigma fd^2}{n-1}} = \sqrt{\frac{2,887.14}{353}} = 2.86 \text{ weeks}$$

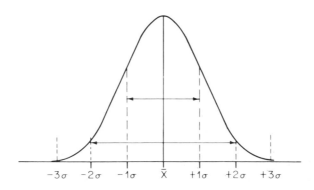

$$-3\sigma \quad -2\sigma \quad -1\sigma \quad \bar{X} \quad +1\sigma \quad +2\sigma \quad +3\sigma$$

FIG 12-9 Diagram showing for a normal distribution the proportion of the population that should fall within one standard deviation, or two, on each side of the mean.

about two-thirds of the flock began laying between $\bar{x} - 1\sigma$ and $\bar{x} + 1\sigma$, or between 23.7 and 29.5 weeks, and 95 per cent began between 20.9 and 32.3 weeks.

Admittedly, one could reach almost the same conclusion (without any standard deviation) by studying the frequencies in various classes as shown in Table 12-3. However, by using a calculator and a different formula, one can determine the standard deviation without sorting the population into classes. In such cases, as here, the standard deviation would give us a good idea of the variability in the thing measured. It is particularly useful, along with the mean, for comparing different groups with respect to any one character. In Fig. 12-8, the tall curve would have a comparatively small standard deviation and the other curve a big one. If we know the standard deviation and the mean for a population, we can visualize its curve of frequency distributions without drawing it.

Coefficient of Variation

This useful measure of variability is derived from the standard deviation and the mean. It is used to determine which of two entirely different things is more variable than the other. If we should want to know whether body weight is more variable in elephants than in a mouse colony, the range from least to greatest would tell us nothing, nor would the standard deviations, because the means for the two measures are so vastly different. The coefficient of variation (C.V.) relates the variation to the mean and measures it on a familiar scale—that of percentage. The formula is simply:

$$\text{C.V.} = \frac{\text{S.D.} \times 100}{\text{Mean}} = \frac{100\sigma}{\bar{x}}$$

Unfortunately, although weights of mice are available in quantity, those of elephants are not. As examples of the use of the coefficient of variability, some of Asmundson's measurements on 707 hens' eggs will serve well:

	Mean	Coefficient of Variation
Length in millimetres	56.9 mm.	5.3 per cent
Weight in grams	53.6 g.	10.1 per cent
Weight of shell, grams	5.6 g.	14.4 per cent

These figures tell us that these eggs were almost twice as variable in weight as in length. Although weight of shell was only a tenth of that for whole eggs, it was much more variable.

The Standard Error

When appraising any population by the mean and standard deviation for the sample that is measured, we must ask ourselves, "How reliable is the mean? If we measured another sample from the same population, how close would its mean come to that for the first?" Looking back at Fig. 12-2, we can easily see how a dozen different samples of 40 pullets each, although all drawn from the same flock of 354 Leghorns, could give us a dozen different means, none of which need be the 26.6 weeks that was determined for the whole lot. Similarly, even those 354 birds may not have yielded the same mean as a flock of 1,000 might have done. Statisticians tell us that the populations measured, no matter how large, should be regarded only as samples of still larger theoretical populations.

Obviously, the reliability of the mean depends in part upon (a) the number of determinations from which it is derived, but also upon (b) the amount of variation in the population. A good measure of the reliability of a mean is provided by its *standard error*, which is based on both of those factors and determined as follows:

$$\text{S.E. of the mean, or S.E.}\bar{x} = \frac{\sigma}{\sqrt{n}}$$

The standard error of *one* mean is equivalent to the standard deviation of a *whole series* of means that might be determined from corresponding samples of the same population. It prescribes limits, therefore, within which would fall about two-thirds of the means for any samples similar in size. Accordingly, by giving the mean we have determined, plus or minus its standard error, we tell something not only about the thing measured but also about the reliability of the sample.

If we apply all this to the sample with which we have sought to elucidate these statistical manoeuvres, the standard error for mean age at first egg for our Leghorns is found to be:

$$\text{S.E.}\bar{x} = \frac{\sigma}{\sqrt{n}} = \frac{2.86}{\sqrt{354}} = 0.15$$

Having determined it, we give the mean as 26.6 ± 0.15 weeks. Another sample like the flock studied would yield a mean between 26.45 and 26.75 weeks in about two-thirds of the cases, and one beyond those limits in only about one-third of similar trials. About 95 per cent of such samples would fall within twice the standard error on both sides of the mean, *i.e.*, between 26.3 and 26.9 weeks. In this example, the standard error of the mean is very small, but that is attributable in large part to the fact that the sample was big. If ages at first egg had been recorded for only 50 birds, instead of 354, the standard error would have been much greater and the mean correspondingly less reliable.

Another measure of reliability is the *probable* error, which is $0.6745 \dfrac{\sigma}{\sqrt{n}}$. It is used less now than in former years. Whereas the standard error of a mean gives limits on both sides within which about two-thirds of the distribution would fall, those set by the probable error would contain only half.

The reader should recognize that one useful function of these several statistics is to give a concise, mathematical description of the population to which they apply. In our example, that description with respect to age at first egg can be summarized as follows:

$$n = 354; \quad \bar{x} = 26.6 \pm 0.15 \text{ weeks}; \quad \sigma = 2.86 \text{ weeks}; \quad \text{C.V.} = 10.7 \text{ per cent}$$

Standard errors of two independent means can be used to determine whether or not the difference between those means is statistically significant. Standard errors can also be computed for constants other than the mean. As one purpose of this book is to present principles and applications of genetics with an indispensable minimum of mathematics, a detailed exposition of the various statistical methods useful to biologists is not justifiable here. Our introduction to quantitative inheritance is concluded, therefore, with the hope that sufficient unto this chapter are the statistics therein.

Problems

12-1 If the range in expression of some quantitative character depends on only five pairs of alleles that exert equal cumulative effects, when a cross is made between the two extreme types, what proportion of the F_2 generation would you expect to show either of the extremes?

12-2 Draw a histogram similar to that in Fig. 12-1, but showing the distribution expected in F_2 from F_1 parents heterozygous for four polygenes with equal cumulative effects.

Castle crossed large and small breeds of rabbits, essentially Flemish Giant and Polish respectively, and then backcrossed the F_1 to both parental races. At 275 days of age, weights of females were distributed as is shown in Table 12-4. The next eight questions deal with those data.

12-3 For each of the five different populations given in the table, work out the mean, the standard deviation, the standard error of the mean, and the coefficient of variation. (Don't throw away these figures; we may need them in a later chapter.)

12-4 From comparisons of the means and distributions for the F_1 and two backcrosses with those of the parental races, would you conclude that the inheritance of size in this cross conformed to that expected for a quantitative character?

12-5 Compare the means in the F_1 and backcrosses with the mid-parent values for the crosses that produced them. Is there any evidence that some genes increasing size tend to be dominant, or recessive, to those depressing it?

12-6 From your figures for $\bar{x} \pm$ S.E., which means would you consider the more reliable, and why?

12-7 Did you expect σ for the Flemish Giants to be greater than that for the Polish? Why?

12-8 The range for the Polish females covered only 600 grams, while that for the Flemish Giants was 1,400. What do the coefficients of variation tell about the comparative variability of these two parent races?

12-9 Which of the five groups was most uniform? Can you think of any comparable groups of domestic animals, or any farm crops, for which uniformity is characteristic?

12-10 If you were to cross (a) the smallest does in one backcross with bucks comparable to the smallest does of the other, and (b) the largest animals from the two backcrosses, you might expect (from mid-parent values) that mean weights of the progeny would be 22 hectograms or less from (a) and 36.5 from (b). How can you reconcile those estimates with the fact that all of them should have the same proportion of Polish "blood"?

12-11 Schneider and Dunn found that coefficients of variation for body weight in their White Leghorns ranged from 12 to 18 per cent, whereas those for bone measurements were only 3 to 4.5 per cent. Why should bodies vary more than bones, and which would give the better measures of size?

12-12 If you have time and energy for additional studies of normal variation, weigh or measure 50 eggs, acorns, or classmates, and determine $\bar{x}$, σ, S.E., and C.V. If you do weigh eggs, compare results with Asmundson's figures given in this chapter.

Selected References

CASTLE, W. E. 1931. Size inheritance in rabbits; the backcross to the large parent race. *J. Exper. Zool.* **60:** 325–38. (Source of the data in Table 12-4; also gives weights at earlier ages and at 365 days.)

TABLE 12-4

Weights of Female Rabbits at 275 Days

(1) Weight (hectograms)	(2) Polish	(3) Backcross to Polish	(4) Weight (hectograms)	(5) F₁ Generation	(6) Backcross to Flemish Giant	(7) Weight (hectograms)	(8) Backcross to Flemish Giant	(9) Flemish Giant
12	3		27	1		42	2	1
13	11	1	28	1		43	2	1
14	13	1	29	2		44	1	1
15	8	6	30	2		45		
16	3	10	31	4	2	46	1	3
17	2	18	32	5	1	47		8
18		17	33	2	5	48		4
19		21	34	1	3	49		1
20		24	35		4	50		3
21		19	36		6	51		
22		15	37		3	52		11
23		10	38		2	53		4
24		5	39		3	54		2
25		4	40		3	55		2
26		2	41		3	56		1
27		1						
(Continued in Col. 4)			(Continued in Col. 7)		(Continued in Col. 8)			

Source: Data of W.E. Castle, 1931.

FALCONER, D. S. 1960. *Introduction to Quantitative Genetics*. New York. The Ronald Press Co. (See particularly statistical analyses of quantitative traits in Chapters 6 to 8, and other chapters for theoretical considerations of responses to be expected with different kinds of selection.)

HUTT, F. B. 1959. Sex-linked dwarfism in the fowl. *J. Heredity* **50**: 209–21. (More information about the effects on growth, productivity, and other quantitative characters of the mutation shown in Figs. 12-3 and 12-4.)

JOHANSSON, I. 1961. *Genetic Aspects of Dairy Cattle Breeding*. Urbana, Ill. Univ. of Illinois Press. (Chapter 8 surveys in readable style present knowledge of two very important quantitative characters: yield of milk and butterfat.)

PUNNETT, R.C. 1923. *Heredity in Poultry*. London. Macmillan & Co., Ltd. (With details of the cross shown in Fig. 12-5 and other studies of quantitative traits.)

ROBERTS, E., L. E. CARD, W. E. SHAKLEE, and N. F. WATERS. 1952. Inheritance of egg weight. *Poultry Sci.* **31**: 870–75. (Details of the crosses shown in Fig. 12-7.)

SNEDECOR, G.W., and COCHRAN, W.G. 1980. *Statistical Methods*. 7th ed. Ames. Iowa State Coll. Press. (A good place to find statistical procedures worked out in helpful detail, mostly with examples of their application to biological material.)

WRIEDT, C. 1930. *Heredity in Live Stock*. London. Macmillan & Co., Ltd. (Contains *inter alia* further details about the cross between Red Danish cattle and Jerseys, one aspect of which is shown in Fig. 12-6.)

Modifiers, Penetrance, and Pleiotropy

From the earlier chapters of this book, some readers may have the impression that in general one gene produces one character, except in special cases where two or more pairs of alleles interact to produce phenotypes different from that of any single pair. However, in Chapter 12 we found that some characters are dependent on the cumulative action of many genes. We come now to some special cases. These include modification of simple traits by many genes, Mendelian characters that do not segregate in Mendelian ratios because of low *penetrance*, and conditions in which a single *pleiotropic* gene affects not just one structure, or function, but several of them at once.

Modifying Genes

This classification is reserved for polygenes of a special kind—those that modify the expression of some simple trait, usually one dependent for its expression on a single pair of genes. Although the modifiers cause a range in the expression of that character, they seldom succeed in obliterating entirely the basic phenotype, the background on which their effects are displayed.

Modifiers have something in common with polygenes that affect quantitative characters. Their numbers are usually unknown, and both kinds are amenable to selection in either a plus direction or a minus one. The difference is that the effects of modifiers are manifested only in a suitable genotype, and their presence and effects are not evident in other genotypes. To a limited extent this same restriction applies to some quantitative characters. We cannot measure production of milk in bulls, or of eggs in roosters, but, apart entirely from such sex-limited traits, most quantitative characters are not dependent for their expression on the presence of some simple genotype affecting form or color.

A familiar example of effects of modifying genes is provided by the range in white spotting to be seen in Ayrshires or in Holstein-Friesians. Both breeds are homozygous for a gene that causes white spotting, but modifying genes are responsible for the range from mostly colored to mostly white (Fig. 13-1). Fortunately, the breed associations accept almost the entire range, and breeders who care to do so can accumulate modifiers for darker or lighter animals as they wish. The phenomenal records of all three cows in our illustration show clearly that productivity is quite independent of the degree of white spotting.

White spotting is not a breed characteristic of the Aberdeen Angus. Undoubtedly, they carry modifiers that could affect the degree of white spotting if there were any at all. In the absence of the spotting

FIG. 13-1 Three Holstein Friesians showing different degrees of white spotting induced by modifiers acting on one genotype. *Top:* Locust-Glen Laird P. Matt, born 1971, a leading lifetime producer, with 119,720 lbs. milk and 4,993 lbs. fat in five lactations. *Middle:* Zairview Arlinda Polly, born 1969, first aged cow and grand champion at Illinois State Fair, 1977, with 111,960 lbs. milk and 4,165 lb fat in five lactations. *Bottom:* Metcalf's Elevation Sylvia, born 1974, with a record of 28,800 lb milk and 1,069 lb fat in her second lactation. (Photographs and records by courtesy of the Holstein-Friesian Association of America.)

genotype the presence of any modifiers for the particular pattern is undetectable.

Sometimes selection of modifiers is necessary to maintain an important breed characteristic. The attractive white belting in the shoulder region of Hampshire swine (Fig. 13-2) is caused by a gene that is dominant to *self-color*, or solid color, in this case, black. Hampshire breeders know that the belt has nothing to do with growth, viability, or capacity for feed conversion, but consider it, as one of them writes, "a million-dollar trademark." To keep that trademark patented, rules for registration of Hampshires specify that the animal must be black, with a white belt encircling the body entirely and including both front legs. The belt may be narrow or wide, regular or irregular, but not so wide that more than two-thirds of the body is white. There can be white on the hind legs, as long as it does not extend up on the ham. Tips of the tail and the snout can be white, but other white on the head bars an animal from registration.

Presumably, most pure Hampshires are homozygous for the gene that causes belting, but constant selection is necessary to maintain the pattern desired, and to prevent the white from increasing. Obviously, the regulations for registration are an important part of that selection. The breeder does not select for white belting, because the major gene for that pattern is already present in all his pigs. He does select for the modifying genes which help to maintain that pattern within the bounds that he and his associates have set up to define the standards for their breed.

White spotting shows great variation in every species in which it is found. C. C. Little attributed spotting in Beagles to their being homozygous for a recessive gene *sp*, one of a series of multiple alleles. Within that genotype he established no fewer than ten arbitrary grades of spotting. These he ascribed to effects of modifying genes, with "plus" modifiers increasing black and "minus" modifiers decreasing it.

Modified Frizzling in the Fowl

For an example of a single gene that does a commendable measure of modification all by itself we return to the Frizzle fowls discussed and illustrated in Chapter 3. The beautifully curled plumage of the heterozygote (*Ff*), which is the kind that wins prizes at poultry shows, can be flattened down to a slight ruffling of the feathers by the modifying gene *mf* when homozygous (Fig. 13-3). In some birds of the genotype *Ff mfmf*, the presence of *F* would pass undetected unless specially sought by someone who knows what to look for. In the

FIG. 13-2 Magnifier Ann, a prize-winning Hampshire gilt showing the white band considered by Hampshire breeders a "million-dollar trademark." See the text for its specifications; selection for modifying genes is necessary to meet them. (Courtesy of Hampshire Swine Registry.)

FIG. 13-3 The modified, heterozygous Frizzle. Compare with the unmodified heterozygote in Fig. 3-2. (From F. B. Hutt in *J. Genetics*.)

homozygous state, this modifying gene suppresses frizzling to a marked degree in all birds *Ff* and in about 40 per cent of those that are *FF*.

This modifier was found in eleven different breeds and in most non-frizzled birds that were not derived from frizzled stock. It seems clear that to perfect the exhibition type shown in Fig. 3-2, the breeder must have had to eliminate this modifying gene, and possibly others.

Fisher's "Evolution of Dominance"

R. A. Fisher suggested that, under natural conditions, any dominant deleterious mutation in wild animals is more likely to be preserved and passed to the next generation by those that carry modifying genes which suppress its effects. By such continued natural selection and accumulation of modifiers over hundreds of generations, such characters could eventually be suppressed in most of the population to the extent that they would be recognizable only in the few individuals that might lack the modifying genes. A mutation originally dominant could thus eventually become recessive to the prevailing wild type.

The gene modifying frizzling provides a case that supports Fisher's theory nicely. From the side-effects of frizzling mentioned in Chapter 3, it is clear that the gene *F* would lessen considerably the chance of survival in nature for any bird carrying it. The single modifier, *mf*, by restoring the plumage to a state almost normal, practically eliminates the extra hazard imposed by *F*. Unmodified frizzled birds cannot fly to the roosts, and usually huddle together on the floor at night. In the wild, they could more easily be picked off by predators than could birds that perch in trees. The modified frizzles can fly, and they roost just the same as other fowls. Obviously, they could survive better in the wild than could their unmodified relatives.

It is easy to see, therefore, how this particular modifying gene would be preserved and accumulated in any wild population exposed to natural selection. It is even conceivable that its presence in eleven different breeds of domestic fowls in the twentieth century may have resulted from a widespread distribution of it among the ancestors of our domestic fowls, long before jungle fowls were scattered to the four corners of the world.

Experimental Accumulation of Modifiers

It is all very well to say that the three degrees of white spotting shown in Fig. 13-1, and the ten degrees of it in the Beagles, result from the action of modifying genes on a single character, but actual proof of that statement could come only from an experiment. If the breeder of Holstein-Friesians could be persuaded to develop two different herds, to select in one for blacker cows and in the other for whiter ones, and to continue that selection for some six or eight generations, the results should show what can be done by the accumulation of modifying genes. Obviously, such an experiment is not very practicable with animals as big as cattle. It would be somewhat easier with Beagles, but there is little point in trying it with that species either. The results of such selection were determined long ago with a species that multiplies fast enough to facilitate exactly such experiments—the rat. In the earlier years of genetics, the rather wide range of expression for some characters led to considerable discussion of the question whether or not the causative gene had become "contaminated" by association with its allele, or with other genes, in crosses. It is easy to see how such a question could arise when a cross of a white-spotted rat or rabbit with another of solid color yielded in an F_2 generation some spotted animals that showed much less white than their spotted grandparent. A few of them might even resemble the self-colored grandparent more than the spotted one.

To answer that question, Castle selected in two lines for and against the hooded pattern in rats. The hooded rat at the mid-point of his 14 grades between extremes showed black color over the head and shoulders and a black dorsal stripe extending back to the tail. Elsewhere it was white. After 20 generations of selection for modifiers, the plus strain was predominantly black, and the minus strain correspondingly white. The feasibility of accumulating modifiers in either direction was proven. From matings of the extreme types with wild, unhooded rats, Castle obtained in the F_2 generations hooded rats less divergent from the average hooded pattern than their grandparents had been. Obviously, the accumulation of modifiers had not "contaminated" the basic gene for the hooded pattern.

While results of some experiments with rats are not as applicable to other species (including our own) as some nutritionists seem to think, Castle's findings about selection for modifiers of the hooded pattern in rats have been confirmed in similar selection for white spotting in mice, rabbits, and guinea pigs. There is every reason to believe that they apply also to that same character in cattle and in Beagles.

Similar accumulation of modifiers is undoubtedly part of the process by which breeders seek to attain perfection of their animals with respect to fancy points and breed characteristics, particularly those related to colors and patterns. All Barred Plymouth Rocks have the same gene, B, for barring, but some have clear, sharp barring, while in others it is so indistinct as to appear smoky. When the show standards of the 1920's called for development of what were essentially two varieties of Barred Plymouth Rocks, one with wide white bars and the other with narrow ones, the breeders produced them almost as if to order. All the birds had the same barring gene; the difference was solely in the modifiers.

The efficacy of selection to accumulate modifying genes was also demonstrated by Dunn and Landauer in their studies of hereditary rumplessness in the fowl. Heterozygotes for the dominant gene (*Rp*) causing that condition sometimes show an intermediate stage in which the tail feathers are reduced in number, but are not entirely eliminated as in completely rumpless birds. By selecting for this type and outcrossing to unrelated stock in quest of modifiers reducing rumplessness, the proportion of intermediates among all rumpless progeny from parents that were both intermediates was raised in a few years from 42 to 88 per cent. As those matings regularly produced approximately 3 rumpless : 1 normal, and the proportion of heterozygotes among the rumpless should not have been more than two-thirds, it seemed evident that the modifiers accumulated were by the last year suppressing the condition not only in all the heterozygotes, but in some of the homozygotes as well.

Rumpless fowls have poor fertility, but it has been proven that the intermediates are less handicapped in that respect than their completely rumpless siblings. Modifiers inducing the intermediate form therefore have what the students of evolution call *survival value*— they are conducive to survival of the species. Such modifiers, like the ones reducing frizzling, are exactly the type that would be accumulated in nature, over thousands of years, by the process of natural selection.

Incomplete Penetrance

Many years ago, the Russian geneticist, Timoféeff-Ressovsky, then working in Berlin, found that in different lines of Drosophila a recessive mutation affecting venation of the wings was not manifested by all the flies homozygous for the causative gene. In different lines that had been inbred to a point at which they should have been homozygous, not only for the gene under study but for all others as well, the proportion showing the mutation varied from 41 to 100 per cent. Timoféeff-Ressovsky described these lines as differing in the *penetrance* of the character.

Reasons for incomplete penetrance, *i.e.*, the failure of a mutation to be manifested by some individuals having the genotype that ordinarily produces the character, are not fully understood. The best interpretation is probably that, during development, the balance between efforts of the gene attempting to induce the character and of others seeking to promote normal development is so delicate that in some individuals the *threshold*, or minimum of activity necessary for manifestation of the character, is not reached. The fact that a gene can thus be thwarted from achieving its destiny may be compared, perhaps, with the similar plight of a person who, having been "bumped" off a plane by a higher priority, fails to reach his appointed destination. We leave that problem for others to study. The important thing to be remembered here is that some animals not showing some hereditary trait may have the same genotype with respect to that trait as do those that show it. Here, as in other walks of life, things are not always what they seem. Proof of such genetic dissembling can come only from breeding tests. A good example of one such test is described in the next section.

Low Penetrance in Pigtailed Mice

Crew and Auerbach studied a recessive mutation in the mouse which caused varying degrees of curling in the tail. These ranged from single or double spirals (Fig. 13-4) through kinks (one or more) down to slight local thickenings. The lowest grades were not evident until several weeks after birth. Appropriately, the mutation was dubbed pigtail.

Breeding experiments showed that pigtail was completely recessive, but the proportions in the various populations that showed the character were far below those expected with any ordinary, simple recessive mutation (Table 13-1). Normal mice from parents both pigtailed, which should therefore have been homozygous for the gene, showed no trace of the character (Fig. 13-4) but bred almost the same as those that did. It is not surprising that such normal mice, when mated *inter se*, should show lower penetrance of the character in their progeny (16.7 per cent) than did progeny from parents both pigtailed. Penetrance of this character could not be rated higher than 24 per cent.

Incomplete Penetrance in Domestic Animals

Breeding experiments of the kind just cited are more difficult to conduct with larger animals, but incomplete penetrance may account for some of the irregularities encountered in genetic studies with domestic animals. One example should suffice to illustrate the point.

Eriksson (1943) studied a peculiar reduction of fertility in Swedish Highland cattle caused by hypoplasia of the gonads. Both sexes

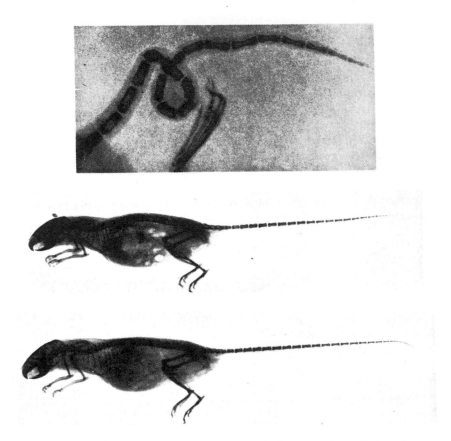

FIG. 13-4   Top: X-ray photograph showing marked expression of pigtail mutation in the mouse. Middle: skeleton of normal control female. Bottom: skeleton
of phenotypically normal female known to be genetically pigtail. (From F.A.E.
Crew and C. Auerbach in *J. Genetics*.)

were affected. The abnormality was usually unilateral, and the ratio of
animals affected on the left, right, or both sides was about 82:3:15,
respectively. It was concluded that the defect was caused by a
recessive, autosomal gene with incomplete penetrance. The proportions of homozygotes clinically classifiable as abnormal (*i.e.*, the
penetrance of the character) were estimated to be about 43 per cent in
males and 57 per cent in females.

   After a genetic basis for the abnormality was recognized, appropriate steps were taken to reduce its frequency, and, among large
numbers of the breed examined on farms or at breeding centres, the
proportion showing genital hypoplasia dropped from about 25 per cent
in 1935 to 8 per cent in 1942.

TABLE 13-1

Matings Showing Low Penetrance of the Pigtail Mutation in the Mouse

Parents	Progeny	
	Number	Pigtail (per cent)
Normal × pigtail	234	0
F$_1$ × F$_1$	311	1.6
Backcross, F$_1$ × pigtail	142	4.2
Pigtail × pigtail	219	23.8
Pigtail × normal from parents both pigtail	117	20.4
Normal × normal, both from pigtail parents	143	16.7

Source: Data of Crew and Auerbach, 1941.

Later investigations showed that the frequency of hypoplasia is highest in white cattle and decreases as areas of black pigment increase. Cattle with a fair amount of black do not have any hypoplasia. Evidently, the genetic basis is more complex than Eriksson had thought.

To prove beyond question that aberrant ratios or other genetic inconsistencies result from incomplete penetrance, it would be necessary to show by breeding tests, as with the pigtailed mice, that animals not showing the phenotype have the same genotype (with respect to the character under study) as those which do. In some cases where low penetrance might be suspected, there may be difficulty in recognizing animals that have only a trace of the character. Conversely, with some traits that are clearly hereditary but for which analyses have as yet failed to show the exact genetic basis, that basis may not be an undetermined number of interacting genes, but one or more with varying degrees of penetrance. (For examples, see Problems 13-4 and 13-5.)

The degree of penetrance may be influenced greatly by the environment. In Drosophila, several different genes can induce a condition called "abnormal abdomen" in which the black bands are very irregular. Flies homozygous for one of these genes will all show abnormal abdomens when produced from fresh, moist food. As the culture gets older and dryer, the proportion of flies manifesting the character decreases, and eventually those produced from dry food are indistinguishable from normal flies.

In highly inbred lines, like those in which Timoféeff-Ressovsky studied incomplete penetrance, the degree of that penetrance remained at a fairly constant figure for each line. Selection to change it in any such highly homozygous stock would be relatively ineffective. However, apart from special laboratory strains, most stocks of animals are highly heterozygous. Among these the degree of penetrance can be greatly altered by selection. Presumably, that selection plays entirely on modifying genes.

Expressivity

In many cases, individuals that all have the same genetic variation will vary widely in the *expressivity*, or degree to which it is manifested. This term (also coined by Timoféeff-Ressovsky) is quite different from penetrance. The latter measures the proportion of a group that shows the character, but expressivity refers to the variation in the character among those which show it. For example, the expressivity in cattle of recessive syndactyly, commonly called mule-foot, varies from complete expression on all four feet, to cases with only one forefoot affected. Sometimes a line of fusion of the digits is evident.

A good example of a trait that varies both in penetrance and expressivity is polydactyly in the fowl. As heterozygotes frequently show no sign of that mutation, it is sometimes described as incompletely dominant. Landauer even found some birds that were proven by breeding tests to be homozygous, but which had no trace of polydactyly. Among fowls that do show it, the expression varies from an elongated single hallux (Digit 1), with extra phalanges, to a double hallux, or one partially double. Sometimes there is only a bifurcation of the claw. One foot may show one of these forms of polydactyly, the other not.

Similar ranges are common in the expression of other genetic characters. The three distinctly different grades of epithelial defects in Jerseys, Holstein-Friesians, and Ayrshires may possibly be caused by different genes, but it seems more likely that they may all result from the same gene, with its degree of expression modified in the three breeds by all the other genes that differentiate them.

Pleiotropy in the Mouse

The gene A^y, although very bad for any zygote that might hope to become a mouse, is very useful to geneticists. In the heterozygotes, which are the only ones capable of showing its effects, A^y demonstrates that a single gene can have multiple effects. It reduces black pigment to a few granules, so that the coat is bright yellow, or even orange. The eyes remain black. It lowers the rate of metabolism so much that yellow mice are conspicuously more obese than their non-yellow litter-mates. Apart from that effect on size, the yellow mice have slightly bigger skeletons. They are less susceptible to spontaneous mammary carcinoma than their non-yellow littermates. Some of these multiple, or *pleiotropic*, effects may seem at first glance to be quite unrelated, but it is probable that the whole syndrome arises from some abnormality of cellular physiology caused by the gene A^y.

Ordinarily, one would not expect genes modifying color of the hair to have any effect on size, but, from a series of extensive backcrosses, Castle found that some simple recessive mutations do have exactly such pleiotropic effects (Table 13-2).

His data show that both b and d increase size of homozygotes and that in combination their effects are cumulative. Two kinds of dilution—pink-eye and pallid—reduce size, as does leaden, but a and c have no effect. Oddly enough, b and ln in combination do not counteract each other, but reduce size more than does $lnln$ in black mice. Castle found evidence that brown color goes with larger size in rats and rabbits, as in mice.

Pleiotropic effects of genes for coat color are known in at least one important domestic animal—the mink. Keeler's investigations showed that the genes causing Moyle Buff, Silverblu, and Pastel tend to make

TABLE 13-2

Changes in Body Size (Percentages) Induced by Certain Genotypes for Color in Mice

Genotype	Phenotype	Weight	Length of Body	Length of Tail
bb	Brown	+ 4.27	+1.51	+1.30
dd	Dilute	+ 2.10	+0.90	+2.64
$bb\ dd$	Dilute brown	+ 5.81	+2.70	+3.89
$bb\ Lnln$	Brown, not leaden	+ 1.07	+0.71	+0.55
$Bb\ lnln$	Black, leaden	− 3.64	−0.61	−2.94
$bb\ lnln$	Brown, leaden	− 5.47	−1.00	−3.42
$bb\ Papa$	Brown, not pallid[a]	− 0.70	−0.36	−0.07
$Bb\ papa$	(Black) pallid[a]	− 5.53	−1.89	−4.04
$bb\ papa$	(Brown) pallid[a]	− 5.90	−2.46	−3.29
pp	Pink-eyed dilution	− 1.01	−0.14	−0.72
aa	Non-agouti	0	0	0
cc	Albino	0	0	0
$A^ya\ ♂\ ♂$	Yellow	+33.00	+2.60	+1.50
$A^ya\ ♀\ ♀$	Yellow	+62.00	+4.90	+0.20

[a]Known erstwhile as Roberts' pink-eye, or pink-eye-2.
Source: Data of Castle, 1941.

minks with those colors bigger than the normal dark kind. Conversely, animals showing the colors known as Breath of Spring, Aleutian, Palomino, and Ambergold are smaller than normal. It is not surprising that Breath of Spring minks are small, because they are heterozygous for an incompletely dominant gene that is lethal to homozygotes. All the other mutants mentioned are recessive. Keeler found that when two of these genes affecting size were present in the same mink, their combined effects were greater than that of either one alone, but not equal to the sum of the two increments induced by the genes when acting separately.

One must not think of pleiotropic effects as being limited mostly to modifications of size. A mutation of the *p* locus in the mouse was found by Hollander *et al.* to cause (when homozygous) not only the usual pink-eyed dilution but also poor growth, uncoordinated behavior, difficulty in chewing, abnormal wearing of the teeth, inadequate maternal care and premature sterility in females, and sterility in males. Not a bad show for one little gene.

Although some genes cause various abnormalities of behavior, or of physiology, not all of these can be considered as pleiotropic effects. For example, animals deprived by some gene of their normal insulation of hairs or feathers frequently are subnormal in viability, have higher rates of metabolism, and show other adaptations to their affliction. These are really secondary effects. In its original usage, the term *pleiotropy* was intended to include other effects not obviously related to the most conspicuous one by which the mutation was recognized.

Lethal Indigestion in Grey Karakuls

Offhand, there would seem to be no reason why some grey sheep should have more digestive troubles than other greys, or black ones, but they do. It has been known for many years that most adult grey Karakul (Shiraz) sheep are heterozygous for that color. When mated *inter se*, they produce lambs in the ratio of 3 grey:1 black, and about one-third of the greys die when four to nine months old. These are the homozygous greys (Fig. 13-5). A few homozygotes may live to reproduce, but they are clearly exceptions.

Nel and Louw found that when the stricken grey lamb begins to weaken, it develops a pot-belly, becomes emaciated, and eventually is too weak to stand. On post-mortem examination, the abomasum is found to be greatly enlarged, and filled with comparatively dry, caked, fibrous material. Contents of the rumen, which is usually distended, are soft, fine, and gaseous. The large and small intestines are empty. In comparisons of homozygous greys with normal lambs, the liver,

FIG. 13-5 Homozygous grey Karakul lamb before onset of digestive disorder. The genotype was indicated by white tongue and chin, light grey face, and white patches on the ears. (Courtesy of J. A. Nel and Experimental Farm, Neudam, South West Africa.)

heart, kidneys, and spleen of the former were much smaller than normal, but the rumen was almost twice as large as it should have been.

Homozygous greys cannot be easily distinguished from heterozygotes at birth by color of the fleece, but Nel found that most of the former could then be identified by their white tongues. Controlled data on the proportion of homozygotes that survives to breeding age are not available, but evidence from different sources indicates that a few, at least, do survive to reproduce. One such ram kept by Nel and Louw had shown typical symptoms at an early age but lived to sire 12 lambs from black ewes. The fact that all of these were grey provided further proof that their sire was homozygous. In this case, one might refer to the digestive disorder as a pleiotropic effect of the gene for grey color of the fleece.

The Screw-neck Pastel Minks

An odd pleiotropic effect was studied by Shackelford and Cole in pastel minks. These have less dark melanin in the fur, eyes, nose, lips, and claws than standard dark minks, so that they look brown rather than black. The difference depends on a single pair of alleles.

About 25 per cent of pastel minks tilt the head to one side or the other. When they turn around, the head is thrown back over the shoulders and some minks turn complete somersaults. A few can scarcely stand, but the condition is not lethal. Symptoms tend to be less severe as the minks get older. Minks of other colors do not show this peculiar behavior.

This peculiar behavior is known to breeders as "screw neck." By selection against it, its frequency has now been reduced in some herds of pastel minks to about 10 per cent.

The anatomical basis for the defect was studied by Erway and Mitchell (1973) who used a swimming test to facilitate identification of minks affected only moderately or slightly. They found that the screw-neck behavior is associated with reduction or absence of otoliths in the inner ear. In normal animals, these structures serve as receptors of gravity and thus ensure maintenance of equilibrium.

Most important for mink raisers, Erway and Mitchell found that supplementing the diet of prospective mothers with about 0.1 per cent of manganese sulphate during the gestation period resulted in better development of otoliths in the progeny and also in elimination of the screw-neck behavior from them.

The screw-neck pastel minks are of interest because they exemplify not only a pleiotropic association of color with an anatomical defect, but also suppression of a genetic weakness by a favorable environment in the form of manganese sulphate. Let us not forget, however, that there are many genetic defects against which even the best environment is of little or no avail.

Pleiotropy in Smooth-tongued Calves

De Groot (1942) studied in Friesian cattle a remarkable syndrome of which the characteristic most easily recognized is a comparatively smooth tongue. It results from a reduction in size of the *papillae filiformes* on the surface of the tongue. The condition was proven to be a simple, autosomal recessive character. Matings between animals known to carry the causative gene yielded a ratio of 90 normal : 38 smooth, and six calves from parents both affected all had the smooth-tongue syndrome.

Pleiotropic effects of the causative gene include defective development of the long hairs, which are shorter and thinner than normal. The under-hair is dense, and in spots the coat appears velvety. The skin seems to hang in vertical folds. The calves appear unthrifty, partly because of dirty coats resulting from a slight diarrhoea, and commonly show eczema. To some extent this last is caused by the pathological condition itself. Some of it results from excessive saliva-tion and remnants of the cud which become plastered on the skin. The eczema causes itching, rubbing, and even superficial sores. The mucous membrane of the tongue is delicate and easily injured; the horns are not as hard as in normal animals. Red blood cells are smaller than normal, and the iron content of the blood serum is abnormally low. Treatment with iron sulphate improves the condition of the calves and brings the red blood cells up to normal size.

In this case, as in many others, subsequent study revealed that the basic abnormality is not the symptom most conspicuous, or first-recognized, from which the condition got its name. De Groot concluded that the underlying defect is an abnormality of blood chemistry which causes a form of hypochromic anemia. The assorted symptoms of the syndrome provide a good example of manifold effects of one gene.

Genetic Intersexuality in Polled Goats

Goat breeders were perplexed for many years by two problems beset-ting their pet species: (a) a deficiency of females, and (b) a compara-tively high frequency of intersexes. Among British Saanens, Paget found the proportion of kids recognizable as intersexes at birth to be 14.8 per cent in 1943. As the intersexes are sterile, they are a total loss in milch goats. They are not comparable to freemartins in cattle, because they occur as often in single births as in twins or triplets. That there is some genetic basis for the intersexes is indicated by the fact that some goats produce them consistently, and others not at all.

A solution to both problems was suggested by Asdell, who had noted that the intersexes were more numerous in populations showing the greatest deficiency of females. If all the intersexes were regarded as originally genetic females, the sex ratios came close to the expected 1:1. The validity of this concept is now recognized. Among Saanen kids recorded in six districts of Japan, Kondo found the recorded sexes to be as follows:

Males: 2,317

Females: 2,111

Intersexes: 201

Addition of the intersexes to the females brings their number as close to that for males as one could hope to get.

Goat raisers prefer polled goats to horned ones. The reasons for that preference are well known to anyone who has learned the hard way not to present his or her derrière to any innocently ruminating, masculine representative of *Capra hircus*. In this species, as in cattle, the polled and horned conditions behave as simple alleles, with the latter recessive. It would seem easy, therefore, to eliminate the horns entirely. However, as Asdell pointed out, it is only among the polled goats that one finds intersexes. If any horned intersexes occur, they are rare, and of some type other than the one that causes the big problem in this species. He suggested that a gene for intersexuality is very closely linked with that for the polled condition, and that goats homozygous for that gene are intersexes if females, but not if males. Accordingly, by using only polled goats as breeding stock in the effort to increase that type in their herds, goat breeders were selecting automatically (albeit unknowingly) to increase also the frequency of intersexes.

It is now clear that Asdell's interpretation is correct, except, perhaps for one point. If two closely linked genes are involved, with *P* (polled) and *h* (intersex) in one chromosome and *p* (horned) and *H* (normal) in another, unless those genes are completely linked, cross-overs should occur and thus eventually produce goats of the genotypes *PP HH* or *pp hh*. The former would produce only polled, normal kids, and no intersexes. The latter could be only horned bucks or intersexes. As neither of these genotypes has apparently ever been found, the alternative interpretation seems more likely, *i.e.* that one gene, *P*, is responsible for both conditions. Intersexuality can thus be considered a pleiotropic effect of that gene. Homozygotes are polled in both sexes, but intersexes only if females.

The extensive data of Eaton (Table 13-3) from herds of two breeds at Beltsville fit both theories equally well, but unless cross-overs are found, it would seem simpler to apply Occam's razor, and to let gene *P* bear the responsibility alone. That does not solve completely the great goat mystery. Homozygous polled bucks, which should be just as common as intersexes, are apparently rare. One such male used by Eaton sired 66 kids, all polled (Table 13-3). It has been suggested that many of them may be the ones that are sterile because of stenosis of the epididymis, a condition not uncommon in bucks.

This case of pleiotropy provides one more example of incompatibility between preferences of the breeders and the genetic facts of life. Obviously, although polled animals are preferred, the breeders should not mate together two of that kind. Since fertile does with horns are preferable to sterile intersexes without them, it will be better to mate polled × horned. If more homozygous polled and fertile bucks could be found, matings of these to horned does should yield the maximum numbers of polled fertile daughters, but, apart from the difficulty of

TABLE 13-3
Segregation of Polled, Horned, and Intersexual Goats

| Type of Mating ♂ × ♀ | Progeny | | | | |
| | Males | | Females | | Intersexes |
	Polled	Horned	Polled	Horned	Polled
1. Polled, Pp × polled, Pp					
In Saanens	107	27	64	31	31
In Toggenburgs	117	42	76	41	36
Totals	224	69	140	72	67
2. Polled, Pp × horned, pp					
In Saanens	18	15	23	11	0
In Toggenburgs	45	52	50	47	0
Totals	63	67	73	58	
3. [a]Polled, PP × polled, Pp	24	0	14	0	12
4. [a]Polled, PP × horned, pp	10	0	6	0	0

[a]The same buck, a Saanen, was used in Matings 3 and 4.
Source: Data of Eaton, 1945.

finding such males, horned females would still have to be produced for that kind of mating.

Finally, it is to be hoped that these few examples of genes with pleiotropic effects will suffice, not only to dispel any illusions that one gene can affect only one character, but also to keep animal breeders alert and on the watch for similar cases that have yet to be recognized.

Problems

13-1 Can you state clearly the distinction between polygenes and modifiers?

13-2 What have the two kinds of genes in common?

13-3 How would you conduct an experiment with Beagles to show whether or not the variation of white spotting in that breed is caused by modifying genes?

13-4 In sheep, prognathia (extension of the incisors beyond the dental pad) and brachygnathia (short lower jaw) are not uncommon defects. Nordby found the frequency of such "overshot" or "undershot" jaws to be about 1.4 per cent in 7,000 Rambouillet lambs. Among those from parents both affected, the incidence was 16.4 per cent. Are these abnormalities inherited? If so, in what way?

13-5 Sometimes in animals the hairs of one region slope in directions opposite to normal. Breeders of guinea pigs call such things rosettes, and treasure them in special breeds. Breeders of other pigs call them whorls, or swirls, and declare them to be disqualifications in show-rings. Craft found the proportion of market hogs with whorls to be only 1.4 per cent. Mating affected pigs together produced offspring among which the incidence was 59 per cent in Craft's Poland Chinas and 46 per cent in another breed studied by Nordby. How would you proceed to clarify the genetic basis for the abnormality?

13-6 Some investigators think that whorls in swine result from action of two pairs of genes. How would you determine to what extent the defect might be influenced by (a) modifying genes? (b) low penetrance?

13-7 From what kind of evidence could you determine whether two associated variations are induced by two closely linked genes or are pleiotropic effects of one gene?

13-8 How many cases do you know in mammals in which a specific color or pattern reveals that the animal showing it is heterozygous for a lethal gene?

13-9 Work out the numbers expected in the various matings of Table 13-3 to see whether or not the results observed support the genetic basis for intersexes given in the text.

13-10 If the intersexes in goats should result from action of a gene h closely linked with P (which eliminates horns), what phenotypes and genotypes would be necessary to establish proof of two such genes, and by what breeding tests would you find them?

13-11 A mutation in the fowl, called "ragged wing" because it eliminated or shortened some of the big flight feathers, appeared to be an autosomal recessive trait because from reciprocal crosses of ragged-wing × normal, 187 F_1 fowls were normal and only three had ragged wings. (Some of the normal parents could have carried the causative gene.) However, when the ragged-wings were mated *inter se*, of the 606 progeny, only half showed the character; the rest were normal.

Assuming that a single gene is responsible, what genetic basis can you suggest for the ragged-wing character in terms of penetrance?

13-12 From backcrosses of F_1 heterozygotes to birds with ragged wing, the progeny were 285 normal to 88 with ragged wing. Do those results fit your answer to Problem 13-11?

Selected References

ASDELL, S. A. 1944. The genetic sex of intersexual goats and a probable linkage with the gene for hornlessness. *Science* **99:** 124. (Original key to solution of the mystery of the intersexual goats.)

CASTLE, W. E. 1941. Influence of certain color mutations on body size in mice, rats, and rabbits. *Genetics* **26:** 177–91. (Details of his experiments on pleiotropic effects of genes for color on body size.)

CREW, F. A. E., and C. AUERBACH. 1941. "Pigtail," a hereditary tail abnormality in the house mouse, *Mus musculus*. *J. Genetics* **41:** 267–74. (A good example of experimental analysis of incomplete penetrance; source of data and illustrations in this chapter.)

DE GROOT, T. 1942. The heredity of smooth tongue, with special references to cattle. *Genetica* **23:** 221–46. (Detailed descriptions of manifold effects of this abnormality together with the evidence of its hereditary nature.)

EATON, O. N. 1945. The relation between polled and hermaphroditic characters in dairy goats. *Genetics* **30:** 51–61. (Source of the data in Table 13-3; a good analysis of extensive data from the goat herds of the U.S. Department of Agriculture.)

ERIKSSON, K. 1943. *Hereditary Forms of Sterility in Cattle: Biological and Genetical Investigations*. Lund. Hakan Ohlssons Boktryckeri. (Detailed description of the hereditary genital hypoplasia in Swedish Highland cattle, with extensive data on its genetic basis.)

ERWAY, L. C., and S. E. MITCHELL. 1973. Prevention of otolith defect in pastel mink by manganese supplementation. *J. Heredity* **64:** 110–119.

FISHER, R. A. 1930. *The Genetical Theory of Natural Selection*. Oxford. The Clarendon Press. (See Chapter 3 for his theory of the evolution of dominance under natural conditions.)

HOLLANDER, W. F., J. H. D. BRYAN, and J. W. GOWEN. 1960. Pleiotropic effects of a mutant at the *p* locus from X-irradiated mice. *Genetics* **45:** 413–18. (Account of remarkable pleiotropic effects of one mutation.)

KEELER, C. 1961. The detection and interaction of body size factors among ranch-bred mink. *Bull. Georgia Acad. Sci.* **19:** 22–65. (Effects on size of various genes for color.)

KONDO, K. 1952. Studies on intersexuality in milk goats. *Jap. J. Genetics* **27:** 131–41. (Extensive data on the incidence of intersexes in goats in Japan, with evidence that they are genetically females; in English.)

NEL, J. A., and D. J. LOUW. 1953. The lethal factor in grey Karakul sheep. *Farming in S. Africa* **28:** 169–72. (A concise summary of studies of this condition at the experiment station at Neudam, near Windhoek, in South West Africa, with references to investigation of the same condition elsewhere. See also Rept. of Exper. Farms, S.W.A., for 1957.)

NORDBY, J. E. 1932. Inheritance of whorls in the hair of swine. *J. Heredity* **23:** 397–404. (A good paper to read before proceeding on any large scale with Problems 13-5 and 13-6.)

SHACKELFORD, R. M., and L. J. COLE. 1947. "Screw neck" in the pastel color phase of ranch-bred mink. *J. Heredity* **38:** 203–9. (Detailed account of the peculiar behavior associated with the pastel color mutation in this species.)

Extra-nuclear Transmission and Maternal Influences

After thirteen chapters devoted to an exposition of the wonders wrought by genes and to their role in the "destiny that shapes our ends," it may be appropriate to consider a few examples in which the effects of genes are somewhat indirect, or even obscured by other forces acting on the organism. After all, cells contain many things other than genes, and, even if most hereditary variation can be accounted for by genes in the nucleus, we should not close our eyes to the possibility that in the cytoplasm there may also be forces affecting inheritance.

If we accept what many geneticists now believe, that the DNA of the genes gives the orders and the RNA in the cytoplasm dutifully does the work, we might still wonder (perhaps from what happens in other communities) whether or not any fractious RNA could successfully revolt against duly constituted authority and set up a little independent republic acting on its own. If the dissident group could manage to include a fertile representative in every germ cell, there could then be a cytoplasmic force of some kind exerting its influence, however small, on heredity. While male gametes contain mostly nuclear material, female gametes carry much cytoplasm (as, for example, in the yolk of a hen's egg) and there is plenty of room for agents independent of genes to be carried along from one generation to the next. And, just to make all this even more plausible, there is the further fact that in some viruses there is no DNA; hence the RNA must "carry the genetic information," *i.e.*, determine the inheritance in those forms.

A few cases are known in which cytoplasmic forces do determine the inheritance, and even act independently of genes. Most of these discovered thus far are in plants, but there is a good example in an animal—a rather lowly animal, to be sure, but, nevertheless, an animal. It is the ciliated protozoan, Paramecium. There are also some interesting cases in which ultimate fate of the organism is determined (a) by pathogenic agents transmitted from one generation to the next, (b) by interactions between genotypes of mother and fetus, or (c) solely by influence of the mother, regardless of the genotypes of the sire or the offspring. Consideration of some of these special situations at this stage may help us to maintain both a balanced viewpoint toward forces that make organisms what they are, and a watchful vigil for other similar cases yet to be discovered.

Plastids and Plasmagenes

Plastids are small cytoplasmic bodies, commonly round, but sometimes lobed or irregular, which produce specific substances, principally starch or pigment. They are common in the cells of higher plants, where, in the presence of light, they produce chlorophyll. Such plastids are called chloroplasts. Other plastids have other functions, but all of them have two characteristic distinctions:

1. They are areas of localized chemical activity.
2. They grow, divide, and multiply on their own, independently of other cellular inclusions.

All plastids are believed to develop from previously existing plastids.

In most cases the activities of plastids seem to be controlled by the genotype of the cell. In various plants, mutations (in genes) have been found which eliminate plastids, or prevent them from forming chlorophyll, and thus produce albinotic white seedlings. Other mutations allow production of chlorophyll in subnormal amounts, with correspondingly pale green leaves in *virescent* seedlings.

In some plants the leaves are variegated, *i.e.*, green but irregularly spotted or blotched with patches of pale green or white. The areas of the plant thus affected may be small, or large enough to include whole branches. It has been found that plastids of the pale or white leaves lack chlorophyll, and those of green ones contain it, while those of variegated leaves have both kinds. The situation elucidated by Correns in careful studies with the four o'clock, *Mirabilis jalapa*, is apparently found in many other cases of variegation as well.

When flowers on branches bearing only green leaves were fertilized separately with pollen from flowers on green, pale, or variegated branches, the type of pollen proved to have no effect on the progeny. All had only green leaves. Similarly, flowers on pale branches produced only pale offspring, regardless of the kind of pollen. Flowers on variegated branches gave rise to green, variegated, or pale plants, but the type of offspring did not depend upon the source of the pollen which fertilized the flowers. It is clear that genes in the pollen had no effect whatever on the kinds of plastids produced. These corresponded in every case strictly to the kind present in the female parent. The inheritance was completely *matroclinous* (resembling the mother) and, moreover, was determined not by maternal genes, but by maternal plastids.

This case, and others of the same kind in other variegated plants, can therefore be ascribed to extra-nuclear transmission, or, more simply, to cytoplasmic inheritance. To distinguish between particles in chromosomes that determine inheritance and particles in the cytoplasm that do somewhat the same, and can reproduce themselves, the latter are sometimes designated as *plasmagenes*. Two other terms coined for the same purpose are the *genome*, by which is meant the entire complex of genes in the chromosomes, and the *plasmon*, a term with equally broad coverage meaning all the inheritance that may be transmitted through the cytoplasm. The fact that the genome seems to do about 99 per cent of the business should not blind our eyes to the fact that in some cases the plasmon may have important effects. Presumably more of these have yet to be recognized.

Plasmagenes in Killer Paramecia

A common one-celled animal in stagnant water is Paramecium, a protozoan belonging to the class Infusoria, and sometimes called the "slipper animalcule" because in profile it resembles a slipper. Some strains of Paramecium produce a substance, *paramecin*, which is secreted into the water and is toxic to individuals of other strains that are unable to produce the same substance. Sonneborn and his associates at Indiana University, who discovered these remarkable properties, designated the producers of paramecin as "killers" and those unable to withstand it as "sensitive."

The toxic material is produced by particles called collectively *kappa*. Like genes, these *kappa* particles can reproduce themselves under suitable conditions. Unlike genes, they are large enough to be seen and counted under the microscope when suitably stained. Like genes, again, they have been found to contain DNA. As the *kappa* particles are in the cytoplasm, and are apparently normal, self-reproducing elements in some strains, they seem to qualify as fully accredited plasmagenes.

There is one important difference between these *kappa* particles and the plastids of plants. Survival of the former is dependent on the genotype (nuclear) of the cell; persistence and behavior of the plastids can be independent of the genotype. *Kappa* particles persist only in animals carrying the gene K, whether KK or Kk, but those genotypes by themselves cannot cause production of paramecin. Accordingly, individuals carrying K but lacking *kappa* particles are just as sensitive to the toxin as are those lacking K. Conversely, if *kappa* particles are introduced to animals of the genotype kk, those particles do not multiply, and are lost after a few cell divisions. These somewhat complex inter-relationships of genes and plasmagenes will be better understood after assimilating the facts as they are arranged in Table 14-1.

Apart from occasional conjugation, when an interchange of nuclear material between two individuals serves the useful function of re-invigorating decadent clones, these animals multiply by simple fission. When the rate of division of killers is accelerated (by subtle techniques known to specialists in Paramecium), the multiplication of

TABLE 14-1
Interactions of Genes and Plasmagenes in Paramecium

Genes in the Nucleus	Plasmagenes in the Cytoplasm	Result of Interaction, or Combined Phenotype
KK or Kk	kappa particles present	Killer, secretes paramecin
KK or Kk	no kappa particles	Sensitive to paramecin
kk	kappa particles present	Loses kappa soon
kk	no kappa particles	Sensitive to paramecin

kappa particles cannot keep up with multiplication of the animals, and eventually some of the latter come through without any kappa whatever. Killers can thus be tamed and reduced to the innocuous rank of sensitive animals, but, as they carry the gene K, they remain potential killers.[1] If a single kappa particle is now introduced to one of these, and the rate of fission is retarded to let that particle multiply, the sensitive Paramecium can thus be restored to the malevolent status of his ancestors. Life among simple, one-celled animals is evidently not as simple as one might suppose.

Maternal Transmission of Pathogenic Agents

It would be easy to regard the kappa particles, not as plasmagenes, but as some kind of pathogenic agent, perhaps an unusually large virus. One might then say that Paramecia carrying the gene K are resistant to the pathogen, or, more accurately, that they tolerate it. Those of the genotype kk will not do so, and would then really be more resistant than the ones that do. At the same time, they are susceptible to the toxin produced by the kappa particles in other individuals.

Some cases of maternal transmission of pathogens are less complicated. Hens infected with Salmonella pullorum frequently pass that bacterium through the yolk of the egg to the chick that hatches from it. Such infected chicks may live or die, depending on their genes for resistance or susceptibility and on environmental conditions. In any case, they usually manage to infect other chicks brooded with them, most of which are likely to have been free of S. pullorum when hatched. The resulting disease, once graphically described as bacil-

---

[1]We would like to write that they remain killers at heart, but that is just the sort of bloomer that book reviewers delight in exposing; as these Paramecia have no hearts, one can only say that they remain potential killers in genotype.

lary white diarrhoea, is now euphemistically labelled as pullorum disease, but the chicks get diarrhoea just the same as in the era of more primitive terminology. Most of them survive the critical period of the first three weeks of life, and live to lay and to reproduce. Many of the survivors vanquish the pathogen entirely, but some continue to harbor it as adults, and these can again transmit it through their eggs to the next generation.

Here is maternal transmission, hence *matroclinous* "inheritance," but of a kind quite different from the plastids and the *kappa* particles. The bacterium is certainly transmitted from one generation to the next in the cytoplasm, but it is not a normal constituent of that cytoplasm, as in the killer clones of Paramecium, or both normal and necessary as are the plastids of green plants. It is an infectious agent—not a plasmagene. Furthermore, although the bacterium can be transmitted through the egg, most infected chicks do not get the organism that way; rather, they pick it up from the droppings of others.

At the Jackson Memorial Laboratory in Bar Harbor, Maine, J.J. Bittner studied two strains of mice: in one the incidence of mammary cancer was comparatively high, and the other was characterized by low incidence of that disease. Because the two strains differed consistently in successive generations with respect to the frequency of the condition, it was logical to assume that one was genetically highly susceptible and the other highly resistant.

Not content with assumptions, Bittner carried out an interesting experiment. Newborn mice of the susceptible strain were removed from their mothers right at birth, without ever sucking milk from them, and were transferred to foster mothers of the resistant strain. Subsequently they developed no more mammary cancer than was normal for the resistant stock.

Conversely, newborn mice of the resistant strain proved to be susceptible when nursed only by mothers of the susceptible line. It was clear that what caused the difference between the two strains in the incidence of mammary cancer was not a matter of genotypes, but of some carcinogenic agent transmitted through the milk by mothers of the strain with high incidence of that disease. That agent is still generally designated as a "milk-factor," for, although it behaves in some respects as a virus, proof that it is one is apparently not yet conclusive. To the best of the authors' knowledge, the existence of such a milk-factor causing mammary cancer has been demonstrated only in mice.

All of this does not mean that there cannot be genetic differences in susceptibility to cancer or to any other disease, but that is a subject for a later chapter. The important point to be made here is that although both pullorum disease in the chick and mammary cancer in the mouse can be transmitted from mother to offspring, neither disease is actually inherited. Both result from mechanical transmission of an infective agent.

Hemolytic Disease in the Horse

An entirely different kind of maternal influence is that responsible for hemolytic disease (destruction of red blood cells) in horses, swine, and man. It occurs because of incompatibility between the genotype of the mother and that of the fetus. There is no infectious agent, and nothing resembling cytoplasmic inheritance. The disease results from the fact that antigens of red blood cells of certain genotypes in the fetus (after passing through the placenta into the maternal circulation) induce the formation of antibodies against those fetal blood cells in mothers of certain other genotypes.

The A system of blood groups is implicated in a large proportion of cases of hemolytic disease in horses. Let us assume that the allele responsible for the antigen which causes production of antibodies is designated as A. Then the gene A in a fetus might be responsible for stimulation of antibodies in a mare of genotype aa (not, of course, in mares of genotypes AA or Aa, since they would not produce antibodies against antigens which they themselves have on their red cells).

Suppose that an aa mare carries an Aa foal sired by an AA stallion. Usually no harm results if it is the first time she has had such a foal, but a second by the same stallion is likely to have trouble. It will appear healthy at birth, but, if left with the mother, might show symptoms of jaundice within 12 to 48 hours, and be dead within three to four days. Symptoms include a yellowish tinge in the sclerotic coat of the eye (best seen when the eyelid is turned back), lethargy, and an erythrocyte count that is far below normal.

The reason why the first pregnancy causes no trouble with genotypes like those given above is that it takes considerable time (more than one pregnancy) for the antigens of Aa fetuses to raise the concentration (or $titre$) of antibodies in the mare up to the level at which they can seriously damage the red blood cells of the foal. The reason why the foal is healthy at birth, but so acutely ill soon afterward, is that the mother's antibodies are concentrated in her first milk—the colostrum. A foal usually drinks the colostrum soon after birth. In nature, its survival would ordinarily be dependent on getting that particularly rich source of nutrients. In this special situation, however, with the first drink of milk there begins a fatal destruction of red blood cells which ends as described above. The yellow color of the tissues, commonly called jaundice, or icterus, comes from the pigments of the destroyed cells.

Proof of this situation was established by Bruner and his associates at the University of Kentucky, and by others elsewhere. Stricken foals in which the disease had not progressed beyond the point of no return could be saved by giving them blood transfusions and transferring them to foster mothers. In cases where tests of the mare's blood before birth of the foal had revealed a high titre of antibodies (and, hence, prospective trouble), transfer of the foals to foster nurses at birth effectively forestalled any hemolysis. Better yet, potential trouble thus detected before it could start was prevented by transferring the foal at birth to a foster mother for 24 to 36 hours, milking out the mare's colostrum by hand, and then returning the foal to its own mother. This is the procedure most commonly used to prevent hemolytic disease. Once the antibodies in the colostrum are eliminated, there are apparently not enough of them in the subsequent milk to cause trouble. Moreover, Bruner found that after 36 hours the foal cannot resorb such maternal antibodies from its own digestive tract.

Alert readers will have foreseen an important point not yet mentioned. The example given above dealt with foals sired by a stallion homozygous for the dominant allele, A. If the sire were heterozygous, only half his offspring would get that gene from him. Without it, the other half must be aa, when carried by mares of that genotype, and a fetus of that genotype cannot cause formation of new antibodies. Foals of that kind are therefore in no danger, even if their mothers had previously produced foals that developed hemolytic disease (Table 14-2). Consequently, an aa mare could, by chance, have two or three foals, or even more, before developing a titre of antibodies high enough to endanger the next one. As some stallions must be Aa, and others aa, it is not surprising that Franks should have found first hemolytic foals to be produced more commonly in their mothers' fourth or seventh years of breeding than earlier.

Students ask why trouble of this sort is not more frequent in foals. As Table 14-2 shows, only two of the nine possible combinations of genotypes in sires and dams can yield hemolytic foals. Mares of the genotypes AA or Aa cannot form antibodies against cells of their own type. This does not mean that two-ninths of *all matings* can have trouble, but only that there are nine possible combinations, two of which can be serious. The proportion of all matings that these constitute will depend on the frequency of aa mares in the population. Another reason for the trouble being less common than it might be is that mares which have had one hemolytic foal are more likely to be bred subsequently to other stallions believed to be better risks. A major

TABLE 14-2
Risks of Hemolytic Disease in All Kinds of Matings Involving Two Alleles.
One of Which, A, is Responsible for an Antigenic Factor Causing Hemolytic
Disease

Genotype of Sire	Genotype of Dam	Genotypic Ratio in Progeny			Proportion of Progeny Endangered
		AA	Aa	aa	
AA	aa	—	all	—	all[a]
Aa	aa	—	1	1	half
aa	aa	—	—	all	none
AA	Aa	1	1	—	none
Aa	Aa	1	2	1	none
aa	Aa	—	1	1	none
AA	AA	all	—	—	none
Aa	AA	1	1	—	none
aa	AA	—	all	—	none

[a] After the first one.

factor in forestalling trouble is undoubtedly the use of blood tests
during pregnancy to determine whether or not the mare's titre of
antibodies is rising to a dangerous level. As those antibodies are just as
effective against blood of the sire as against that of his foal, the sire's
blood can be used for such tests.

To simplify the explanation of this interesting incompatibility of
genotypes, our example has been written as if only a single pair of
alleles were involved, but that is not the case. The A system is known
to be a system involving multiple alleles; in our example, we considered these alleles having an antigenic factor which might stimulate
antibody formation as one group of alleles, which we designated as A
without considering possible variations. The alternative alleles not
involved in hemolytic disease can then be designated as a. Since not
all cases of hemolytic disease are due to differences between mare and
fetus in genotypes in the A system, other blood-group systems should
also be considered in assessing possible risks of hemolytic disease.

Hemolytic Disease in Pigs and People

In pigs the course of hemolytic disease is almost the same as in foals.
The piglets are normal at birth, may show symptoms within six hours,
and usually all in the litter are dead within three days. In some litters
there is variation in the severity of the disease, and Buxton et al. found

that a few of those lightly affected eventually recovered. As in the foal, jaundice, lassitude, and anemia are characteristic. The basic cause is an incompatibility of fetal and maternal genotypes very similar to that just described in the horse. A number of different blood-group systems may be involved. Again the antibodies in the colostrum are responsible.

The cases described by Doll and Brown are of interest because the sow involved (a Duroc Jersey) had six litters, the last five of which were all by the same boar. Her first four litters were all healthy. In the fifth litter there were 13 piglets, all of which died within 9 to 12 hours after birth. In the sixth litter there were 11 piglets alive at birth, but all were dead within 12 hours.

Recognition of hemolytic disease and the basis for it in foals and piglets was facilitated by the fact that the same condition had earlier been identified and accounted for in man. Serologists classify us as Rh+ or Rh−, but genetic studies have made it clear that the Rh+ class includes both homozygotes and heterozygotes. If there were only one pair of alleles involved, and if we use for them the symbols *Rh* and *rh*, then Rh+ people can be *Rh Rh* or *Rh rh*, and Rh− people are *rh rh*. As we learned in Chapter 11, there is a series of multiple alleles at the *Rh* locus, and the exact number of genes in that series is probably not yet known. The proportion of *rh rh* people is approximately 15 per cent in white people of North America.

In man the *Rh rh* fetus carried by an *rh rh* mother causes production of maternal antibodies, but, unlike the situation in the horse and in swine, those antibodies are able to get through the placenta and thus begin hemolysis before the baby is born. The resulting condition is called *erythroblastosis fetalis*, because, with so many erythrocytes destroyed as they mature, there is a high frequency (in the blood of the fetus) of immature blood cells, or erythroblasts. It is frequently fatal before birth, or soon after. Now that the basis for it is understood, afflicted infants can be saved by transfusions soon after birth with Rh− blood. Moreover, the need for such treatment can be anticipated by tests during pregnancy to determine the titre of antibodies in the blood of the mother.

It is beyond the scope of this book to give any detailed account of the working of the *Rh* alleles in man. Readers will find further details in Stern's useful book, cited at the end of this chapter.

It is important to recognize that the foals, piglets, and babies afflicted with hemolytic disease would escape that trouble if born to mothers of other genotypes. Similarly, mothers who make antibodies against their own offspring do not do so for all of them, but only for those of certain genotypes. It is not either genotype alone that makes trouble, but the incompatibility between specific genotypes.

Blocked Riboflavin, a Maternal Effect in the Fowl

We come now to a different kind of maternal influence, one in which the fate of the developing embryo, regardless of its own genotype, depends on the genotype of the mother.

The vitamin, riboflavin, is necessary for normal development of the chick embryo. When hens are not laying, the amount of that substance in their blood is at a comparatively low level, but in laying hens it is increased about fifty-fold. That increase can be effected only if the diet contains adequate riboflavin. Because such diets are essential if the eggs produced are to hatch well, it is now standard procedure to ensure that there is a high level of riboflavin (over 300 micrograms per 100 grams of feed) in diets for flocks that supply eggs to hatcheries. Hens apparently need less riboflavin to live and lay well than they do for reproduction. Accordingly, mashes compounded for ordinary laying flocks usually contain less riboflavin than those used for breeding flocks.

Hens have been found which, even on diets very high in riboflavin, are unable to transfer that substance to their blood when they lay. When eggs of such hens are incubated, they are found to be fertile like those of other hens, but every embryo dies. Few survive beyond 14 days of incubation.

Investigation of such abnormal hens by Maw and his associates at Pennsylvania State University revealed that, when enough supplementary riboflavin was injected into the air cells of the deficient eggs early in incubation, hatchability of the embryos within them was fully restored to the normal rate. Genetic studies with the birds thus hatched and their descendants showed that the underlying trouble was caused by a simple recessive, autosomal gene, *rd*. Hens homozygous for that gene, although themselves healthy and able to lay well, produced eggs so deficient in riboflavin that none of them could maintain an embryo through the normal period of incubation. Subsequent studies showed that the gene is incompletely dominant. Heterozygous hens put in their eggs amounts of riboflavin that are subnormal but apparently still adequate for normal development of the embryo (Table 14-3).

Eventually it was found that hens homozygous for the causative gene (*rd*) lacked a protein that binds riboflavin. For details, see Buss (1969). In this most unusual case, the genotype of the embryo does not determine its fate. The genotype of the mother does not affect her own well-being, but it is lethal to all her embryos and thus deprives her of any hope of posterity! To appreciate this tale fully, one should break open a fresh egg in a glass dish set on pure white paper. It will be noticed that the albumen has a slightly greenish or yellowish tinge.

TABLE 14-3

Effects of the Gene *Rd* on Levels of Riboflavin (micrograms per gram) in Hens of Three Genotypes

Where Measured	Rd Rd	Rd rd	rd rd
In blood of hens not laying	0.008	0.008	0.008
In blood of laying hens	0.434	0.272	0.008
In egg yolk	4.30	2.50	0.41

Source: Data of Buss *et al.*

That comes from riboflavin. Most of it is put in the yolk, but there is usually enough in the albumen to color it faintly. Maw found that with practice he could easily identify the *rdrd* hens in his experimental flock by the complete absence of color in the albumen of their eggs.

Body Size in Shire × Shetland Crosses

Still another kind of maternal influence is illustrated by crosses between breeds of animals that differ greatly in size. The reciprocal crosses between Shetland ponies and Shires made by Walton and Hammond provide a good example. Artificial insemination was used to effect them. Weights in kilograms of some of the animals concerned were as follows:

	Shire	Shetland
Birth weights of purebred foals	59.4, 82.6	Average for 4: 19.6
Adult weights of mares that produced crosses	797	185, 229
Birth weights of crossbred foals	53.5, 45.3	17.2, 13.6, 22.7

Although only one Shire mare produced two foals by the Shetland sire, and two Shetland mares had three foals in the reciprocal cross, the results of this experiment were very striking. Some idea of the disparity in size between the two parent breeds and between the two kinds of F_1 progeny is conveyed by the accompanying illustration (Fig. 14-1).

Crossbred foals from the Shire mare were about three times as heavy at birth as those from the Shetland mares. For three of the latter, average weight was 17.8 kg., a figure even below that for four purebred Shetlands. Apart from the possible role of sex-linked genes

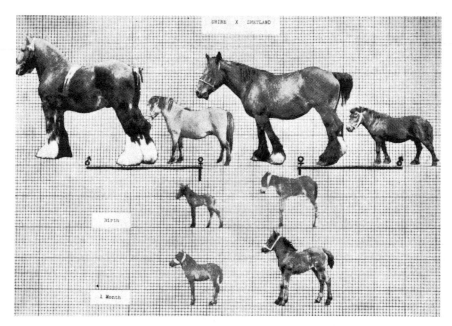

FIG. 14-1 Reciprocal crosses between Shires and Shetland ponies, with F_1 from the Shire mares thrice the size of those from Shetlands. (From A. Walton and J. Hammond in *Proc. Roy. Soc. London.*)

affecting size, the two kinds of F_1 crossbreds should have had similar genotypes with respect to size. Regardless of that, size at birth was clearly determined by the mother.

After birth the F_1 foals from the Shetland mares grew more rapidly than pure Shetlands, but at three years of age they had not caught up to the crossbreds from the Shire mares, and there seemed little chance of their ever doing so.

Here is a maternal influence on the expression of genotypes that is different from all the other examples cited earlier. There is no need to invoke any cytoplasmic inheritance; there is no transmission of infectious agents, no specific antagonism between genotypes of dam and offspring. Maternal regulation of fetal growth is a normal process, even though the mechanisms effecting it are not fully understood. Although mares and other mothers may skimp occasionally on size, they usually do not wrap up the goods in packages bigger than they can deliver.

Corresponding limitations on size at birth have been found in reciprocal crosses between large and small breeds of rabbits, cattle, and sheep. A similar maternal limitation on size, but one apparently not permanent, is imposed by the size of a hen's egg. Even though the chick hatched therefrom can sometimes claim a sire twice as big as its dam, the original size of the chick is limited completely by the size of the egg shell in which it is incubated. After hatching, chicks thus restricted grow rapidly, but there is evidence that some of their initial handicap may persist for eight weeks at least. For that reason, growers of broilers do not want to start with small chicks.

Mothers Have Limits

After these several examples showing how expression of a genotype can be limited by maternal influences, it may be well to remind the reader that in many cases genes exert their effects completely untrammeled by maternal environment or any other. This was demonstrated in one way long ago by investigators who transplanted ovaries or fertilized eggs from rodents of one color to foster mothers of another. In such cases, even though the foster mothers were in control of the developing embryos until birth, the little guinea pigs or mice showed their own genotypes (colors)—not those of the foster mothers.

Such complete control by the genotypes might, perhaps, be expected more with genes affecting color or structure than with those determining physiological processes, but that these last can also be resistant to any maternal influence was clearly demonstrated in an experiment with mice by Runner and Gates. In mice there is a simple recessive mutation, *ob*, which causes homozygotes to be unusually obese. At ten months of age such mice weigh over three times as much as their normal litter-mates. Obese mice of both sexes are sterile, but that sterility can be overcome by environmental modification. The females can be made to ovulate by administering gonadotropic hormones, and males can be kept trim and fertile by restricting their feed. Runner and Gates bred a black heterozygous male (*Obob*) to a black obese female that had been thus made to ovulate, and transferred the fertilized eggs to a female mouse homozygous for recessive pink-eyed dilution. At the same time, the pink-eyed mouse was mated to a pink-eyed male.

The resultant litter (Fig. 14-2) showed no effect whatever of the foster mother on her somewhat unnatural progeny. Among the five black mice, all from the transplanted eggs, the ratio of 3 obese:2 normal was as close as one could come to the equal numbers expected, and the obese mice were in every way true to type. Conversely, these had no effect whatever on the three pink-eyed dilute offspring which the mother may have considered more properly her own.

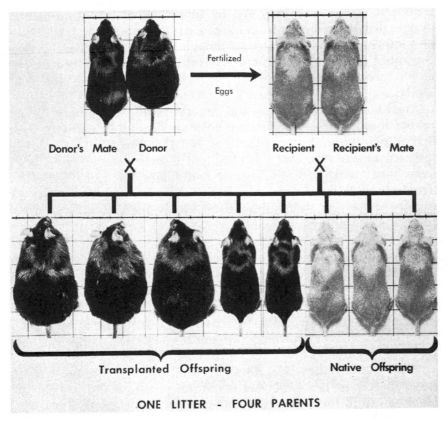

FIG. 14-2 Parents and foster mother (above) of the eight mice in one litter (below), showing absence of maternal influence on color and obesity. (From M. N. Runner and A. Gates in *J. Heredity.*)

Telegony

There is no evidence to warrant the persistent belief in *telegony*, the influence of a previous impregnation on a later one. Even today many a dog breeder worries over the possibility that some promising bitch, having had an illegitimate litter by some disreputable mongrel, might thereby be ruined for production of purebred stock. She isn't. However undesirable the results of such canine promiscuity may be, telegony is not among them. In earlier years a belief in telegony was widespread among breeders of other domestic animals, and some breed associations even refused registration to animals whose dams had previously been bred to males of other breeds.

Oddly enough, while the principle involved can be quickly tested with mice, rats, or rabbits, the most famous investigations of telegony are probably the "Penycuik experiments" of Cossar Ewart with horses. His use of that species for what bacterial geneticists of the twentieth century might view as somewhat leisurely experiments of the Victorian era was dictated by the case that caused them. The trouble began with Lord Morton's mare, an almost pure Arabian, which produced a hybrid by a quagga (a type of zebra) and later foaled by a pure Arabian stallion two colts that were not only conspicuously striped on the legs and elsewhere, but also showed a mane much like that of the quagga. The case was reported to the Royal Society of London as indisputable evidence of telegony.

In Ewart's experiments eight mares altogether bred to a Burchell zebra stallion produced 13 hybrid progeny and later (by stallions of their own species) 18 foals. Without going into details (which were nicely summarized by Crew) it will suffice to say that no evidence for telegony was found, but it was shown, among other things, that some horses can produce striped colts without ever having seen a zebra.

Ewart's results at Penycuik in Scotland were subsequently confirmed in Russia by Ivanov. Let no modern geneticist comfortably immured with his murmuring mice in some urban laboratory cast a supercilious eye on the Penycuik experiments and others like them. It is desirable that breeders of domestic animals—from honey-bees to horses—should understand the workings of heredity in their favorite species. Those who have helped them gain such an understanding know that any stockman who had seen the striped filly from Lord Morton's mare would never have had his faith in telegony shaken by any mouse, but only by normal, unstriped colts from mares that had earlier produced striped hybrids.

Concluding Comments

The rarity of true cytoplasmic inheritance in animals is indicated by the fact that we had to resort to plastids and Paramecium for examples. There are probably other cases in animals waiting to be recognized. On the other hand, transmission of infectious agents from mother to offspring is not unusual, but, whether the pathogen is put into the cytoplasm of a hen's egg even before its fertilization, or acquired by a young mammal at birth, or soon after, such extranuclear maternal influences should not be confused with true inheritance. Sometimes, as with hemolytic disease, disaster results—not from the genotype of parent or offspring alone, but because one is incompatible with the other. The genotype of the mother may cause maternal effects that suppress the genotype of the progeny. Such suppression may be complete, as for the chick embryos that die because they are deprived of riboflavin, or partial, as in the F_1 foals from the Shire × Shetland cross, whose potential size cannot be realized if they are born to Shetland mothers.

If these examples have made us realize that, although genes accomplish marvellous wonders, and many are sufficiently powerful to cause death, they are sometimes suppressed or overwhelmed by forces stronger than their own, this chapter will have served its useful purpose.

Problems

Most fruit flies are highly resistant to CO_2, but one strain is known in which all the flies can be killed by low concentrations of that gas. When females of the sensitive strain are crossed to males of resistant ones, all the progeny are sensitive. If the F_1 females are backcrossed to resistant males, again all the progeny are sensitive.

14-1 How would you account for the sensitivity to CO_2?

14-2 When males of the sensitive race are crossed with resistant females, some sensitive offspring are produced but the proportion varies. Can this be reconciled with your answer to Problem 14-1?

14-3 Sensitivity is not spread by contact but the substance causing it can be inoculated into eggs of resistant stocks, which then produce sensitive flies. It is inactivated by high temperatures. What kind of agent could it be?

In the fresh-water snail, Limnaea, some strains have clockwise or *dextral* coiling of the shell, and others are coiled counter-clockwise, or *sinistrally*. Boycott and his associates found that, in reciprocal crosses between the two kinds, the F_1 snails always showed the same type of coiling as in the female parent. As this snail is *hermaphroditic* (*i.e.*, producing both male and female functioning gametes), it was possible to get F_2 populations by isolating the F_1 snails individually and thus ensuring their self-fertilization. In the resulting F_2 populations, from both reciprocal crosses, all snails were dextral. When these were self-fertilized to produce an F_3 generation, each F_2 snail produced progeny that were all alike, but (from both original crosses) about three yielded only dextral coilers to every one that produced only sinistral coilers.

14-4 Recognizing that the F_3 snails revealed the F_2 genotypes, what does that $3:1$ ratio reveal about the genetic basis for dextral and sinistral coiling in this species?

14-5 How then can you account for the matroclinous inheritance shown by the two F_1 generations from the reciprocal crosses? Would it be expected if the type of coiling were determined not by the genotype of the snail in the shell, but by that of the (maternal) snail which produced the oöcyte from which that snail developed?

14-6 Assuming that the oöcyte has thus had its subsequent pattern of coiling impressed upon it at a very early stage (actually by the four-cell stage of development), are we dealing here with cytoplasmic inheritance, or with a maternal influence?

14-7 On this basis, why are all the F_2 shells coiled dextrally, even though, as the F_3 generation showed, about a quarter of the F_2 snails were homozygous for sinistral curling?

14-8 The intersexuality in goats described in the previous chapter has been incorrectly described as hermaphroditism. Can you describe the difference between an intersex and a true hermaphrodite, as in these snails?

14-9 What diseases in domestic animals other than pullorum disease can be transmitted from dam to offspring before, during, or after birth?

14-10 Would you consider hemolytic disease in the foal as being inherited, or the result of maternal influence, or attributable to neither? Think it over.

14-11 Experienced serologists with adequate reagents can now identify blood groups in horses, but cannot tell whether an animal carrying a dominant allele is homozygous or heterozygous for that allele. If a mare that does not carry a specific antigen has a foal which also lacks that antigen by a stallion which carries it, what is thus revealed about the genotype of that stallion?

14-12 Bruner found some cases in which a mare produced two successive foals by the same stallion, and the second one showed severe hemolytic disease. What would such a record suggest about the genotype of the stallion?

14-13 What is the chance that such a stallion could still be heterozygous for the causative antigen even if it occurred in his second foal by a negative mare?

14-14 What does that chance become if his third is also hemolytic?

14-15 Assuming (for purposes of this problem only) that the nine different kinds of mating shown in Table 14-2 occur with equal frequency and that one fetus carrying the antigen is enough to raise a mare's titre of antibodies to a level dangerous to her next such foal, allow eight foals to each of the nine kinds of mating and estimate what proportion of them all could show hemolytic disease.

14-16 Rh− women may be sensitized not only by an Rh+ fetus, but also if given transfusions of Rh+ blood. Accordingly, transfusions of that kind should be avoided. Is there any risk of similarly sensitizing female pigs given whole blood to immunize them against hog cholera?

14-17 Would it be easier to demonstrate the effects of plasmagenes than of genes, or *vice versa*?

14-18 Since maternal influences, like sex-linked genes, can cause differences in reciprocal crosses, how would you distinguish in such crosses (or in other matings to clarify the situation) whether the observed differences result from the one cause or the other?

14-19 If hens of the genotype *rdrd* cannot normally have any offspring because of the deficiency of riboflavin in such hens' eggs, would that gene automatically be eliminated?

14-20 If your answer to Problem 14-19 is "yes," how would you explain the fact that hens can be homozygous for *rd*?

14-21 Can you think of any maternal limitation of size other than the examples cited in this chapter? Are twins as big at birth as animals born singly? Do big litters have puppies or piglets just as large as in small ones?

Selected References

BRUNER, D. W., E. F. HULL, and E. R. DOLL. 1948. The relation of blood factors to icterus in foals. *Amer. J. Vet. Res.* **9:** 237–42. (Report of cases; how to anticipate and save them.)

BUSS, E. G. 1969. Genetic interference in the egg transfer, utilisation and requirement of riboflavin by the avian embryo. In *The Fertility and Hatchability of the Hen's Egg*, ed. by T. C. Carter and B. M. Freeman. Edinburgh. Oliver and Boyd. pp. 109–116. (Review of many years' study of the blocked riboflavin.)

BUSS, E. G., R. V. BOUCHER, and A. J. G. MAW. 1959. Physiological characteristics associated with a mutant gene in chickens that causes a deficiency of riboflavin. 1. Eggs and embryos. *Poultry Sci.* **38:** 1192. (See also report by the same authors on p. 1190 of the same volume.)

BUXTON, J. C., N. H. BROOKSBANK, and R. R. A. COOMBS. 1955. Haemolytic disease of newborn pigs caused by maternal iso-immunisation. *Brit. Vet. J.* **111:** 463–73. (A good account, including studies of the blood of sows and piglets.)

CASPARI, E. 1948. Cytoplasmic inheritance. *Advances in Genet.* **2:** 1–66. (A good review, with extensive list of references.)

CREW, F. A. E. 1925. *Animal Genetics*. Edinburgh. Oliver and Boyd. (See Chapter 12 for Ewart's experiments, and the rest of the book for much that had to be omitted from this one.)

DOLL, E. R., and R. G. BROWN. 1954. Isohemolytic disease of newborn pigs. *Cornell Vet.* **44:** 86–93. (Further details of case cited in context.)

EWART, J. C. 1899. *The Penycuik Experiments*. London. A. & C. Black. (Results in crosses between horses and Burchell's zebra, with disproof of telegony.)

MAW, A. J. G. 1954. Inherited riboflavin deficiency in chicken eggs. *Poultry Sci.* **33:** 216–17. (Genetic study with chicks rescued by supplementary riboflavin put in eggs.)

RUNNER, M. N., and A. GATES. 1954. Sterile, obese mothers. *J. Heredity* **45:** 51–55. (Details about obese mice and the litter shown in Fig. 14-2.)

SONNEBORN, T. M. 1949. Beyond the gene. *Amer. Scientist* **37:** 33–59. (Plasmagenes in Paramecium made easy to understand by a very readable account.)

STERN, C. 1973. *Principles of Human Genetics*. 3rd ed. San Francisco. W. H. Freeman & Co. (A good condensation of what is known about the *Rh* alleles and erythroblastosis, but the discussion is scattered. See the index.)

WALTON, A., and J. HAMMOND. 1938. The maternal effects on growth and conformation in Shire horse-Shetland pony crosses. *Proc. Roy. Soc. London*, Ser. B, No. 840, **125:** 311–35. (Details of reciprocal crosses between these two breeds.)

Heredity and Environment

In the good old days before television put the debating societies out of business, no subject for debate was more popular than the hardy perennial, "Which is the more important in shaping the individual: heredity or environment?" In fewer words, the question was sometimes designated as the "nature-nurture problem." The purpose of this chapter is not to answer the old question. For geneticists there never was any question. Few of us got very far in our studies before discovering that the environment provides many complications and that, in some cases, it seems to swamp the genes entirely. We who are concerned with increasing the productivity and viability of domestic animals have a special reason for considering always the environment. Most of the things that we value are quantitative characters, and these seem to be particularly subject to environmental modification.

We have already considered some environmental modifications of genetic traits, although we were careful not to label them specifically as such. In the previous chapter we had several examples. Thus, some of Bittner's mice were genetically susceptible to the mysterious milk-factor, but a few were resistant. That milk-factor was a part of their environment. The hemolytic foals can be saved if put in one environment (a foster mother), but will perish if left in another (their own mother). Similarly, the riboflavin added by Maw to his deficient eggs provided for the embryos therein an environment in which they could develop normally. Without it they would perish. Finally, we saw that in the Shetland × Shire cross F_1 foals born from one maternal environment were about three times as big as those from the other one. In this chapter we shall consider a few more examples to illustrate some common kinds of interactions between environment and heredity.

The Diet

No examples are needed to remind us of the importance of an adequate quantity of food, whether for ourselves or for our animals. We may not have thought of our diet as a part of the environment, but it is an inescapable part of that environment. It can be favorable for some purposes and unfavorable for others. For the broilers and pigs that we are trying to get to market at the earliest possible age, the more feed we can get into them the better. For ourselves, at least for those of us who are trying to keep our waistlines under control, big meals constitute an unfavorable environment.

Similarly, there is not much question about the importance of the quality of the diet as well as its quantity. Our nutritionists have done so well in showing us the assorted afflictions besetting animals that do not get their vitamins or minerals that we all know the importance of having the diet adequately balanced to permit normal health and productivity. It is not realized by most people that animals differ genetically in their dietary requirements, and particularly in their ability to withstand deficiencies of some important constituent. For example, White Leghorns thrive on a diet containing only 30 parts of manganese per million, but some chicks of Rhode Island Reds and other heavy breeds may need as much as 50 parts per million to prevent perosis, a condition in which the "hock" joint becomes enlarged and causes bad lameness. Most young animals require for normal growth not only a proper balance of calcium and phosphorus, but also vitamin D. However, when the diet is deficient in any of these important ingredients, the pigs or chickens getting that diet do not all come down with symptoms at once. The most susceptible show it first, and those genetically resistant show it later, or even not at all. Chicks are more favorable for experimental studies of requirement of specific nutrients than are larger animals. Genetic differences have been demonstrated in their requirements of thiamine, vitamin D, and vitamin E, and of two amino acids, methionine and arginine. If we consider a deficiency of any one of these nutrients as an unfavorable environment, obviously the genetically resistant chicks are better able to cope with such adversity than are the others.

In earlier chapters we considered genes for yellow fat in sheep and rabbits. Yellow fat is normal for most breeds of chickens, but there is a gene, W, which restricts yellow to the visceral fat so that it is not deposited in the skin. Accordingly, birds carrying W have white shanks and a white beak. In some European countries, where this white skin is a characteristic of Sussex, Houdans, and other breeds, birds with white skin are preferred in the markets. In North America most breeds are ww and, as a result, they have yellow shanks and (before they begin to lay) a bright yellow beak. American markets prefer birds with yellow skin. As laying proceeds, the yellow pigment is gradually lost because the hens put more of it in the eggs than they extract from the feed, so they use up their storage in the skin.

Most poultry feeds in North America have enough carotinoid pigments in them to let growing birds (and those not laying) accumulate the xanthophyll that causes the yellow color. Diets can be made up, however, which lack those pigments, and hens thus fed eventually lose all yellow from the shanks and skin and produce yolks that are almost colorless. On such a diet, birds thus become indistinguishable, so far as one can tell from the shanks and beak, from those that are made white by the gene W. Such an environmental modification is a comparatively harmless one, for the birds seem to get along just as well without the xanthophyll in their skin as with it.

Geneticists have found a number of cases in which environmental forces of one kind or another have produced forms indistinguishable from those caused by genes. These are called *phenocopies*. In a broad sense, we could say that the birds made white-skinned by a diet deficient in carotinoid pigments are phenocopies of those made white by W.

Physical State of Food

Even when the diet is fully adequate with respect to nutrients and ample in amount, its effects can differ greatly according to its physical state. At Michigan State University, Hunt and his associates bred rats of two different strains, one highly susceptible to decay of the teeth and the other highly resistant. Susceptibility was measured by the time required (on the diet that induced dental caries) to the first sign of decay. Contrary to what dentists tell us about our own teeth, finely ground feed apparently does not bother the rats' teeth at all. It is coarsely ground feed that gives them trouble, as the following results with rats of the susceptible strain showed:

	Caries Time
7th generation, on coarse feed	35 days
8th generation, on feed fine as flour	125 days
9th generation, on coarse feed	30 days

After seven generations of selection, Hunt's rats were clearly so susceptible to dental decay that they showed lesions at four to five weeks after going on coarse feed. When rats of the same strain were given the same feed exactly, but finely ground, no tooth decay appeared until 18 weeks, and even then only four of the 57 rats tested showed any decay. Here is an environmental difference that few might suspect until it had been demonstrated.

Even the difference between dry and moist feed can have an effect on the viability of the animals getting it. Sturkie studied a bizarre mutation in the fowl called apterylosis because the pterylae, or feather tracts, were greatly reduced (Fig. 15-1). On dry feed, mortality among these chicks to 10 days of age was 54 per cent, but among those given the same feed moistened to make it more palatable only 22 per cent died. The difference resulted from the fact that, without their

FIG. 15-1 Chick showing apterylosis. (Courtesy of P. D. Sturkie.)

protective covering of down, those chicks needed a high intake of feed to maintain body temperature and normal functions. Dry feed is not as palatable as wet feed. Those getting only dry feed did not eat enough of it to meet requirements, but those getting the more palatable mixture did so; hence more of them survived. When these last were again put on dry feed at 10 days, or at 30 days, their mortality rates immediately rose.

Effects of Temperature

Farm animals react in most cases to changes in temperature much the same as do humans. It is not so generally realized that there are genetic differences among animals in their ability to withstand extremes of heat and cold. Perhaps that it because the reactions are not always easy to measure in quantitative terms. A few examples will make the point clear. Zebu cattle of various breeds are common in Asia and Africa, and are now being used in various parts of the world, particularly in the tropics, to produce new breeds that can maintain useful productivity at relatively high temperatures. The remarkable ability of zebu cattle to retain their bovine complacency when cattle of European origin have been completely upset by high temperatures was well illustrated in some tests made by Rhoad in Brazil. He compared pure zebus, pure Holstein-Friesians, and eight F_1 animals from the cross between those two breeds (Fig. 15-2). Even at 23° C. the

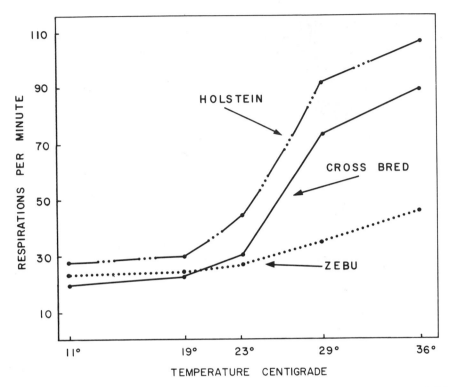

FIG. 15-2 Differing responses of three kinds of cattle to stress of high temperature, as measured by respirations per minute. (From A. O. Rhoad in *J. Agric. Science*, by permission of Cambridge Univ. Press.)

respiration rate in the European cattle and in the crosses had begun to rise. At 36° C. the Holstein-Friesians were in real distress and were panting at the rate of 107 respirations per minute. At the same time, the zebus were doing a comfortable 46, and the F_1's were intermediate. This little experiment merely measured in quantitative terms what was already well known, namely, that the zebus, as a breed, are characterized by a remarkable ability to withstand high temperatures. That ability is a breed characteristic, and hence just as much a hereditary characteristic as is the great hump on the shoulders of the zebu cattle.

Similar differences among breeds in ability to withstand high temperatures are found in fowls. This was demonstrated at Cornell some years ago, when for three days in July temperatures rose to levels unprecedented during the previous eleven years. During the whole heat spell, which lasted some six days, mortality from heat prostration alone was three times as high in heavy breeds as in Leghorns.

At Davis in California, Asmundson and Hinshaw investigated in Bronze turkeys a peculiar abnormality which they called pendulous crop. In affected birds the crop becomes filled with fluid and greatly distended (Fig. 15-3). A great many of them die, sometimes from pneumonia or from wounds in the enormous crop which subsequently becomes infected. Even the survivors are unthrifty, and few of them can be marketed.

Because the condition was found only in Mammoth Bronze turkeys, and never in the Bourbon Reds raised on the same premises, the condition was clearly a genetic character. Additional proof of that was seen when, by selective breeding, a strain was developed in which over two-thirds of the turkeys from afflicted parents developed pendulous crops. However, it was noticed that the pendulous crops appeared in poults at 8–16 weeks of age after excessive drinking during spells of hot weather. In order to test further the possible influence of the environment, poults of the susceptible strain were divided into two groups, one of which was sent from Davis to Tomales, a place only two miles from the ocean. The results of this experiment showed clearly that in the cooler atmosphere at Tomales the genetic susceptibility to pendulous crop was never manifested (Table 15-1).

Here is an interesting and typical case of interaction between genetic forces and environmental influences. Any raiser of Mammoth Bronze turkeys at Davis who experienced this difficulty in his own

FIG. 15-3   Pendulous crop, induced by a combination of genetic and environmental influences. (From W. R. Hinshaw and V. S. Asmundson in *J. Amer. Vet. Med. Assoc.* **88** (N.S. 41): 154-65.)

TABLE 15-1

Effects of Differing Environments on the Incidence of Pendulous Crop in Turkeys of the Same Stock

Place	Maximum Daily Temperature (average)	Highest Temperature Recorded	Poults Exposed (number)	Incidence of Pendulous Crop (per cent)
Davis	92.5° F.	109° F.	67	67.2
Tomales	74° F.	92° F.	83	0.0

Source: Data of Asmundson and Hinshaw.

flock, but noticed that his neighbor's Bourbon Reds across the road were having no trouble whatever, would, of course, blame the stock. However, when he made his claim for redress to the turkey breeder who had supplied the poults, that harassed individual could probably reply, "No, it can't be the fault of the stock, because I sent some from the same hatch to Jim Smith down at Tomales and he never had a bit of trouble." The same condition has been noticed in some dry areas in Australia. It is clear that breeders supplying poults for the hot environment should breed strains that can endure the high temperatures without trouble. It is also clear that enough genetic variation is available to facilitate such selection.

Ordinarily, colors of animals don't seem to be influenced much by temperatures, but that they can be so influenced is demonstrated by Laura Kaufman's Himalayan rabbits shown in Fig. 15-4. Normally, Himalayans should have black pigment only on the nose, ears, feet, and tip of the tail. However, when a spot is shaved in the white area of such a rabbit and the animal is then put in a rather low temperature while the new hair is being grown, the new hairs come in black. Conversely, a shaved ear wrapped in bandages to keep it warm grows white hairs where black grew before.

Effects of Sunlight

Evidence that strong sunlight affects some genotypes differently from others can perhaps be most easily seen at any bathing beach in mid-summer, where sunburned blondes and tanned but unburned brunettes attest the fact that differing genotypes do not react alike to some environments. Domestic animals have the same problem. In areas where there is intense sunlight, dark skin provides a protection not available to the white-skinned animals. Sensitivity to sunshine is

FIG. 15-4 Changes from normal color caused by variations in temperature. The rabbit on the left, after being shaved on its left side and put in a cold room, grew black hair where white had been before. The other rabbit had its right ear warmed with bandages, and grew white hairs where black had been before. (Courtesy of Dr. Laura Kaufman, Lublin, Poland.)

sometimes aggravated by eating certain plants, such as buckwheat, St. John's Wort, and some species of the genus Senecio. It is also aggravated by some mutations that occur in cattle, sheep, and swine, which will be considered later. The genetically sensitive animals can be maintained in good health so long as they are not exposed to the sun. In fact, that difference in the environment can make all the difference between life and death.

Another influence of light on a genetic character is shown in Table 15-2. Here it is not the intensity of sunlight that counts, but the duration of the day. It will be noted that Strain C matured five or six days earlier than Strain K in both years (a genetic difference), that all mid-season and late hatches were retarded, and that some environmental effect caused unusually slow maturity in late-hatched pullets of both strains in 1953. The successively older mean ages at first egg for pullets hatched in mid-season or later have really nothing to do with the date of hatch, except insofar as that date determines when they will be somatically mature and able to lay about five months later. Shortening days in September provide a handicap that retards mean age at first egg for the late-hatched bird.

Apart entirely from the effect of the amount of light on age at first

TABLE 15-2

Variations in Age at First Egg in White Leghorn Pullets, Showing Influences
of Strain, Date of Hatching, and Years

		Mean Age at First Egg in Days, for Pullets Hatched		
		---	---	---
Year	Strain	March 3–10	March 24– April 2	April 21–29
1953	K	164	+ 8[a]	+30[a]
1953	C	159	+ 7	+28
1954	K	161	+ 9	+16
1954	C	155	+14	+15

[a]Number of days later than for hatches of March 3–10.
Source: Data of Hutt and Cole, Cornell University.

egg is its effect on the rate of egg production. As every poultryman
knows, late-hatched pullets and other malingerers can be forced into
egg production by providing enough artificial light to lengthen their
working day to about 13 or 14 hours.

Effects of Pathogens and Allergens

Pathogenic organisms are just as much part of the environment as are
the diet, the housing conditions, and the temperatures. Resistance to
them depends on many things, including the severity of the exposure
or dosage, the genotype of the animal attacked, and also the degree to
which environmental conditions are favorable. Chicks infected with
Salmonella pullorum will die if chilled during the first ten days or so
after hatching, but similarly infected chicks can nearly all be safely
reared if they are brooded at 36°C. or higher. Of all this, more anon.

Animals that are genetically highly susceptible to disease when
born or hatched may become more resistant at later ages. If fowls
genetically susceptible to leukosis and Marek's disease can be isolated
until five months of age, and then exposed, they are then relatively
safe (Fig. 15-5). In this case we might consider the causative viruses
as a part of the external environment and the age of the bird as a part
of its internal environment. With other pathogens the situation is
exactly reversed. Young chicks are comparatively resistant to coccidia, but are highly susceptible as adults. The recommendation of
former years that everything be done to protect the chicks from
infection has proven completely futile, and one modern control is to
expose them when very young to small doses that will allow them to

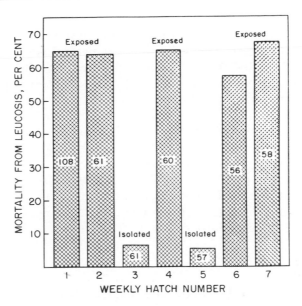

FIG. 15-5 Effects of differing environments on incidence of leukosis to 500 days of age in White Leghorn females of genetically susceptible stock. Hatches 3 and 5 were isolated at hatching, but returned to infected premises at about 150 days of age. The other hatches (all of the same stock at those isolated) were kept on the infected premises. Numbers of birds per hatch are shown in columns. (From R. K. Cole and F. B. Hutt in *Poultry Sci.* **30**: 205–12, 1951.)

develop immunity. As there are genetic differences in susceptibility to coccidia in chicks, this case provides another example of interaction among genes, pathogen, and age.

An inescapable part of the pathogenic environment for some millions of people is provided by the assorted allergens to which they are sensitive. Those who flee to the north woods to escape the ragweed pollen, or take anti-histamines to stop their runny noses when the timothy is in flower, may not think of their affliction as an interaction between their genotype and the environment, but sensitivity to allergens is caused in most cases, if not all, by one's genes.

Other pathogens include the carcinogens. Those who still refuse to admit that cigarettes can cause lung cancer in man are usually able to produce some confirmed addict who has managed to live to a ripe old age in spite of his persistent smoking. A point overlooked in all the publicity about lung cancer is the fact that undoubtedly some people are much more resistant to it than others. Resistant people could afford to take chances with an environment that includes carcinogens, but susceptible ones cannot. A helpful solution might be to find some means of identifying the two kinds at an early age.

Still other pathogens are the toxins. As the house flies have shown us, every population contains some individuals that are more resistant than others to certain toxins, in their case, to DDT. Those who subject the flies to DDT really do that species a service by eliminating those most susceptible and permitting those most resistant to multiply. This they have done with a will.

Somewhat similarly, the rabbits in various parts of the world are gradually conquering the virus of myxomatosis with which people in Australia and elsewhere sought to vanquish them. Evidence is accumulating to show that in Australia the rabbits are becoming more resistant with each passing year. While part of that rise in resistance may be attributable to the development of attenuated strains of the virus, there is ample evidence that over a period of years the resistance of wild rabbits to standard doses of the pathogen has steadily increased.

Effects of Management

A typically complex interaction of genetic and environmental forces is illustrated by a kind of arthritis that occurs in turkeys at 12 to 24 weeks of age. Affected animals become lame in one leg, or both, and the inter-tarsal joint becomes enlarged. The following conditions have been found to affect the frequency of the condition:

1. It occurs chiefly in males, seldom in females.
2. There are significant differences among sire families in the incidence of the condition; hence there is a genetic basis for it (Fig. 15-6).
3. It is influenced to some extent by the diet, as it is slightly less severe when the amount of inorganic phosphorus is maintained at the right level. It is apparently reduced somewhat by feeding extra vitamin E.
4. The arthritis is about four times as common in turkeys raised on slatted floors as in those left on litter.
5. Its frequency is much higher in some hatches than in others for reasons impossible to discern (Fig. 15-6).

Obviously, here is a condition which could be reduced to some extent by selective breeding, perhaps also by improving the diet; but while nutritionists seek for trace elements that might lessen the trouble, it is clear that the turkey raiser's best protection, if he has susceptible stock, is to put it on the floor, or on the ground, and not to

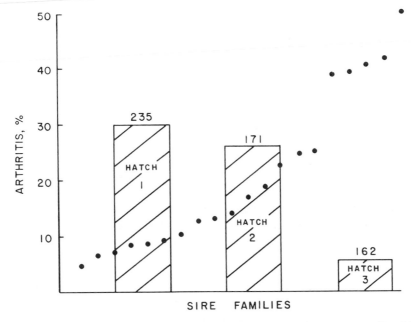

FIG. 15-6 Incidence of arthritis in male turkeys, showing inexplicable environmental differences among hatches of the same stock, and highly significant genetic differences among sire families. Each dot represents one sire family and shows the incidence of arthritis therein during the first two hatches. (From data of A. S. Johnson in *Poultry Sci.*)

try to rear it on wire or on slatted floors. Can we assign the cause to heredity or to environment? Obviously, both are important factors.

Twins

It is clear that there are many cases in which it is difficult to assess the relative importance of heredity and environment. In man and in cattle there is a very precise tool for doing just that. It is called the *twin-study method.* Before proceeding to see how it works, we should know a little more about twins. Of these there are two kinds:

Monozygotic, or identical, twins resulting from the division of an egg after its fertilization; commonly designated, for short, as MZ.

Dizygotic, fraternal, or dissimilar, twins, resulting from the separate fertilization of two eggs. Monozygotic twins must always be of the same sex, but the dizygotic twins can obviously be of either sex, and the combinations should appear in the ratio of approximately 1 ♂ ♂ : 2 ♂ ♀ : 1 ♀ ♀. They are commonly called DZ twins.

The frequency of human twins varies somewhat from one country to another. From Greulich's extensive data from 21 countries it would appear that the frequency of twins in some 120 million births was about 1 in 86.2. In the United States, among Caucasians the frequency is about 1 in 88.6, and in Negroes 1 in about 71. Obviously, the department store that lures the customers for layettes with the assurance that, if there should be twins, a second layette will be provided free of charge, will have to make good on its promise only about once in 87 sales. In man there seems to be operating a very peculiar law whereby, when the frequency of twins is 1 in n, the frequency of triplets is about 1 in n^2, and that of quadruplet births becomes about 1 in n^3.

By what is called Weinberg's differential method, it is possible to estimate what proportion of twin births is monozygotic and what proportion dizygotic. Since in DZ twins the ratio of combinations of the sexes is about $1:2:1$, there must be just as many pairs with the same sex as there are with both sexes. Accordingly, in large numbers the amount by which same-sexed twins exceed pairs of unlike sexes should be represented by the monozygotic twins. In such numbers the frequencies in North America data for man work out about as follows:

	♂ ♂	♂ ♀	♀ ♀
Dizygotic twins	1	2	1
Monozygotic twins	1	—	1
	$\overline{2}$	$\overline{2}$	$\overline{2}$

In other words, the proportion of identical twins among all twin births is approximately one-third. Among them, the ratio of pairs both boys to pairs both girls is not exactly $1:1$, because the sex ratio in man at birth is not exactly $1:1$. It is closer to about 104 ♂ ♂ : 100 ♀ ♀. However, taking the round numbers given above, it is seen that the chance of being monozygotic for any pair of twins of the same sex among all such pairs is approximately 1 in 2, and for any pairs of opposite sexes, it is 0.

From extensive data compiled by Johansson, the frequency of twins in dairy cattle was found to be about 1.88 per cent of all births, and the corresponding figure for beef cattle only 0.44. Johansson calculated by Weinberg's differential method that the frequency of monozygotic twins in cattle is about 6 per cent of all twin births.

One problem confronting all those who use twins for various studies is that of identifying with certainty which pairs are monozygotic and which dizygotic. Special techniques have been worked out in man, and others in cattle, to facilitate the identifications, but we shall not be able to go into them here. Readers interested will find them reviewed in the publications by Stern and Hancock cited at the end of this chapter.

The Twin-Study Method

The value of this method lies in the fact that the monozygotic twins should be identical with respect to their genes, except insofar as there may have been any mutations since the two parted to go their separate ways. This does not apply to the dizygotic twins, which are no more alike genetically than brothers and sisters born separately. Accordingly, the procedure and reasoning in the twin-study method are as follows:

1. Ensure that adequate representative samples of both MZ and DZ twins are available for study.
2. Determine in them the incidence of the particular trait to be investigated.
3. If that trait is determined chiefly by heredity, then the *concordance*, or agreement of one twin with its partner, should be comparatively high in MZ twins. Dizygotic twins would agree no more than sibs.
4. If the trait is determined chiefly by the environment, then monozygotic twins should show concordance no higher than that for dizygotic twins of the same sex.
5. The concordance in monozygotic twins becomes specially significant if, as frequently happens, the two members were reared apart.

Some findings with the twin-study method, as applied to man, are given in Table 15-3. These are only samples from many such studies. (For others see Stern and Kallmann, as cited.) It is clear that some human afflictions which many people considered to be caused solely by the environment—for example, tuberculosis, schizophrenia, and poliomyelitis—are really to a high degree dependent upon genetic

TABLE 15-3

Concordance in Human Twins With Respect to Various Conditions

Condition Studied	Investigators	Dizygotic Twins		Monozygotic Twins	
		Pairs (number)	Concordant (per cent)	Pairs (number)	Concordant (per cent)
Tuberculosis	Kallmann and Reisner	230	25.6	78	87.3
Poliomyelitis	Herndon and Jennings	33	6.1	14	35.7
Diabetes	Lemser	29	20.7	18	83.3
Schizophrenia (U.S.)	Kallmann[a]	685	14.5	268	86.2
Schizophrenia (England)	Slater[a]	115	14.0	41	76.0
Epilepsy	Conrad[a]	130	3.1	30	66.6

[a]Cited in the book by Kallmann listed at the end of this chapter.

susceptiblity to them. The data of Herndon and Jennings suggested that susceptibility to poliomyelitis might depend upon a single recessive gene with a penetrance in homozygotes of about 36 per cent, but studies with still larger numbers of twins would be needed to prove that point.

Unfortunately, the twin-study method has not yet been used on any extensive scale to determine the comparative importance of heredity and environment in causing disease in cattle. One investigator has noted that monozygotic twins agree more than dizygotic twins in the incidence of bloating. At Wiad in Sweden, among 104 head of cattle that had been fed mouldy hay, one died of a mycotic infection at 1 day over 35 months of age. As it happened, she was one of a pair of monozygotic twins. Exactly two weeks after the first twin showed symptoms, her mate did exactly the same, and she eventually died just two weeks after the other one. On examination, the same infiltration of moulds in the mucous membrane of the intestine was evident. These two MZ twin heifers were the only ones infected among 104 animals that ate the mouldy hay. While one swallow does not make a summer, or one pair of twins tell a complete story, it is difficult to account for this case on any basis other than genetic susceptibility of those twin heifers to mycotic infection.

Twins for Other Experiments

Monozygotic twins in cattle are particularly useful for experiments to determine the effects of different diets, different therapy, or any other modification of the environment. Since both members of a pair have the same genotype, by using one as a control and giving the other the treatment, the experimenter is able to determine the effects of that treatment without any complication from differing genetic susceptibilities such as are found in animals not twins.

The pair of MZ Ayrshire heifers shown in Fig. 15-7 was used to demonstrate the value of a special feed formula for starting calves. The two twins, Candy and Sandy, started life at 103 and 97 pounds, respectively. At five months their weights were 315 and 318. The control twin, Candy, got five quarts of whole milk for the first 90 days, but Sandy, who had been raised at much less expense on the special formula, weighed every bit as much.

Some investigators who have made similarly controlled experiments with larger numbers of twins have estimated that for many kinds of experimentation a single monozygotic twin as a control is worth some 10 to 25 other calves, or even more (see Hancock). Monozygotic twins are apparently remarkably uniform with respect to the time in grazing, in loafing, lying down, and in the distances walked. One can only regret that such ideal material for experiments is not common, is difficult to identify with accuracy, and apparently is not available in species other than cattle (or man). If domestic animals

FIG. 15-7 Identical Ayrshire twins, Sandy and Candy, discussed in text. (Courtesy of Dorothy Long, Dawnwood Farms, Amenia, New York.)

would only produce monozygotic quadruplets, a trick that is routine for the armadillo, answers to problems in animal husbandry would come faster and more easily, but who wants to eat an armadillo?

Twin Studies in Dairy Cattle

Utilization of twins for study of economically important, inherited traits of dairy cattle was begun in 1937 by Bonnier at Wiad, Sweden, and continued for many years. Similar studies have been conducted by other investigators in New Zealand, Scotland, and Sweden. Their findings were thoroughly reviewed by Johansson (1968). One of them in which there was remarkable agreement in two countries was that the correlation between milk yields of the two sisters of a pair of MZ twins was about 0.84 to 0.88, while the corresponding figure for DZ twins in the same herds ranged from 0.56 to 0.68.

These intra-pair correlations in twins were found to be much higher than in pairs of full sisters, half-sisters, and unrelated controls. As Rendel and Johansson (1966) have shown, these latter groups lack the contemporaneity which characterizes twins. Part of the high correlations within pairs of DZ twins, and the still higher ones for MZ twins, may therefore be ascribed to the fact that both members of such twins had shared the same environment since birth.

The inevitable influence of environment was also demonstrated by tests at the Ruakura Station in New Zealand in which pairs of MZ twins were split and the two sisters raised in two different environments through the first lactation period. In one environment (selected farms), milk production was known to be high; in the other, it was comparatively low. As was to be expected, in the latter group the environment triumphed over the genes. It suppressed the innate genetic potentiality for milk production of the calves sent to the low-producing herds, while their sisters in the high-producing herds showed that in a good environment their genes for milk production were fully expressed.

Johansson pointed out that problems with MZ twins have caused some research stations to forego their use and to do their tests with DZ twins instead. It is difficult to collect enough MZ twins within the short time desirable for precise comparisons. They are expensive, and some are found on accurate diagnosis not to be MZ. Finally, if one becomes incapacitated, the other is automatically out of the test. The DZ twins are easier to collect in good numbers, and intra-pair correlations between them are remarkably high.

Relation of Environment to Selection

Most breeders selecting among domestic animals for improvement in quantitative characters of economic value, such as milk production and egg production, would try to maintain their animals under test in conditions as nearly ideal as possible. Only thus can one get maximum expression of such traits. Sometimes, however, it may pay the breeder to test his animals under conditions of stress. This applies particularly to poultry. Most of the chickens raised in this country, whether for egg production or for broilers, are no longer bred by the people who raise them, but are produced by a few dozen specialized big breeders. The wise ones know that their products will be subjected by their customers to various kinds of stress including disease. It may pay the breeder, therefore, to test his populations under conditions of exposure to such stress. Any genetic resistance to disease developed on the breeder's premises could be a valuable asset to that breeder if his stock should perform better than others under adverse conditions.

Breeds for Special Environments

Under this heading it is necessary only to remind the reader that breeders of domestic animals the world over have tried to get into their favorite breeds the best possible collection of genes to adapt those breeds to the local environment. The Santa Gertrudis cattle developed on the King Ranch in Texas are approximately three-eighths zebu and five-eighths Shorthorn (Fig. 15-8). Because of their zebu blood, they are particularly well qualified to withstand hot climates. At the other extreme, Highland cattle with their unusually heavy coat of hair are particularly adapted to a cold winter and are famous for their hardiness (Fig. 15-9). Many similar examples could be cited. Independently, in Canada and in Russia, poultry breeders developed breeds with walnut combs, a type less likely to be frozen in winter than the upright single combs of other breeds. Unfortunately, economic qualities were not improved at the same time; hence the Canadian Chantecler has almost disappeared in spite of its built-in genetic adaptation to Canadian winters.

Similarly, zebus of the Sahiwal type were crossed with Holstein-Friesians in Jamaica to produce eventually the Hope Holsteins, a breed better qualified than pure Holstein-Friesians to produce milk under tropical conditions. They contain approximately 25 percent zebu blood. Zebus of other breeds have been utilized similarly elsewhere.

FIG. 15-8 Santa Gertrudis I, one of the founders of the Santa Gertrudis breed, showing well the good conformation of that breed for beef, and the shoulder hump of his zebu ancestors that contributed their ability to tolerate heat. (Courtesy of R. J. Kleberg of the King Ranch, Texas.)

It is not by accident that the breeds of dogs variously known as Mexican Hairless, Persian or Turkish Hairless were developed nearer to the equator than to the North Pole. When next the reader sees one of these animals at a dog show, he might remember that its genes do not fit it for northern climes and that the blanket it wears when not on the show bench is a man-made device by which the little dog is enabled to cope with an adverse environment.

Problems

15-1 In a large number of bovine twins, Johansson found that sexes to be: ♂♂ : 732; ♂♀ : 1,357; ♀♀ : 799. Assuming for convenience that the sex ratio in cattle is 1 : 1, and applying Weinberg's differential method, what proportion of these pairs was likely to have been monozygotic?

15-2 If the finding in the previous question were generally applicable, then, if your neighbor's Guernsey should produce a pair of twins, what are the odds against their being monozygotic?

FIG. 15-9 Highland calf showing in its heavy coat one of the breed characteristics by which this hardy breed is enabled to range in the open through Scottish winters. (Courtesy of Mary MacLennan Cheska, Ellensburg, Washington.)

15-3 If the twins of Problem 15-2 prove to be both heifers, what then are the odds against their being monozygotic? (To answer this one, first determine the proportion of such pairs expected among DZ twins.)

15-4 In many pairs of twins in Swedish sheep, Johansson found the sexes to be: $\male\,\male$: 1,164; $\male\,\female$: 2,685; $\female\,\female$: 1,239. With the same procedure as for Problem 15-1 determine the probable frequency of MZ twins in sheep.

15-5 What hereditary characters in domestic animals can you think of that are fully expressed regardless of the environment after birth or hatching?

15-6 What specific environmental stress factors do you know in addition to the examples mentioned in this chapter, and what traits of domestic animals do they affect?

15-7 Can you think of any specific practices in management of domestic animals that are followed to aid those animals in coping with an adverse environment?

15-8 Rumplessness in the fowl can be caused by a dominant gene, *Rp*, but it can also be induced (a) by changing the temperature of the incubator abruptly during the first week of incubation, (b) by shaking the eggs severely before incubation, or (c) by injecting insulin before incubation. Do these phenocopies and the environmental stresses that produced them suggest any way in which *Rp* might induce genetic rumplessness? Remember that the embryos have developed for about 24 hours by the time the egg is laid.

15-9 Would you expect the phenocopies induced by these experimental treatments (in Problem 15-8) to transmit rumplessness as do the birds that carry *Rp*?

15-10 The white-skinned chickens that some European markets prefer can be produced by feeding special diets low in carotinoid pigment, or by using the gene *W* and eliminating its allele, *w*, which (when homozygous) induces yellow skin. Considering that tests of genotypes *WW* and *Ww* are simple and that the dietary treatment is not perfect, is the genetic control, or the environmental one, to be preferred?

15-11 By what kinds of test-crosses would you eliminate the gene *w* from a flock that carries it? What kinds of males and females would you use in such tests?

15-12 Would it be necessary to continue such testing in each generation? If not, how soon would further testing become unnecessary?

Selected References

ASMUNDSON, V. S., and W. R. HINSHAW. 1938. On the inheritance of pendulous crop in turkeys (Meleagris gallopavo). *Poultry Sci.* **17:** 276–85. (Genetic and environmental influences causing pendulous crop.)

HANCOCK, J. 1954. Monozygotic twins in cattle. *Advances in Genet.* **6:** 41–81. (With special reference to their identification and use in controlled experiments.)

HERNDON, C. N., and R. G. JENNINGS. 1951. A twin-family study of susceptibility to poliomyelitis. *Amer. J. Human Genet.* **3:** 17–46. (With evidence that susceptibility is genetic in origin.)

HUTT, F. B. 1958. *Genetic Resistance to Disease in Domestic Animals.* Ithaca, N.Y.: Cornell Univ. Press. London: Constable & Co. Ltd. (See Chapter 3: On coping with the environment.)

JOHANSSON, I., and J. RENDEL. 1968. *Genetics and Animal Breeding.* Edinburgh. Oliver and Boyd. (See Chapter 5 for twin studies and the rest of the book for everything else. If you cannot find it, try for the Swedish edition published in 1963 by L. T.'s Forlag, Stockholm.)

JOHNSON, A. S. 1956. Incidence of an abnormal hock condition in male turkeys as influenced by genetic differences and by hatch. *Poultry Sci.* **35:** 790–92. (Source of data shown graphically in Fig. 15-6.)

KALLMANN, F. J. 1953. *Heredity in Health and Mental Disorder*. New York. W. W. Norton & Co., Inc. (With many examples of the use of the twin-study method.)

RENDEL, J., and I. JOHANSSON. 1966. A study of the variation in cattle twins and pairs of single-born animals. IV. The effect of contemporaneity and some other factors on heritability estimates for milk yield and fat percentage during the first lactation. Z. *Tierz. ZüchtBiol.* **83:**56–71. (The title tells what is in it.)

RHOAD, A. O. 1936. The influence of environmental temperature on the respiratory rhythm of dairy cattle in the tropics. *J. Agric. Sci.* **26:** 36–44. (Source of data shown in Fig. 15-2 and discussed in this chapter.)

STERN, C. 1973. *Principles of Human Genetics*. 3rd ed. San Francisco. W. H. Freeman & Co. (See Chapters 25, 26, 27 for concise summary of information about twins, the twin-study method, and what it has revealed about the influence of heredity and environment on physical and mental traits in man.)

STURKIE, P. D. 1942. A new type of autosomal nakedness in the domestic fowl. *J. Heredity* **33:** 202–8. (Details about the kind of chicks shown in Fig. 15-1 and environmental influences affecting their viability.)

CHAPTER 4

Selection and Changing Populations

In breeding to improve domestic animals, whether for more gallons of milk, thicker egg shells, finer wool, or any other objective of economic value, the breeder is really selecting from the genes available in his animals those that are most likely to work toward attainment of his ideal. Part of that process is the elimination of genes with undesirable effects—the lethals, and others that induce serious defects. Some of the selection is directed toward the maintenance of breed standards, like the belting of Hampshire pigs, the rose comb of Wyandottes, or the lack of horns in Polled Herefords. Most of it, however, is directed toward the accumulation of quantitative genes and modifying genes that affect polygenic characters, particularly to induce more or better milk, eggs, meat, and wool.

There are thick books on this subject. In this chapter we can only mention some of the problems involved, and the procedures followed to overcome those problems. The basic principles are much the same, whether we are breeding for a heavier fleece, a bigger pail of milk, or a heavier basket of eggs.

Long before man began striving to increase the productivity of his captive animals, another force was hard at work to make them better. It is still doing so. That other force is natural selection, or if you like, the survival of the fittest. Contrary to what some breeders of domestic animals seem to think, natural selection is not something that operates only in jungles. It operates also, albeit less severely and often imperceptibly, on all our flocks and herds, even those in the best of environments. Let us consider a few cases.

Natural Selection in *Biston betularia*

Of the Peppered Moth, *Biston betularia,* two distinct forms are known in England. The original type is whitish but carries a pattern of many thin black lines and dots that caused its common name. A simple dominant mutation makes a black form known as *B. b. carbonaria.* The latter type, first collected about 1850, has increased so greatly in certain areas of Great Britain that in some of them it is the predominant form, and the paler moths are rare. Elsewhere, particularly in the south-western parts of England, *carbonaria* is not found, but the typical form is abundant. It is now clearly established that the dark mutant predominates in areas that are heavily industrialized, but is not found in counties predominantly rural. Thus, 90 per cent of the moths of this species in Glasgow are *carbonaria,* but in northern Scotland none of that color is found.

As Kettlewell has clearly demonstrated in a remarkable series of investigations, the reason for predominance of the *carbonaria* form in

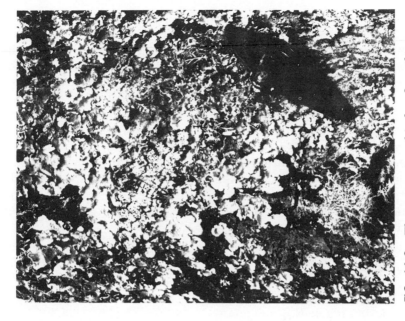

FIG. 16-2 The same two kinds of moths on a tree covered with lichen in Dorset. (Courtesy of H.B.D. Kettlewell.)

FIG. 16-1 Typical *Biston betularia* (above) and the melanic *carbonaria* form (below) at rest on the dark bark of a lichen-free tree near Birmingham. (Courtesy of H.B.D. Kettlewell.)

the industrial areas is that the trees there, on the trunks and branches of which the moth normally rests, are blackened by the smoke. As a result, the black moths are not easily seen, but the paler, typical ones are conspicuous. The latter are eaten by birds, but more of the *carbonaria* mutants escape the notice of predators, and thus live to reproduce their kind (Fig. 16-1). By contrast, in rural areas, where the tree trunks are light grey and often covered with lichens, the situation is reversed. There it is the pale form that most often escapes detection and *carbonaria* that falls to the predators (Fig. 16-2).

These facts were determined not only by watching the predaceous birds at work, but also by releasing known numbers of marked moths of both kinds, and then determining somewhat later the proportion of each kind that could be recaptured. Recovery rates (per cent) were as follows:

	Pale Form	*Carbonaria*
Industrial area near Birmingham	25.0	53.3
Rural area in Dorset	12.5	6.3

Clearly, natural selection gave the black moths a 2 to 1 advantage over the others in the industrial area, but a 1 to 2 handicap in the rural one. If we assume that contributions of the two types to the next generation were of somewhat the same order, it is easy to understand how even a century of such selection could result in the predominance of one kind in one environment, and of the other elsewhere.

The Fast-working Oysters of Malpeque Bay

Many people fail to realize that natural selection is at work because they do not live long enough to see its effects in the comparatively large animals with which they are familiar. In an average lifetime we are not likely to see more than four generations of our own species, or 25 generations in cattle, and neither of these now gives natural selection much material to work on, or a free hand to do it. It is different with the oysters, many of which have the capacity to bring into their watery world up to 60 million potential little oysters in a year. The species needs them all, for, apart entirely from the comparatively small proportion of each crop that goes down the gullets of gourmets, life for oysters is fraught with many hazards.

On the east coast of North America, one such hazard is Malpeque disease, a mysterious sickness not yet fully understood, which, about 1915, wiped out nearly all the oysters in the hitherto productive waters of Malpeque Bay, on the north shore of Prince Edward Island, in the Gulf of St. Lawrence. Diseased oysters had characteristic symptoms, including flabbiness of the big adductor muscle, cessation of growth, failure to spawn, and yellowish-green pustules up to 0.5 cm. in diameter. By 1922, discouraged oystermen had shifted operations to other areas, but eventually the disease spread not only around the island, but also to near-by beds on the shore of the mainland.

In 1925, persistent fishermen got four barrels of oysters at Malpeque Bay. Next year there were 14. From then on, the yields increased, and by 1930 only about one oyster per thousand showed any signs of the disease. By 1940, yields were far above these obtained before the disaster struck (Fig. 16-3). Transfers of the hardy oysters of

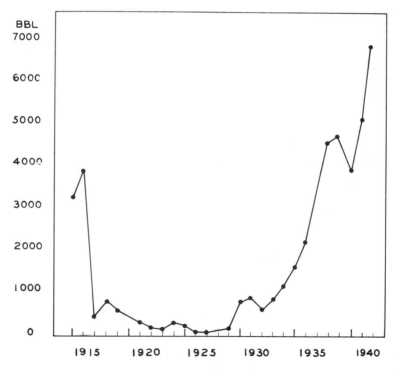

FIG. 16-3 Yields of oysters over 27 years in Prince County, Prince Edward Island, which includes Malpeque Bay, showing the rapid decline after disease appeared in 1915, and subsequent recovery after 1928, as natural selection led to production of genetically resistant oysters. Because the statistical area included some districts not stricken, the figures do not show the full damage in Malpeque Bay, where no oysters whatever were obtained in 1922. (From A.W.H. Needler and R. R. Logie in *Trans. Roy. Soc. Canada.*)

Malpeque Bay to other areas where the disease had recently struck showed that the former were now highly resistant. The big yields at Malpeque resulted, not from disappearance of the unknown pathogen, but from genetic resistance to it gradually acquired under the selective pressure of natural selection. To attain that resistance, the fast-working oysters of Malpeque Bay needed only about a dozen years. For later news of these oysters, see p. 563.

Natural Selection; Other Examples

Similar rapid response under natural selection has been demonstrated by other species capable of rapid multiplication in large numbers. Perhaps the most familiar example is that of the irrepressible house flies which, in various parts of the world, have become resistant to DDT.

Natural selection is not just something that happens to insects and oysters, but its effects are harder to see in larger animals. The prolific rabbits of Australia and elsewhere, once thought by wishful thinkers to have been vanquished by myxomatosis, are gradually becoming resistant to the virus causing that disease. That process would proceed at a more rapid rate if the rabbits could be assured of regular exposure of each generation, but, in Australia at least, conditions of drought sometimes result in too few mosquitoes to spread the virus; hence many rabbits do not get the exposure necessary for efficient operation of natural selection. Even rabbits have their limitations, but, at the rate at which they are now raising their genetic resistance to myxomatosis, it seems probable that, within the lifetime of most readers of this book, the Australians will have to deal with rabbits that may not be any bigger or fiercer than those of yesteryear, but will be genetically better equipped for survival than their ancestors ever were.

Sturkie's apterylosis (Fig. 15-1) was fatal within 15 days for 58 per cent of the affected chicks of the first two generations. Thereafter, with no selection on his part, but with survival (and multiplication) of the fittest, viability increased, and in the seventh generation mortality was only 6 per cent. At the same time chicks of the later generations became better covered with feathers. Here was natural selection operating to accumulate modifying genes that helped to suppress a deleterious dominant mutation.

Similar examples of long-continued natural selection are harder to find in larger animals—but chiefly because most of us do not live long enough to see them. The remarkable tolerance of zebu cattle to heat probably results from natural selection in hot climates over centuries. Similarly, the problem of protecting cattle in Africa from trypanosomes carried by tsetse flies is aggravated by the fact that the wild animals have developed genetic resistance to trypanosomiasis by centuries of natural selection, and thus maintain a reservoir of infection.

A well-known case is provided by the sheep raised for years on the lush pastures of the Romney Marsh in Kent. Under conditions conducive to maximum infestation with trichostrongyle worms, and with natural selection multiplying those most resistant, the Romney Marsh sheep as a breed are more resistant to such parasites than breeds developed where exposure was less severe. This fact is attested, not only by comparisons with other breeds in England, but also by controlled tests in faraway California.

Any observant biologist will see natural selection at work whenever disease or any other adverse environment strikes any flock or herd. Frequently some animals are killed, and others prevented from reproducing, or so reduced in productivity that they are eliminated from the breeding of future generations. In such cases, always some animals prove more fit than others.

Natural Selection and Evolution

Under this heading we can do no more here than to point out that mutations and the resulting variation provide the material upon which natural selection plays to bring about changes in species. There is no "evolution theory." Evolution is a fact, but there are many theories about the relative importance of forces that direct or influence it. These include the migration of animals to new environments, polymorphism, isolation of small groups geographically, or genetically (by combinations that make some hybrids sterile), inbreeding in isolated communities, genetic *drift*, or chance elimination or retention of genes in small samples of a species that become isolated but may later mutiply greatly, and others. For good surveys of this absorbing field, the reader should consult the original masterpiece: Darwin's *Origin of Species,* and later writings in this field, some of which are cited at the end of this chapter.

Mass Selection

Artificial selection can be roughly classified as either *mass selection* or *progeny-testing*. The former is the selection (to reproduce) of the individuals considered from their appearance or performance to be excellent and of the type desired. It is essentially selection based on phenotypes—a belief in the dictum that "like begets like." It is often extended to include consideration of pedigrees, in the general belief that illustrious ancestors are preferable to those unknown or non-descript. This is all to the good when the pedigree gives the performance of those ancestors, but it is no guarantee that their descendants will maintain the family tradition.

To some extent like does beget like. Cows beget calves, not foals, and purebred animals usually breed true for breed characteristics. However, as every poultry breeder knows, daughters of 300-egg hens do not all lay 300 eggs; only a disappointing minority does so. There are limits beyond which there is little assurance that like will beget like. Similarly, there is a limit to what can be accomplished by mass selection. The usual history is that mass selection is effective in raising productivity to certain levels, but beyond those levels further progress can be attained only by progeny-testing.

The truth of this statement has been attested by several cases in which selection for the same objective was continued for many years. For example, at Cornell University, mass selection long ago to raise annual egg production per hen in White Leghorns did so (with help from an improving environment) during the first 15 generations, but failed to do so thereafter. At the same time, in another strain maintained in the same environment, attempts to *lower* egg production by mass selection were completely ineffective. Productivity remained fairly constant and, after 22 generations of selection, was still about 100 eggs per bird. When mass selection was replaced by progeny-testing, productivity dropped quickly and in five generations was down to 40 eggs per bird.

While thus recognizing the limits of mass selection, it is important to recognize some of the things accomplished by that procedure. For one thing, in every domesticated animal, the important breeds have been built up almost solely by mass selection. To get some idea of how much selection breeders have made from the genes affecting size and conformation, one should go to any good dog show or poultry show. It is sometimes difficult to believe that the extremes on exhibit there could belong to the same species (Fig. 16-4), but they have been differentiated by mass selection.

Apart from the establishment of breeds, mass selection has been effective in raising and maintaining high yields, particularly in animals raised primarily for meat. It is easier to change conformation by selection than to raise the yield of milk, butterfat, eggs, or wool. Several of the breeds of beef cattle were brought to a high degree of perfection solely by mass selection. When the turkey breeders decided to change from high, sharp keels to the broad-breasted type, the trick was done within two decades, mostly by mass selection. (To be sure,

FIG. 16-4 The St. Bernard and the Chihuahua show what breeders have done by mass selection to differentiate breeds. The St. Bernard seems not amused. (Courtesy of Gaines Dog Research Center.)

their product is so short-legged and so awkward that it can be reproduced efficiently only by artificial insemination, but the customers like it and come back for more.)

The Fetish of the Pedigree

Until disillusionment dawned, the belief in the value of pedigrees was a fetish with many animal breeders. Most of them have now learned that pride of names in a pedigree carries no assurance of equal pride in performance of the progeny. As a typical example among many, consider the case of Leland Sarcastic, a Holstein-Friesian bull used in the herd of the University of Missouri in 1907–1908. He was the son of Sarcastic Lad,[1] Grand Champion at the World's Fair in St. Louis in 1904. What more could one ask? As it turned out, his daughters produced 221.5 lbs less fat than their dams, and it took 20 years to build the herd back up to the former level of production (Fig. 16-5). Most modern pedigrees of dairy cattle include records of performance, which are more reliable than names alone.

Nowadays, sires used by associations for artificial insemination of dairy cattle consistently beget daughters that excel in milk production the daughters of bulls kept in the owners' herds. That difference results—not from any superiority of artificial insemination over natural service—but from the fact that sires are not used extensively in the A.I. circuits unless they have been proven by preliminary progeny tests to beget high-yielding daughters. Disasters like that shown in Fig. 16-5 are thus avoided, and productivity is steadily raised.

Limitations of the pedigree, even of one based on performance of near relatives, are indicated in another way by the records of three White Leghorns, all full brothers, used successively in cockerel tests with the same pen of females (Table 16-1).

Although the pedigrees for these full brothers had to be identical, it is clear from performance of their daughters that they were not transmitting the same genes to their offspring. U 2's daughters exceeded average egg production of their flock by 26 eggs; those of the other two sires were 33 and 20 below their controls. U 14's daughters seemed most resistant to neoplasms (chiefly leukosis), but died of other causes. Daughters of U 26 were comparatively susceptible to leukosis. Actually, U 2 was the best sire found in several years of cockerel testing. His full brothers have yielded similar disappointing results, but in some cases two brothers prove equally good.

Progeny-testing

The difference between mass selection and progeny-testing is that, whereas the former attempts to evaluate the genotype (and therefore the worth for breeding) by the individual's phenotype, the latter evaluates the genotype by adding information on the phenotypes of an individual's offspring. Progeny tests are particularly important for

[1] We met this old boy back in Chapter 9, where he was indicted for spreading far and wide an undesirable lethal gene.

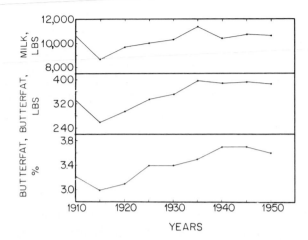

FIG. 16-5 Depression of yields caused in one dairy herd by a poor sire, with very slow recovery therefrom. (From R. C. Laben and H. A. Herman in *Missouri Agric. Exper. Sta. Res. Bull.* **459,** 1950.)

sires for such traits as milk or egg production, because they themselves cannot produce milk or eggs, and as the old adage goes, "The sire is more than half the herd." Any cock exerts 15 times as much influence on the next generation as does any one of the 15 hens to which he is mated. When a single bull can sire over 25,000 calves in a year through a centre for artificial insemination, the importance of the sire needs no further emphasis.

Details of procedure for conducting and utilizing progeny tests

TABLE 16-1
Differing Progeny Tests (in White Leghorns) of Three Full Brothers, All Mated Successively to the Same Females; Daughters Tested to 500 Days of Age

Item	First Brother, U 2	Second Brother, U 14	Third Brother, U 26
Daughters tested, number	51	40	34
Eggs per daughter, number	226	173	168
Eggs per bird, flock average[a]	200	206	188
Mortality, per cent	17.6	25.0	23.5
Died of neoplasms, per cent	3.9	0.0	14.7

[a] For all birds of the same hatch, for comparison.
Source: Data of Hutt and Cole, Cornell University.

need not be given here. Although the methods differ with varying objectives in different species, the principles are the same in all. They are:

1. To select the most promising young sires on the basis of pedigree, appearance, and every other possible means by which hopeful prospects might be identified,
2. To get from these enough daughters (with tests of their performance) to permit early evaluation of their sires, and
3. To use the sires thus proven best as long as they can reproduce, or until still better ones are found.

All of this is much more easily said than done. How does one select the most promising young sires for testing? Presumably, they should be sons of proven sires and of dams with high records. If the pedigree shows two generations of such ancestors, so much the better. Families of full sisters and of half-sisters that perform well give some reason to hope that their brothers might have good genes to pass along to the next generation, but this is just another way of saying that it is desirable to test sons of proven dams and proven sires. Some poultry breeders tend to overrate the value of such *sibtests* (*i.e.*, by performance of sibs) and *half-sib* tests, and to assume that brothers of good sisters must also be good enough to use extensively without actual progeny tests. Such sib-tests are better than none, and they do give a guide to selection of cockerels for testing, but, as the record in Table 16-1 shows, three full brothers with identical sib-tests can yield remarkably different results in their progeny.

Finally it should be made clear that the progeny-tester can evaluate his desirable animals by their phenotype, their own performance, and their pedigree, just as does the mass selector. In addition, the progeny-tester goes one step further and withholds final evaluation until he has seen how the offspring perform. In some respects we can think of progeny-testing as a test-cross with many genes involved. It is an attempt to appraise the genotype.

Progeny tests of females are feasible in fowls, but less so in large mammals. By mating the same pen of pullets to two or three cockerels in succession in one breeding season, and then testing five or six daughters of each female by each sire, one has a diallel test by which the breeding worth of the females can be evaluated. With such a system of "multiple shifts," appropriate intervals must be left between sires to permit accurate identification of paternity. It is desirable that the dams' families to be compared have approximately equal representation of daughters by each sire. As the second and third shifts are hatched later than the first one, the daughters in all three are automatically subjected to different environments, a factor improving the value of the test.

An indispensable part of good progeny tests of sires is that, for those to be compared, their daughters should be either (a) all compared in a uniform environment, or (b) numerous enough and well enough dispersed to sample several different environments. After belaboring in the previous chapter the importance of interaction between heredity and environment, further emphasis on that score is hardly needed here.

Criticisms of Progeny-testing

Some statistical vilipenders have criticized progeny-testing on the ground that if one must wait to test sires, the interval between generations is thereby lengthened; hence progress in improvement is slower than it might be if only young sires were used. The theory is that the young sires this year should be better than those of last year; hence the former should be used in preference to the latter, and using old sires of four years ago would be a sheer waste of time. So much for theory.

While this may be true enough in unimproved stocks where a little selection should make each generation better than its predecessors, the fact remains that, in the highly improved strains of cattle, fowls, and other animals available today, performance is so high that each new generation is only very slightly better, if at all, than the previous one. This year's outstanding proven sire may be better than any uncovered by progeny-testing in the next three years. At current high levels of milk production and egg production, the important thing is not to get another generation quickly, but to find better sires. Ways of doing so at younger ages are now being studied.

Consistent differences between proven sires and untested cockerels in performance of their daughters were found in White Leghorns at Cornell University. The birds were bred primarily to live and to lay. Over a period of 15 years, the proven sires begot daughters that laid better, with less mortality, than those of the cockerels (Figs. 16-6 and 16-7). There is nothing wrong with untested cockerels (or other sires) except that about 80 per cent of them are not worth using again. The remaining 20 per cent are sometimes worth using as proven sires as long as they can reproduce. So far as the interval between generations is concerned, it is worth noting that, with respect to egg production, the proven sires of 1952 begot better daughters than those from the cockerels in the next five years.

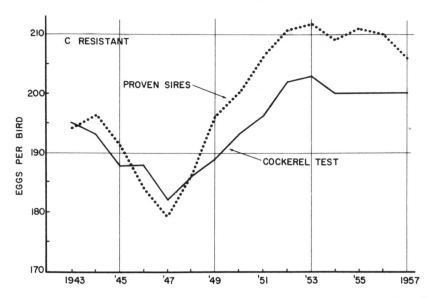

FIG. 16-6 Average egg production to 500 days of age for surviving daughters of proven sires and of cockerels in the C-Resistant White Leghorns undergoing tests, showing superiority of the former during 12 of the 15 years; graphs smoothed by a three-year moving average except for terminal points. (Data of R. K. Cole and F. B. Hutt at Cornell University.)

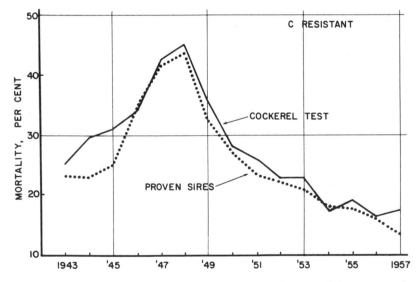

FIG. 16-7 Mortality to 500 days in the White Leghorns of Fig. 16-6, showing consistently better viability in daughters of proven sires than in those of cockerels under test. Each year the proven sires were selected (for viability and egg production of their daughters) from the cockerels tested in previous years. (Data of R. K. Cole and F. B. Hutt at Cornell University.)

A more valid criticism of reliance on progeny-testing is that, as proven sires get older, they tend to become less fertile, and, in some species, difficult to handle. As a result, their effective life is not as long as the breeder would like after he has gone to the trouble and expense of proving them.

Complications in Selection

Any inexperienced geneticist seeing the steadily rising curve showing how selection has increased the number of bristles on the abdomen of fruit flies, or the size of mice, might think that breeding for quantitative characters is simple and rewarding. The geneticist who has tried it with domestic animals knows that breeding for characters of economic value is beset with complications, and is often frustrating. The most common difficulty is that he must select not just for one objective, but for several or many at the same time. Cows must give not only a lot of milk, but also (until recently, at least) a high level of fat in that milk. The poultry breeder can hardly number all his objectives on the fingers of both hands.

If we assume for simplicity that the breeder will use as superior sires the best 25 per cent of those tested, the chance of any one sire's qualifying with respect to one objective is $\frac{1}{4}$. With three independent objectives, such as egg production, viability, and hatchability of fertile eggs, the chance of any one sire's being good in all three becomes $(\frac{1}{4})^3$ or 1 in 64. Even if the breeder is willing to re-use any sire above average, he would have to test 16 to find one that qualified with respect to four different objectives, and that one might die before the second breeding season.

Not all objectives of the breeder are independent. It would be nice if the correlations between them were all positive, so that cows with high yields of milk would also have high levels of fat in that milk, or good layers would lay bigger eggs than poor ones. Unfortunately they don't do so. Many of the correlations are negative, so that while selecting for high productivity in one measure, we must compromise in another.

Because of multiple objectives, and frequent negative correlations between some of them, breeders must test many young sires to find a few good ones. The problem is complicated, in some species at least, by wastage or attrition by death or infertility among those tested before they can be used very long.

Some geneticists, seeking to overcome complications resulting from several objectives, combine them all in one figure, an index, which is purported to measure the net breeding worth of the animal being appraised. Total yield of butterfat by a cow is a simple index. Its value depends not only on how much milk the cow gives, but also on the richness of that milk. Poultrymen talk of a "production index" or a "hens-housed" average, which is the same thing. It gives the average number of eggs laid per pullet housed. If most of them live and lay well, the index goes up, but it goes down if many of them die before the test finishes, and, also, even if they all live but lay poorly. In really complicated indices the various components are weighted according to their heritability and economic value. There are some arguments in favor of indices and some against them, but there is no need to enter that controversy here.

Effectiveness of Selection

Much discussion has been devoted to the comparative effectiveness of different procedures for breeding more productive animals, and to speculation about the limits to which productivity might be raised. The rate of improvement is dependent in large measure upon the degrees of *selection pressure,* a force which can be measured in different ways. If a flock average to 500 days of age is 220 eggs, and the hens used for breeding have an average of 260, the *selection differential* is 40 eggs. Such selection pressure is greater than one could get with breeders at the 240-egg level, and, obviously, if the hens used to produce the next generation were only 220-egg birds, there would be no selection pressure whatever, and one could hope for no higher production except insofar as it might be raised by bettering the environment.

One can also measure the selection pressure by the proportion of one generation used to produce the next. For example, if the hypothetical flock considered in the paragraph above consisted of 1,000 hens, the selection pressure would be greater if one used only the best 100 birds than if the top 300 were used. Similarly, if 20 males are tested, but only the best two of them are used as proven sires to beget the next generation, the selection pressure is greater than if the best five, or eight, were used.

It follows from all this that the amount of selection pressure possible with each generation depends in large measure upon the number of animals available from which the breeder can choose the ones to produce the next generation. If only ten sires are tested each year, and five proven sires are needed to beget the next generation, to get those five the breeder will have to use every male tested that is better than average. There is not much selection pressure in such a program. If he could test 30 sires a year, and use the top five of those, he should make better progress. Better facilities for testing sires have enabled big breeders (including centres for artificial insemination of cattle) to produce better stock than the small-scale operator can hope for.

The effectiveness of selection is also influenced by the degree to which the characters sought are influenced by the environment. If a flock or herd gets a very poor diet, or too little feed, or is stricken with disease, the animals in it cannot show their genetic potential for producing eggs or milk. If every animal lives and thrives, one cannot select for resistance to disease. Even with an optimum environment some of the variation is genetic and some attributable to the environment.

Heritability

In order to predict the rate of improvement in the next generation, statistical geneticists have attempted to determine how much of the total variation of a trait in a population can be attributed to heredity and how much to environment. *Heritability* (h^2) is an estimate of that part of the total variance which can be attributed to the cumulative (additive) action of genes, *i.e.,* the ratio of the additive genetic variance to the total phenotypic variance. The various methods for estimating h^2 are based on a comparison of the increased resemblance for the trait in question among relatives compared to the resemblance among unrelated individuals. The higher the h^2, the greater the increased resemblance of relatives. We will not go into the various statistical methods used to estimate h^2. When h^2 is used to predict improvement, the predicted response to selection (genetic gain) in each generation is calculated as the product of the h^2 times the selection differential. If h^2 is not known, but the response to selection and the selection differential are, h^2 can be estimated by dividing the response by the selection differential. When heritability is so estimated, it is sometimes referred to as *realized* heritability. Heritability estimates are, strictly speaking, only valid for a particular population (herd or flock) at a particular time; they are based on genetic and environmental components of variance which vary from generation to generation and population to population.

Representative estimates of h^2 are given in Table 16-2. In general, h^2 for traits which are measures of productivity (such as production of milk or eggs) are lower than for traits related to conformation (such as body length or mature size) and higher than for traits involving reproductive performance (such as calving interval and number of offspring born). (The higher the h^2, the more an animal tends to "breed true" for the trait in question.)

Heritability estimates can also be used to predict the probable

TABLE 16-2

Representative Estimates of Heritability in Domestic Animals, from Many Sources

Characteristic	Estimate of heritability (h^2)
Dairy cattle	
Mature size	.4
Milk production	.3
Butterfat, per cent	.6
Type score	.2
Beef cattle	
Weaning weight	.3
Weaning conformation score	.3
Yearling weight	.5
Calving interval	.1
Sheep	
Birth weight	.3
Yearling weight	.4
Clean fleece weight	.4
Number of lambs born	.1
Horses	
Heart girth	.3
Height at withers	.6
Trotting speed	.4
Fertility	.1
Swine	
Litter size at weaning	.1
Growth rate	.3
Length of body	.5
Back-fat thickness	.5
Fowl	
Body weight	.4
Egg weight	.6
Egg production	.2
Total mortality	.1

breeding value of an individual. If the average phenotypic value for a herd or flock is equal to the average genotypic value, the phenotypic superiority of an individual above the average times h^2 is an estimate of his genotypic superiority. This estimate of genotypic superiority added to the average of the herd or flock is an estimate of breeding value (sometimes referred to as probable breeding value).

If h^2 is high, the correlation between genotype and phenotype is high, and the phenotype is closer to the breeding value than if heritability is low. With lower heritability, aids to selection such as progeny tests are of increasing importance to determine breeding value. Of course, if progeny tests are being used anyway, the information they provide about breeding values for all traits is available to increase the accuracy of predicting response to selection.

It is important to remember that the aforementioned estimates of breeding values are intended to measure only the *additive* genetic value. Nonadditive gene action is important for some traits, especially for those traits that are related to reproductive performance and viability. To improve performance for such traits, it is important to consider which specific individuals, lines, or breeds combine best to produce the most and the best F_1 progeny. *Special breeding value* which depends upon nonadditive gene action is sometimes distinguished from *general breeding value* which depends upon additive gene action only. We will return to a consideration of combining ability in Chapter 17.

Limitations

If we omit speculation about possible limits to productivity of farm animals, there is one limitation of which animal breeders should be aware, even though it is of less concern to them than to plant breeders. In highly inbred lines, the individuals become increasingly homozygous with respect to all their genes. When complete homozygosity is reached, further selection in such a *pure line* is ineffective, and any variation in the population is attributable only to environmental influences. This is considered further in the following chapter.

Control Populations

When improvement accompanies selection for some character, one might wonder to what extent that change has resulted from the selection practised, and how much of it has resulted from improvement of the environment. The question can be answered either by maintaining an unselected control population, or by selecting simultaneously in two directions. An example of the latter method was provided by MacArthur's mice (Fig. 16-8) in which, after 21 generations of selection, males of the large line were 3.3 times as big as those in the small line. One can think of better health, diets, etc., that might be conducive to good growth and large size, but it is difficult to see how

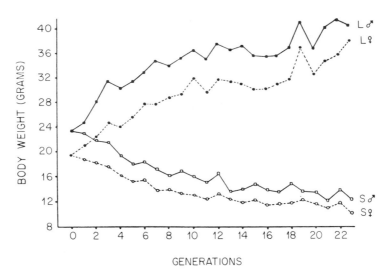

FIG. 16-8 Results of simultaneous selection for large mice in one line and small ones in another. In the 21st generation, males of the large line were 3.3 times as heavy as those in the small line. (From J. W. MacArthur in *Genetics*.)

any such environmental effects could increase size in one line, and, at the same time, decrease it in the other. It is clear, therefore, that MacArthur's selection was effective. The fact that by selection mice could be made either increasingly bigger or progressively smaller provided evidence that body size in the mouse is determined in large part by many genes with small effects.

Control flocks and herds have been much discussed by animal breeders, and at least two such flocks have been established, both in poultry. When unselected individuals in a population are allowed to mate at random, with each one contributing an equal number of offspring to the next generation, the proportions of the various genes in that next generation remain the same as in the previous one. In other words, the population does not change genetically. In such a *random-bred* flock, any change in productivity in succeeding generations would have to be ascribed to the environment. In any one year, the difference between performance of a selected strain and that of the control flock could safely be credited to genetic differences. Gowe was thus able to satisfy himself that the steady improvement of his "New" strain of Leghorns, when it was compared with his control flock, showed that the method of selection practised was effective in raising egg production (Fig. 16-9).

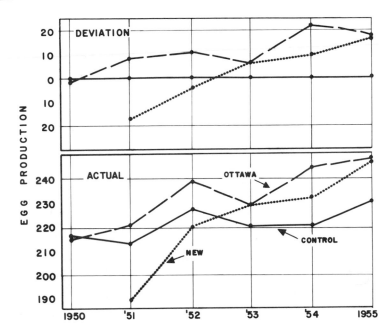

FIG. 16-9 By holding the annual performance of the control strain at zero (above), the deviations from it show that selection had improved the "New" strain but had less effect on the Ottawa strain of fowls. From the lower chart, one could not tell how much of the rise in egg production should be credited to the environment and how much to breeding. (From R. S. Gowe in *Research for Farmers.*)

One difficulty about control strains is that, although they are commonly considered to represent a population breeding at random, elaborate precautions have to be taken to make sure that each male and each female in the control flock does actually contribute equally to the next generation. Differences in viability and in ability to reproduce, along with chance fluctuations in sex ratios, all cause problems when one seeks to maintain uniformity. For large domestic mammals it has been estimated that 50 females and 10 males would be needed to maintain such random-bred controls. Few breeders could afford the facilities and expense needed for such a venture. Gowe maintained his control flock of Leghorns with a little under 200 females and 50 males.

Altogether, the maintenance of such random-bred controls, however desirable, would seem to be an enterprise for co-operative support by several agencies and perhaps by governmental experiment stations, rather than for individual breeders.

Effect of Random Breeding on a Population

To prove the validity of the statement in the previous section that random mating with equal reproduction does not change the proportions (or frequencies) of genes in successive generations, let us consider a very simple example. If an F_2 generation were raised from an original cross of $AA \times aa$, the proportions of the genotypes in that F_2 population would be:

In males: $1AA : 2Aa : 1aa$
In females: $1AA : 2Aa : 1aa$

Let us assume that every male mates with every female and that four offspring are produced from each mating. Results will be as shown in Table 16-3.

It is clear that with this system the ratio of genotypes in F_3 is exactly the same as it was in the F_2 generation. The same would apply in F_4 and succeeding generations. Since each pair of alleles should behave as did our sample pair, A-a, the proportions of all genes in successive generations of random mating should remain unchanged. In actual practice it may be influenced by unequal reproduction of differing genotypes, by natural selection which favors some genotypes at the expense of others, and by chance fluctuations in small populations whereby some genotypes are not reproduced.

If we consider genes rather than genotypes (*i.e.*, the frequency of A and a in each generation) and remember that the genotypes are in the ratio of $1 AA : 2 Aa : 1 aa$, the frequency of A alleles is $(2 + 1)$ in six chromosomes altogether, or ½, or 0.5. Similarly, the frequency of a is $(1 + 2)$ in the six chromosomes, or 0.5, as for A.

Random Breeding, With Elimination of the Recessive

Now let us consider a slightly different system of breeding. If we have random mating as before, but happen to encounter in the F_2 generation some undesirable recessive mutation—perhaps a lethal one—that recessive would be eliminated. Only the homozygous normal and the heterozygotes would be allowed to reproduce.

Let us assume that the recessive to be eliminated is the gene b, which, when homozygous, makes Holstein-Friesians red where black is preferred. Assume further that we start again with an F_2 population in which the genotypes are in the ratio: $1 BB : 2 Bb : 1 bb$. With the bb animals eliminated, we again allow four offspring from all possible matings among the black animals. They should be in the ratio of

TABLE 16-3

Results in the F_3 Generation from Random Mating and Equal Reproduction of F_2 Parents from a Monohybrid Cross

Parents in F_2			Progeny in F_3		
Males	Females	Matings (number)	AA	Aa	aa
1 AA	1 AA	1	4	—	—
	2 Aa	2	4	4	—
	1 aa	1	—	4	—
2 Aa	1 AA	2	4	4	—
	2 Aa	4	4	8	4
	1 aa	2	—	4	4
1 aa	1 AA	1	—	4	—
	2 Aa	2	—	4	4
	1 aa	1	—	—	4
	Totals:		16	32	16
	Ratio of genotypes:		1	2	1

1 *BB*:2 *Bb*, but, of course, we have no way of knowing which are which. Following the procedure shown in Table 16-3, the ratios of genotypes in successive generations prove to be as given in Table 16-4.

On plotting the frequencies of red calves in successive generations one finds that although the red animals were systematically eliminated in each of them, they were still cropping out at the rate of

TABLE 16-4

Ratios of Genotypes in Successive Generations of Random Mating but with Elimination of Recessives, *bb*

Generation	BB	Bb	bb	Proportion bb in Population
F_2	1	2	1	1 in 4
F_3	4	4	1	1 in 9
F_4	9	6	1	1 in 16
F_5	16	8	1	1 in 25
·	·	·	·	·
·	·	·	·	·
·	·	·	·	·
F_n	$(n-1)^2$	$2(n-1)$	1	1 in n^2

one per cent in the tenth generation (Fig. 16-10). More important, among 99 black animals of that generation, 18 per cent still carried gene *b*. As this kind of selection is the same as that resulting with any lethal character, it is clear that, if left to themselves, genes causing such defects can persist long after their presence is first noted.

An important point to note is that, when the red calves are eliminated in each generation, the frequencies of the alleles *B* and *b* do not remain equal. In the tenth generation the genotypic ratio is: 81 *BB* : 18 *Bb* : 1 *bb*; hence the gene frequencies in that population are:

For *B*: (81 × 2) + 18 = 180 in 200, or 90 per cent, or 0.9
For *b*: 18 + 2 = 20 in 200, or 10 per cent, or 0.1

The Hardy-Weinberg Law

We have just seen (Table 16-3) that with random matings the ratio of genotypes in successive generations remains the same when the frequencies of the two alleles of a pair were equal (0.5) to start with. We have also seen that even if two alleles were once equal in frequency, selection (in our case, continuous elimination of one

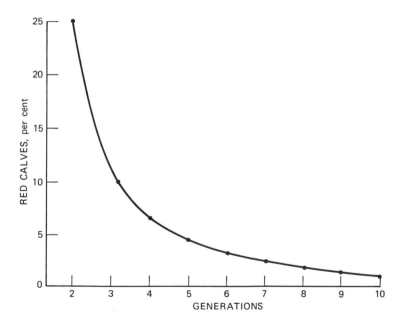

FIG. 16-10 Reduction in the frequency of a recessive character during ten generations of selection in which animals showing the character are excluded from reproduction.

homozygote) can radically change their frequencies in subsequent generations.

Now we consider cases in which the two alleles are not necessarily equal in frequency. So long as mating is at random, their separate, differing frequencies should be maintained in subsequent generations. The principle involved is commonly called the Hardy-Weinberg law, after the two men, Hardy in England and Weinberg in Germany, who recognized it independently in 1908.

Instead of giving gene frequencies numerically as we have been doing in recent paragraphs, they can be designated algebraically as p and q. As all the genes at one particular chromosomal locus in a population must add up to 100 per cent, or 1.0, it follows that $p + q$ must equal 1. Let us assume that we are dealing with a pair of alleles B and b with frequencies of p and q, respectively. We could make a little four-fold Punnett square to work out all possible combinations of each sex being p B and q b. We could find exactly what the Hardy-Weinberg law tells us, that the frequencies of the resulting genotypes are:

$$p^2 \, BB + 2 \, pq \, Bb + q^2 \, bb$$

By substituting the appropriate values for p and q, the relative frequencies of the three genotypes can be determined.

The formula is useful in two distinct ways. In the first place, if a random sample of a population shows that with respect to B and b the three genotypes are in the relative frequencies just given, then we have reason to believe that reproduction is at random, without selection for or against any one genotype. Such a population is said to be in equilibrium. On the other hand, if there are more or fewer of any one type than our formula calls for, one would suspect that there is not random mating, or that there is preferential selection of some kind, perhaps natural selection, which prevents the genotypes from contributing equally to the next generation.

In the second place, with completely recessive traits the heterozygotes cannot be identified, but we might want to know their frequency among the (in this case) indistinguishable $BB + Bb$ genotypes combined. This we can do, so long as the frequency of the bb type is known. If there are $q^2 bb$, we can find q, and hence p, and thus work out the expected frequencies of the BB and Bb genotypes.

London's Cats Obey the Hardy-Weinberg Law

Among 691 cats destroyed in 1947 at three animal clinics in London, Searle found the frequency of orange (*OO* or *O*—), tortoise-shells (*Oo*), and others (*oo* or *o*—), to be as follows:

	Not Yellow	Tortoise-shell	Orange	Total
In females	277	54	7	338
In males	311	—	42	353

Recalling that O and o are sex-linked, we remember that there should be no tortoise-shell males. In this case there was one, but consideration of his kind will be left to a later chapter. With females having two sex chromosomes, and males one, the total number of genes at the O locus in this population was $(388 \times 2) + 353$, or $1,029$.

The frequency of O was $54 + (2 \times 7) + 42$, which is 110, or 10.7 per cent of the total. The frequency of o must, therefore, have been $(100 - 10.7) = 89.3$ per cent.

In terms of p and q, as used in the Hardy-Weinberg formula, these frequencies can be expressed as follows:

$$p = 0.893; \quad q = 0.107; \quad p + q = 1.0$$

Substituting these values in the formula, and multiplying expected frequencies by the number of cats in the sample, one can determine whether or not matings were completely at random with respect to color. As males cannot be heterozygous, the Hardy-Weinberg formula can be fitted only to the 338 females. (The males were useful by enlarging the sample from which gene frequencies were determined.)

Expected frequencies: $p^2 oo + 2\,pq\,Oo + q^2\,OO$
Expected female cats $= 338\,[0.893^2 + 2(0.893 \times 0.107) + 0.107^2]$
$ = 269.5 + 64.6 + 3.9$
or, in whole cats: $\quad$ 269 not orange, 65 tortoise-shell, 4 orange

Comparing observed numbers with those expected, we see a slight deficiency of tortoise-shells, and two or three oranges above expectation, but these deviations could occur by chance.

The frequencies observed are thus in accord with those expected if:

1. All three genotypes are equally viable and fertile.
2. The owners of the cats had either no preferences for colors or no power to enforce them.
3. The feline gentlemen of London do not prefer blondes.

Prospective Veterinarians and the Hardy-Weinberg Law

In the foregoing section, we were able to determine the gene frequencies for O and o in London's cats directly from the phenotypes. It was not necessary to estimate what proportions of those carrying O were homozygous or heterozygous, because the orange cats told us about one genotype and the tortoise-shells about the other. In the many cases where heterozygotes and dominant homozygotes are quite indistinguishable, their frequencies in a population can be calculated from the Hardy-Weinberg law, so long as the frequency of homozygous recessives is known.

Let us try this with an interesting genetic character in man—the ability to taste phenylthiocarbamide, commonly called PTC. For about 70 per cent of North Americans it has a bitter taste, but about 30 per cent of us cannot taste it at all. Genetic studies have indicated that the latter are homozygous for a recessive autosomal gene, t, and that tasters are therefore TT or Tt. The situation may not be quite so simple because there seem to be differences in the sensitivity of some people to different concentrations of PTC.

Over a period of ten years, members of the senior author's class of prospective veterinarians enlivened their Easter vacations by testing ability to taste PTC of their fathers and their mothers, their brothers and their sisters, their uncles—and so on. Among all those, the proportion of non-tasters in ten years was 478 among 1,612, or 29.71 per cent, practically identical with that found in other samples of the population. This means that 70.29 per cent were tasters, but, if a single dominant major gene is responsible for that ability, some of those tasters were TT and others Tt. Let us estimate the probable frequencies of these two genotypes.

Since people do not ordinarily select their mates for the ability to taste PTC, marriages can be assumed to be at random with respect to that character. The frequencies of the three genotypes should therefore be: $p^2\ TT : 2\ pq\ Tt : q^2\ tt$.

Taking q^2 as 0.2971 (from its observed frequency), q (the frequency of t) = $\sqrt{0.2917}$ = 0.545. Therefore p (the frequency of T) = $1 - q$ = 0.455, and, substituting in the Hardy-Weinberg formula, the frequencies of the three genotypes are calculated to be

$$0.2070\ TT : 0.4959\ Tt : 0.2971\ tt$$

So much for theory. How good is it? We have a way to find out. From the pedigrees of PTC-tasting brought back by the students, all families in which both parents and some children had been recorded were sorted into the three types possible with respect to PTC (Table 16-5)

No Mendelian ratios are to be expected in Classes 1 and 2 because the parents include mixed genotypes. Thus, with parents both tasters (Class 1), one, both, or neither could be heterozygous. Theoretically, if inability to taste PTC goes with the genotype tt, parents both of that type (Class 3) would have only non-tasting children. As this fact gradually sank in during the analyses of the data in class, and sidewise glances were being directed toward those who had reported exceptions, it became necessary to point out some of the possible sources of error in classification. If we reject entirely the crutch of illegitimacy on which some geneticists lean heavily when awkward exceptions must be explained, and even allow for the differing sensitivities mentioned earlier, most of the exceptions could be accounted for by the almost incredible zeal of the investigators, some of whom (presumably in the absence of the mothers) had administered bits of the test paper to infants six months old, and estimated results from the ensuing facial expressions or expostulations!

Let us now consider Class 2, the marriages of taster × non-taster. Our earlier calculations have told us that among the tasters the ratio of genotypes should be 0.2070 TT : 0.4959 Tt. If we assume that these proportions should apply even to the comparatively small sample of tasters in Class 2, and allow two children per family, the proportion of non-tasting offspring theoretically expected in this class can be determined as follows:

	Offspring	
Parents	Tasters	Non-tasters
0.2070($TT \times tt$) × 2	0.4140	0
0.4959($Tt \times tt$) ×2	0.4959	0.4959
	0.9099	0.4959

The expected proportion of non-tasters among the offspring in Class 2 is thus found to be 0.4959 / 1.4058 or 35.3 per cent. This is a very good fit to the 36.7 per cent actually found (Table 16-5). That, in turn, means that the proportion of heterozygotes among the tasting parents in the taster × non-taster marriages was accurately estimated from the frequencies of genes (and genotypes) determined by the Hardy-Weinberg formula. The sample from which those frequencies were calculated numbered 1,612, but the sample of tasters in Class 2 on which the expected frequencies of TT and Tt genotypes were tested

TABLE 16-5

Ability to Taste PTC in Families of Veterinary Students at Cornell University

Class	Parents	Tasters (number)	Non-tasters Number	Non-tasters Per cent
1	Taster × taster	654	76	10.4
2	Taster × non-taster	354	205	36.7
3	Both non-tasters	7	98	93.3

numbered only about 124. The fit of observation to theory is therefore all the more remarkable.

Altogether, these findings, including the gene frequencies for T and t, are much the same as those determined by others from different samples of the population. Admittedly, no special satisfaction is to be derived from the fact revealed by PTC and the Hardy-Weinberg formula that prospective parents of prospective veterinarians, like the rest of us, evidently do not consider ability to taste PTC important when a mate is to be selected. In other cases, the formula can be very useful. For example, if a flock, herd, or breed is regularly showing some recessive lethal character or defect, when its frequency is known, one can determine the proportion of the normal animals that carries the unwanted gene, and take appropriate steps to eliminate it.

Populations Differing or Changing

Geneticists who study the genes in different races of man find remarkable variations in the frequencies of many of them, and biologists are much interested to know how such differences have arisen. Racial differences in the frequencies of genes determining blood antigens are particularly conspicuous. Some genotypes are favored by natural selection, thus leading to frequencies of certain genes considerably higher than would be expected with random matings and a population in equilibrium.

Noticing that the recessive gene causing a certain abnormal type of hemoglobin (called S) were higher in areas subject to endemic tertian malaria than elsewhere, Allison postulated that people heterozygous for the hemoglobin S are more resistant to malaria than either homozygote. Homozygotes for the abnormal hemoglobin die of sickle-cell anemia (usually early in life), but because heterozygotes have a selective advantage over the homozygous dominant type, the gene causing the anemia persists in malarial regions at an unusually high level. We cannot be diverted here into those interesting variations in the frequencies of genes in different human populations. For a lucid summary, readers could consult Stern's book. Similar differences are now being found among breeds of domestic animals. Some of these, such as genes affecting color, horns, and capacity to produce milk or eggs, are conspicuous, but others causing biochemical differences in the milk and blood are also being investigated.

Although he may not think in terms of genes, what the breeder of domestic animals actually tries to do is to change the existing frequencies of genes in his flocks or herds in such a way that they will raise the average productivity of his stock. Much of this change necessitates the accumulation of desirable genes influencing quantitative characters. The problem is not so much to make the best animals better, as to increase the proportion of the flock that can come close to the best.

Problems

16-1 Can you think of any other cases like that of *Biston betularia* in which species in the wild have developed colors or patterns that help to protect them against predators?

16-2 Can you think of other examples like those given in this chapter in which natural selection has acted on domestic animals in such a way as to improve the biological fitness of the species?

16-3 What instruments of natural selection (weather, disease, predators, etc.) have you seen operating on domestic animals to lessen the contribution of the unfit to the next generation?

16-4 When you go to a dog show, you must agree with Darwin that variation is much greater under domestication than in the wild, but how do you account for that fact?

16-5 Do you know of any breed of dogs in which the color, coat, or conformation has changed within the last 50 years because of changing preferences of the breeders? This is a good one to discuss with elderly dog fanciers!

16-6 In genetic terms, how would you explain the differences among the three brothers of Table 16-1 in their progeny tests? Remember that the mothers were the same with all three.

16-7 Using the procedures shown in Tables 16-3 and 16-4, work out the expected proportions of heterozygotes in each of ten generations from a monohybrid F_1 if all the recessive homozygotes are eliminated from reproduction. Plot the frequencies (proportions) of heterozygotes as in Fig. 16-11.

16-8 Without using the general formula, work out the frequencies expected for F_6 in Table 16-3.

16-9 Searle found the frequencies of cats in Singapore to be:

	Not Yellow	Tortoise-shell	Orange
In females	63	55	12
In males	74	—	38

Using the same procedure as was followed for those in London, determine the frequencies of the O and o alleles, and whether or not matings with respect to these were at random in Singapore.

16-10 How would you account for the differing freqencies of the O-o alleles in London and Singapore?

16-11 Searle found that 69 per cent of the Singapore cats had kinky tails, an abnormality not seen in the London cats. How could that happen?

16-12 Let us suppose that two ships are wrecked in the South Pacific and that the survivors of one, taking with them their two ship's cats, establish a community on a hitherto catless island. Survivors of the other ship do the same (independently) on another tropical isle, 1,000 miles distant. If the cats taken from one ship were a black tom and an orange female, and from the other ship an orange tom and a black female, what gene frequencies of O and o would you expect on the two islands two years later?

16-13 Having worked out Problem 16-12, do you now see how genetic "drift" (selection of certain genes in small samples, with their subsequent multiplication) might be a factor in evolution?

16-14 In Wentworth's compilation of data on coat colors in Shorthorns, there were 4,169 red, 3,780 roan, and 756 white. Sewall Wright showed that these colors could be accounted for by a single pair of alleles, which we have been designating as N and n.

Assuming that roans are Nn, and, hence, that the frequencies of N and n can be determined directly from those of the phenotypes, (a) find them, (b) work out the ratio of genotypes expected if matings were at random, (c) multiply each by the total number (8,705), and (d) compare resultant expectations with the observed numbers. You could even test how well they fit with χ^2.

16-15 If the answers to Problem 16-14 show that the observed numbers do not fit the expectation with random matings, what are the most likely reasons for the deviations? Apart from the obvious one, there is "white-heifer disease," about which we may hear more later.

16-16 If, in many large herds of Holstein-Friesians combined, the frequency of red calves in one year is one per cent (0.01), what are the frequencies likely to be of the B and b alleles, and what proportion of the blacks is likely to carry red?

16-17 The proportion of Rh— people *(rhrh)* in the whites of North America is 15 percent. Assuming that mates are not selected in man for their blood antigens, what is the chance for the average Rh— girl that her husband will be (a) *rhrh*? (b) *Rhrh*? (c) *RhRh*?

Selected References

DARWIN, C. 1860. *The Origin of Species by Means of Natural Selection, or the Preservation of Favored Races in the Struggle for Life.* 1st ed. New York. D. Appleton & Co., Inc. 6th ed. is reprinted in The Home Library, A. L. Burt Co. Many other printings. (The starting point for students of evolution.)

FISHER, R. A. 1930. *The Genetical Theory of Natural Selection.* Oxford. The Clarendon Press. (A standard authoritative book in this field; five chapters devoted to man.)

GOWE, R. S., A. S. JOHNSON, J. H. DOWNS, R. GIBSON, W. F. MOUNTAIN, J. H. STRAIN, and B. F. TINNEY. 1959. Environment and poultry breeding problems. 4. The value of a random-bred control strain in a selection study. *Poultry Sci.* **38:** 443–62. (With many examples of its use.)

JOHANSSON, I., and J. RENDEL. 1968. *Genetics and Animal Breeding.* Edinburgh and London. Oliver and Boyd. (See Chapters 15, 16, 17 on selection in domestic animals.)

KETTLEWELL, H. B. D. 1956. Further selection experiments on industrial melanism in the Lepidoptera. *Heredity* **10:** 287–301. (Details of the experiments discussed briefly in this chapter.)

NEEDLER, A. W. H., and R. R. LOGIE. 1947. Serious mortalities in Prince Edward Island oysters caused by a contagious disease. *Trans. Roy. Soc. Canada,* 3rd Ser., Sect. V, **41:** 73–89. (With ample evidence that natural selection produced genetically resistant oysters.)

SEARLE, A. G. 1949. Gene frequencies in London's cats. *J. Genetics* **49:** 214–20. (For Singapore's cats, see *ibid* **56:** 111–27, 1959.)

STURKIE, P. D. 1950. Further studies of autosomal nakedness in the domestic fowl. *Amer. Naturalist* **84:** 179–82. (Conspicuous effects of eight generations of natural selection on a character originally lethal.)

WRIGHT, S. 1917. Color inheritance in mammals. VI. Cattle. *J. Heredity* **8:** 521–27. (Analysis of reds, roans, and whites in Shorthorns with data used in Problems 16-14 and 16-15.)

CHAPTER

17

Inbreeding and Hybrid Vigor

To most animal breeders, "inbreeding" is a nasty word. It is desirable that we explore a little to see why it has that reputation, and perhaps to find out whether or not anything good can be said for it. Whether the breeder is trying to produce better cows or better cats, his problem is somewhat the same in most cases. He selects for a desired type or performance, and, after a few generations of such selection, builds up a flock, herd, or kennel in which most of the animals conform to his specifications. By that time many of them are related. The breeder knows about the risks of inbreeding, and he hesitates about what to do. He would like to continue with his own stock, if it were not for the risk that inbreeding might lead to some degeneration of it. If he introduces some new blood, that would overcome the danger of inbreeding, but the new stock might bring with it undesirable genes that could spoil the fruits of all his years of selection. The breeder has heard about a way of inbreeding called *line breeding*, which is said to be less risky than close inbreeding, but he is not sure how to proceed with it in order to get good results without any bad ones.

Increase of Homozygosity With Inbreeding

The general effect of inbreeding is to increase the number of gene pairs that are homozygous. It is this very process that has given inbreeding a bad name. At many of the loci there are lethal genes, or others causing serious defects, as we should know from the previous chapters of this book. Homozygosity at such loci results in defective animals, and the breeder sadly blames himself for having mated together animals that were too closely related.

Let us see how fast the homozygosity is increased with differing degrees of inbreeding. The closest degree of inbreeding, and hence the most rapid approach to complete homozygosity, is found in organisms capable of self-fertilization. There are very few of these in the animal kingdom, and none among domestic animals (except for a special case coming up in the next chapter), but in the plant world there are many. Oats, wheat, peas, beans, barley, and tomatoes are all examples of self-fertilizing plants. It is well to remember at this juncture that most varieties of them are vigorous, even though so highly inbred that they should be completely homozygous.

We can see how rapidly homozygosity is increased in such an organism by considering the hypothetical case of a single pair of alleles in a self-pollinated plant heterozygous at one locus. If such a plant produces four offspring, the ratio in these, if the original heterozygote were Aa, must be $1\ AA:2\ Aa:1\ aa$. In an F_3 generation similarly produced by self-fertilization of each plant in the F_2, and again allowing four offspring from each of those F_2 plants, the ratio becomes $6:4:6$ (Table 17-1). In the first three generations, the proportion of heterozygotes in the population has thus been reduced by one-half in each generation, and, conversely, the proportion of homozygous plants is raised correspondingly. By carrying the process a little farther, we can see that the same result is obtained in each succeeding generation. The proportion of heterozygotes is steadily halved in each, and almost complete homozygosity is attained in a little over ten generations of self-fertilization.

With other degrees of inbreeding the increase in homozygosity proceeds at a somewhat slower rate (Fig. 17-1). Continuous matings of brother × sister for ten generations would result theoretically in a population in which 91.3 per cent of all the originally heterozygous gene loci would be homozygous. It would take nearly 18 generations of such brother-sister matings, which are the closest possible kind of inbreeding in domestic animals, to attain as much homozygosity as one gets in ten generations with a self-fertilizing plant. Sewall Wright calculated that, whereas matings of brother × sister reduce the residual heterozygosity by about 19 per cent in each generation, if a male is mated with his half-sisters, heterozygosity is reduced by only 11 per cent in each generation. This is obviously a much safer process.

TABLE 17-1

Effect of Self-Fertilization in a Plant Originally Monohybrid on Proportion of Homozygotes in Successive Generations (Each Plant is Allowed 4 Progeny)

Generation		AA		Aa		aa	Heterozygotes (per cent)	Homozygotes (per cent)
F_1				1			100	0
F_2		1	:	2	:	1	50	50
F_3		4		0		0		
		2		4		2		
		0		0		4		
		6		4		6	25	75
	or:	3	:	2	:	3		
F_4		24		0		0		
		4		8		4		
		0		0		24		
		28		8		28	12.5	87.5
	or:	7	:	2	:	7		

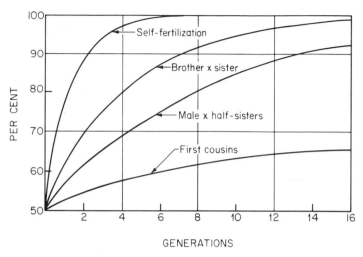

FIG. 17-1 Rates of increase in homozygosity with different degrees of inbreeding. (After S. Wright in *Genetics.*)

Many breeders do not realize that, contrary to common opinion, continued breeding within one flock without introduction of new blood does not necessarily result in rapid inbreeding. The increase in homozygosity in such closed flocks depends largely upon the number of sires used to beget each generation. Wright calculated that when the number of females greatly exceeds the number of males as is usual with most domestic animals, heterozygosity is reduced in each generation by about $1 \div 8N$, when N is the number of males used. Thus, if four males are used per season in some flock or herd, homozygosity is increased by only 3.1 per cent in each generation. If ten sires are used, the figure is reduced to about 1.2 per cent. Gradual approaches to homozygosity like these are not likely to hurt any flock very much if the breeder is careful to exclude from reproduction any animals that are defective, and they may be reduced still more by special care to avoid matings of close relatives. Proof of this fact was provided by the Cornell S strain of White Leghorns. It was reproduced for many years by the annual use of eight males (four proven sires and four untested cockerels), each mated to about 13 females. After 1940, no new blood was introduced. Close inbreeding was avoided. After 30 years as a closed flock, the degree of inbreeding by 1971 was found by Gavora *et al.* (1979) to be a little over 39 per cent. In spite of this, reproduction in the S strain was excellent. In the last two years, average fertility was over 90 per cent, and hatchability of fertile eggs was 89.7 per cent.

General Effects and Risks

The most conspicuous result of close inbreeding is often a lower efficiency of reproduction, together with the appearance of lethal characters or other defects. As the bad genes responsible for these things become homozygous, they are, of course, eliminated. One hears therefore that inbreeding is a method of purifying the stock. Unfortunately, when the bad genes go out, they take a lot of good ones with them, because good genes are inevitably linked with bad ones. The result is that some of the highly inbred lines are poor, and many of them are eliminated by their inability to reproduce. There is usually what some geneticists call *inbreeding depression* of quantitative characters in animals that survive the purifying process. In other words, things like hatchability of eggs, number of eggs laid, yield of milk, litter size, viability, and growth rate of animals are all likely to be depressed. Along with all this there is usually segregation into distinct lines, but increasing uniformity within those lines, as individuals within them gradually acquire more genes common to all.

In spite of the usual inbreeding depression, it has been feasible to breed and maintain highly inbred lines of animals—the products of so many generations of brother-sister matings that they could be considered homozygous at all loci. At the Wistar Institute, Helen Dean King's white rats were put through such a purging of bad genes, and, although many lines were lost, a few came through with remarkable retention of vigor after more than 25 generations of brother-sister matings. The same was done by Wright with guinea pigs, and by others who have developed highly inbred strains of mice. That such mice are indeed highly homozygous has been demonstrated by Ruddle, who has succeeded by special techniques in producing from such lines mice which are the result of zygotes formed by the division of single gametes. Such homozygous mice (which are, of course, homogametic and thus always females) appear to be as vigorous as the other members of their inbred lines.

Although it has been possible to breed highly inbred lines of laboratory mammals, attempts to breed similar highly inbred lines quickly in domestic animals have not been so successful. Continued brother-sister matings in the domestic fowl have led, in most attempts, to extinction of the line within six or eight generations because of failure to reproduce. When inbreeding was carried out using matings that were less close, so that homozygosity was increased more gradually, the reproductive performance in later generations was better.

While inbreeding does help increase uniformity, any animal breeder using it for that purpose should be continuously on guard against reproducing from animals that are lacking in vigor. In dogs and swine small litters would be one early indication of trouble, just as low hatchability of eggs is an early sign that inbreeding in the domestic fowl is too close. In general, the breeder should avoid matings of close relatives and try to keep the level of inbreeding rather low. Finally, we should realize that inbreeding does not by itself create any undesirable genes. It merely reveals those that are latent in the stock.

Pure Lines

We owe this term to the Danish botanist Johanssen. Studying variation in beans, he differentiated 19 lines differing in the average weight of the seed. As the bean plant is self-fertilizing, and because Johanssen kept the progenies in each line separate from all the others, they were undoubtedly homozygous, or what he called *pure lines*. Within each of these there was still considerable variation in the size of seeds, but, when (within lines) single beans differing in size were grown separately, it was found that their progenies did not correspond to the size of the seed planted, but rather to the average for the line. Clearly, the variation in size among the seeds planted was attributable to environmental effects, perhaps, among other things, the position of the seed in the pod. This was confirmed by further attempts to select within the pure lines, all of which were fruitless.

Breeders of domestic animals do not have this problem because, with the exception of laboratory rodents, it is probable that there are no pure lines in domestic animals. However, in closed flocks, in which homozygosity may be somewhat high, occasionally the breeder finds that his selection is less effective than in former years. In other words, he has reached a plateau, above which further progress is difficult. It is sometimes recommended that, when such a stage is reached, the breeder should bring in an infusion of new genes from some other source, or from several, and start over again, but, while this seems sound in theory, the feasibility of thus overcoming a plateau in practice, especially when there are several objectives, has yet to be demonstrated.

Theoretically, it should be feasible to produce pure lines in lower animals that are hermaphroditic, unless, as sometimes happens, there are bars to self-fertilization. In higher animals, genetic uniformity is found in monozygotic twins. In this case, both individuals presumably have the same genes (except for any mutations that may have occurred since they separated), but each is highly heterozygous and neither could breed true as did Johanssen's beans.

Degrees of Inbreeding

Matings of brother × sister, or parent × offspring, constitute close inbreeding. *Line breeding* is a form of inbreeding in which individuals are mated to keep their descendants closely related to some outstanding ancestor or ancestors, but it is ordinarily not close inbreeding. There is not much point in trying to define sharply differing degrees of inbreeding because, as we can perhaps appreciate better from the scale shown in Fig. 17-2, it is as almost as difficult to mark in that scale one point that separates inbreeding from other kinds of matings as it would be to put a point on a thermometer to separate hot from cold.

The degree of inbreeding in any individual (X) may be calculated from a formula devised for that purpose by Wright:

$$F_X = \Sigma\left[\left(\frac{1}{2}\right)^{n + n' + 1}(1 + F_A)\right]$$

in which:

F_X = coefficient of inbreeding of X

n = number of generations from the sire of X back to some ancestor common to both sire and dam

n' = number of generations from the dam of X back to the same common ancestor

Σ = summation (of separate contributions from each different common ancestor)

F_A = coefficient of inbreeding of the common ancestor when that animal is itself inbred

Wright's coefficient is a measure of the degree of relationship between the sire and the dam. That relationship, if there is any, will depend upon how far back in the pedigree those two have a common ancestor. If they have more than one such common ancestor, they are likely to be more closely related than if they have only one, and separate contributions of each must be calculated. These are then added up (Σ) to determine F_X. The ½ part of the formula recognizes that the contribution of an ancestor's genes (blood) to X is halved with each generation that separates them. Finally, the formula becomes less formidable when (as in most cases) the common ancestor is not inbred, and the coefficient is simply $\Sigma\frac{1}{2}^{n + n' + 1}$.

The following examples show how the formula is applied. In each

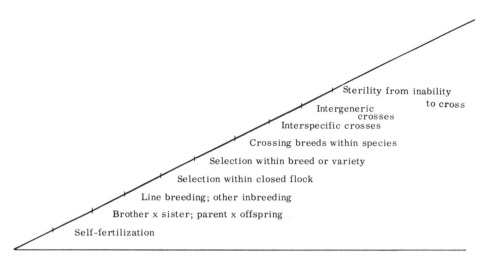

FIG. 17-2 Gradations in the degree of inbreeding with different types of matings.

case the pedigree is shown together with a path (arrow) diagram. To convert correctly a pedigree to a path diagram and then calculate F_X:

1. Draw arrows from parents to offspring (be sure the path diagram indicates that each individual has only two parents, *i.e.*, two arrows point to him).
2. Be sure each individual appears only once in the path diagram.
3. Calculate F_X by counting the paths (generations) from the sire of X back to some ancestor common to the sire and dam of X and then forward to the dam of X. Be sure that no individual appears in any *one* chain of paths contributing to F_X more than once.

I. Mating of Half Sibs to Produce X

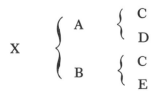

Bracket pedigree

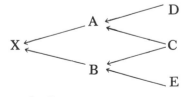

path diagram

One common ancestor: C

$F_x = 1/2^{1+1+1} = (1/2)^3 = 0.125$
Paths: A-C-B
X is 12.5 per cent inbred

II. Mating of Brother × Sister to Produce X

$$X \begin{cases} A \begin{cases} C \\ D \end{cases} \\ B \begin{cases} C \\ D \end{cases} \end{cases}$$

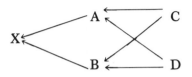

Two common ancestors: C and D

$F_x = 1/2^{1+1+1} + 1/2^{1+1+1} = (1/2)^3 + (1/2)^3 = 1/8 + 1/8 = 1/4 = 0.250$
Paths: A-C-B   A-D-B
X is 25 per cent inbred.

III. Two Generations of Mating Brother × Sister to Produce X

$$X \begin{cases} S \begin{cases} M \begin{cases} A \\ B \end{cases} \\ N \begin{cases} A \\ B \end{cases} \end{cases} \\ D \begin{cases} M \begin{cases} A \\ B \end{cases} \\ N \begin{cases} A \\ B \end{cases} \end{cases} \end{cases}$$

Four common ancestors: M, N, A and B

$F_X = 1/2^{1+1+1} + 1/2^{1+1+1} + 1/2^{2+2+1} + 1/2^{2+2+1} + 1/2^{2+2+1} + 1/2^{2+2+1} =$
Paths: S-M-D   S-N-D   S-M-A-N-D   S-M-B-N-D   S-N-A-M-D   S-N-B-M-D
$1/8 + 1/8 + 1/32 + 1/32 + 1/32 + 1/32 = 3/8 = 0.375$

IV. Matings in Which the Common Ancestor is Himself Inbred

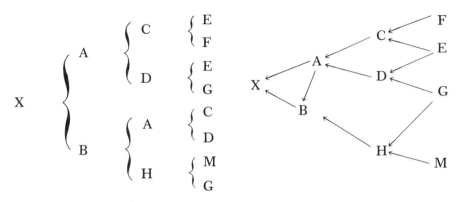

Common ancestor A has the same grandsire (E) on both dam's and sire's side of his pedigree.

$$F_A = 1/2^{1+1+1} = 0.125$$
$$\text{C-E-D}$$

Contribution of A to F_X
$$= (1/2)^{0+1+1} (1+0.125) = (1/2)^2(1.125) = 0.282125$$
$$\text{A-B}$$

Contribution of G to F_X = $\dfrac{1/2^{2+2+1}}{\text{A-D-G-H-B}}$ = 0.03125

$$F_X = \Sigma = 0.3125$$

In determining such coefficients of inbreeding, one must re-member that the numbers of generations back to common ancestors are not counted from the animal for which the coefficient is being computed, but from his sire and his dam. When the sire is himself a common ancestor, as in example IV above, the number of generations back to the common ancestor on the sire's side is 0. The same would apply to the dam if she were to appear on both maternal and paternal sides of the pedigree. It will also be noted in Example IV that, although C and D appear on the dam's side of the pedigree, as well as on the sire's side, calculations are not made for their contributions to the inbreeding of X because those contributions all go through A, and hence are taken into account when his contribution is considered.

Students sometimes forget that an animal may appear twice in the pedigree (as does E in Example IV) without being a *common* ancestor. That adjective means (in this usage) common to both the dam's and the sire's sides of the pedigree.

It is evident that close ancestors affect the degree of inbreeding more than distant ones. For common ancestors back of both parents by 1, 2, 3, or 4 generations the contributions are $(\frac{1}{2})^3$, $(\frac{1}{2})^5$, $(\frac{1}{2})^7$, and $(\frac{1}{2})^9$, respectively. This means that, in a pedigree for any animal (X) showing a common ancestor in the fifth generation behind X, the contribution of such a common ancestor to any inbreeding of X is only

$$\frac{1^{4\,+\,4\,+\,1}}{2} \;, \text{ or } \left(\frac{1}{2}\right)^9, \text{ or } 0.0019$$

To illustrate calculation of F_x from a real pedigree, Wright determined the coefficient of inbreeding of a famous Shorthorn bull, Comet. Comet's pedigree is given in Fig. 17-3 and is converted to a path dia-

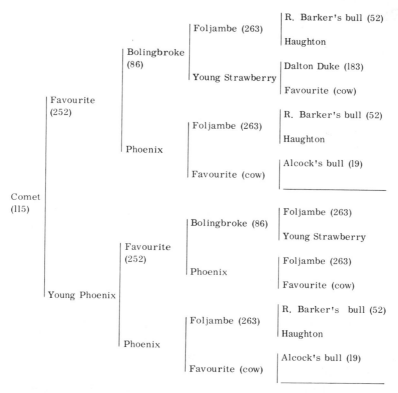

FIG. 17-3   Pedigree of the famous Shorthorn bull, Comet. (As compiled by S. Wright in *J. Heredity.*)

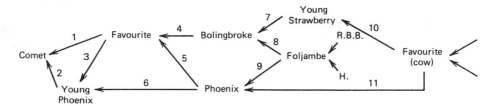

FIG. 17-4 Path diagram of Comet's pedigree.

gram in 17-4. The paths used in calculation of F_x for Comet are numbered for convenience.

The first thing to note in this pedigree is that the nearest common ancestor, Favourite, is himself inbred. We must calculate F_A for him. When this is done, and contributions of all four common ancestors have been determined as in Table 17-2, the coefficient of inbreeding for Comet is found to be 0.4687.

Wright's coefficient measures on a scale between 0 and 1 the reduction by inbreeding of the original heterozygosity in the common ancestor. That figure, in turn, cannot be known exactly, so that the coefficient of inbreeding is a value relative to the population from which the individual originated. Our coefficient of inbreeding of 0.4687 for Comet would mean that his heterozygosity had been reduced by 46.9 per cent below that of the average for his great-great-grandparents.

Breeders who still think that inbreeding belongs in the same class

TABLE 17-2

Calculation of the Coefficient of Inbreeding of Comet

Animal	Common Ancestor of Sire and Dam	F_A	n	n'	Paths	$1/2^{n + n'+1}$
Favourite	Foljambe	—	1	1	8, 9	0.1250
Favourite	Favourite (cow)	—	2	1	7, 10, 11	0.0625
					Coefficient of inbreeding of Favourite:	0.1875
Comet	Favourite	0.1875	0	1	3	0.2969
Comet	Phoenix	—	1	1	5, 6	0.1250
Comet	Foljambe	—	2	2	4, 8, 9, 6	0.0312
Comet	Favourite (cow)	—	3	2	4, 7, 10, 11, 6	0.0156
					Coefficient of inbreeding of Comet:	0.4687

as failure to pass a Wassermann test might be interested to know that, in spite of Comet's high degree of inbreeding, he was considered by Charles Colling as the best bull he had ever bred or seen. Others evidently agreed, for, when the Colling herd was dispersed in 1810, Comet brought a record price of 1,000 guineas—and that in an age when guineas and dollars were still real money.

At one time it was believed that the amount of hybrid vigor to be obtained from different crosses is related to the degrees of inbreeding in the stocks crossed, and hence that the breeder in quest of hybrid vigor should know (from average coefficients for representative samples) the degree of inbreeding in the various stocks available. Later, it became evident that there is no exact relationship between the degree of inbreeding and the amount of hybrid vigor to be obtained, and that the best way to find out how much hybrid vigor one can get in different crosses is to make them. The coefficient of inbreeding has been useful for measuring degrees to which inbreeding has been instrumental in developing prominent breeds or families, and also in attempts to determine just how much inbreeding depression results from just how much inbreeding (Table 17-3).

Hybrid Vigor

In general terms, hybrid vigor can be defined as the extra vigor, exceeding that of both parent stocks, which is frequently shown by hybrids from the crossing of species, breeds, strains, or inbred lines. It

TABLE 17-3

Depressions Estimated for Each 10 Per Cent Rise in Coefficient of Inbreeding

Item Measured	Change with Inbreeding
In swine (Dickerson *et al.*)	
Litter size at birth, piglets	−0.20
Pigs per litter at 154 days, number	−0.44
Weight per pig at 154 days, pounds	−3.44
In the fowl (Shoffner)	
Hatchability of fertile eggs, per cent	−4.4
Age at first egg, days	+6.0[a]
Eggs laid per bird to 500 days, number	−9
Body weight and egg weight at maturity	little effect

[a] Although plus, the retardation of maturity is a depression.

may be expressed as more rapid growth, larger size, increased productivity, better viability, or in other ways. The "Austra-White" hybrids from the cross of Australorp × White Leghorn are more resistant to respiratory diseases than the parent breeds, and this same valuable effect of hybrid vigor is also found even in crosses between different strains of the same breed.

Perhaps the best illustration in agriculture of economically profitable hybrid vigor resulting from suitable crosses is afforded by the hybrid corn (maize) which has done so much to raise yields of that grain in the United States. Among domestic animals, the most familiar example of hybrid vigor is provided by the mule. That worthy creature, although deprived of any hope of posterity by the incompatibility of its maternal and paternal chromosomes, is justly famous for its hardiness and capacity for work. It is even said to be endowed with a superior intelligence which tells the mule when to stop stuffing its stomach—a valuable bit of perspicacity not always found in representatives of another species that is generally believed to outrank the mule in cerebral accomplishments.

Hybrid vigor has been observed in all domestic animals from the largest to the smallest. In crosses of inbred lines of honey-bees, Cale and Gowen found that egg production by hybrid queens excelled that of random controls, and colonies headed by hybrid queens produced significantly more honey than others.

Animal breeders have somewhat more difficulty than plant breeders in reaping the benefits of hybrid vigor. For the former, adequate testing of many crosses is expensive. So is inbreeding to produce special strains for crossing. Hybrid vigor in animals is a fleeting asset, at its best in the F_1 crosses, but usually fading away in any later generations from them. Parent strains of animals must be maintained and fresh crosses made anew for each generation. By contrast, in some species of plants, the hybrids can be kept going indefinitely by such useful forms of asexual reproduction as budding, grafting, and the propagation of tubers or rhizomes.

In a few rare cases, hybrid animals can breed true, producing only hybrids like themselves, but this applies to a limited number of genes. Such organisms have *balanced lethals*, some cases of which are known in Drosophila. In one of these, the affected flies are dihybrids, carrying two linked lethal genes, *Bd* and *l* in the third chromosome, but with *Bd* in one member of the pair and *l* in the other, thus:

$$\frac{Bd\ +}{+\ l}$$

As either chromosome is lethal to any zygote in double dose, from matings of dihybrids together the only viable progeny are dihybrids like the parents. Crossing-over between *Bd* and *l* could break their linkage, but in this case an *inversion* (rearrangement of genes) prevents crossing-over (in females; there is none in males anyway). Consequently, these hybrids breed true. Their state is sometimes designated as *enforced heterozygosity*.

Genetic Basis of Hybrid Vigor

Hybrid vigor is still one of the great enigmas in the field of genetics. There are two distinctly different theories about forces that might be responsible for it. One of these might be called the *theory of dominant genes*. Most mutations are recessive, and recessive genes in general have more bad effects than good ones on viability, growth, and other measures of vigor. In general, dominant genes are more favorable than recessive ones. It is not unreasonable, therefore, to make the generalization that dominant genes raise vigor, whereas recessive genes tend to depress it.

As there are many genes affecting vigor, no animal is likely to carry all of them in the homozygous state. A cross between two animals combines in the progeny representatives of all the dominant genes that were homozygous in either parent, plus about half of those that were heterozygous. Parents distantly related, such as those in two widely different breeds, strains, or inbred lines, have a greater chance of contributing differing dominant genes to their progeny and thus increasing hybrid vigor. We might think of the dominant gene in a heterozygote as protecting that heterozygote against the adverse tendencies of the recessive allele. Considering only five loci in each of two distantly related animals that are crossed, we might get results somewhat as follows:

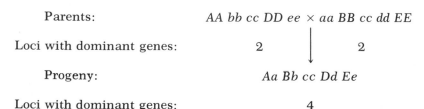

Parents:	*AA bb cc DD ee* × *aa BB cc dd EE*	
Loci with dominant genes:	2 2	
Progeny:	*Aa Bb cc Dd Ee*	
Loci with dominant genes:	4	

In this case, all the recessives except one will be masked by a dominant allele. If our theory is right, such animals should be more vigorous than either parent.

The second theory is that of *heterosis*, or *over-dominance*. It was suggested by Shull many years ago that, since the heterozygous state of a pair of alleles is frequently expressed differently from that of either homozygote, the heterozygote might be more vigorous than either parent. An easy way to visualize this situation, for those of us who like to think in simple terms, is that, in the heterozygote, both genes might be competing with each other to see which will determine the resultant phenotype. In most cases the dominant allele wins, but the many examples of incomplete dominance suggest that the battle is not always one-sided. In some cases, as, for example, in persons of blood group AB, both alleles manage to express their own separate effects. If we have had enough imagination to proceed thus far, it requires little more to suppose that the conflict between the two alleles, presumably going on at many loci rather than just one, might stir up the biochemical processes in the cell enough to make that cell more active, *ergo* more vigorous, than it could otherwise be. By contrast (according to the theory) life in cells with many loci homozygous would be comparatively secure, complacent, and even lethargic—not a state of affairs conducive to vigor.

Shull referred to this condition in the heterozygote as heterosis, a term which later came to be synonymous with hybrid vigor. In recent years others have referred to the same phenomenon as over-dominance. One difficult confronting the devotees of this theory is that comparatively few examples are known in which heterozygosity at one locus produces an individual more vigorous than either parent. A good example of the superiority of a heterozygote in one environment is afforded by the individuals heterozygous for the type of hemoglobin causing sickle-cell anemia, who are thus made more resistant than either homozygote to malaria. The advantage of the heterozygote would be lost in regions free of malaria. Similarly, other genotypes that show heterosis in one environment could fail to do so in another.

Crosses Between Species

A familiar interspecific hybrid is the mule, which results from crosses between the Jack, *Equus asinus*, and the mare, *E. caballus*. Makino and his associates found the diploid chromosome numbers in the horse, the ass, and the mule to be 64, 62, and 63, respectively. Evidently, mules get one haploid set of chromosomes from each parent. Earlier cytological studies have shown that their sterility results from incompatibility of maternal and paternal chromosomes, so that viable germ cells are not formed in meiosis. Reports persist of fertile female mules, but Benirschke *et al.* (1964), after finding that the chromosomes of one supposedly fertile mule were like those of a donkey, reviewed other reported cases and concluded that there is no acceptable record of any animal proven to be both a mule and fertile. Crosses between the zebra and mares (see Chapter 14) are also sterile.

Crosses between cattle and the American bison have not contributed much to agriculture. The resulting "cattalo" usually come from the cross of *Bos taurus* ♂ × *Bison bison* ♀, which causes less dystocia than the reciprocal cross. The female hybrids are fertile but the males are not.

The best example of an interspecific hybrid of great value in agriculture is provided by the many crosses between European cattle (*Bos taurus*) and zebus (*Bos indicus*). The special merits of breeds derived from such crosses, whether they be the Santa Gertrudis of Texas, the Hope Holsteins of Jamaica, or the various other crosses elsewhere, can hardly be attributed to hybrid vigor. Ordinarily, hybrid vigor should not persist for more than one or two generations after a cross is first made except in plant hybrids maintained by asexual reproduction. The breeds of cattle just mentioned more likely owe their special value to the good characters introduced by the zebus, particularly ability to tolerate heat, to repel ticks, and thus to resist tick-borne diseases.

Cross Between Breeds

The merits of crosses between breeds have long been known to animal breeders. It is neither feasible nor necessary here to review the many such crosses which have demonstrated how effectively hybrid vigor can raise the productivity of farm animals. Growth of such hybrids has been shown to excel that of either parent in cattle, swine, and chickens. In various crosses of dairy cattle at the U. S. Department of Agriculture, milk production of crossbred heifers was 20 per cent higher than that expected without a benefit of hybrid vigor. As is shown by the extensive data of Winters *et al.* on breed-crosses in swine, their hybrids not only outnumbered the purebreds in average litter size at birth and at weaning, but also got to market weight faster and on less feed (Table 17-4). For contrast, these advantages of hybrid vigor should be compared with the depression from inbreeding shown in Table 17-3. The value of hybrid vigor from breed-crosses is nowhere better indicated than by the fact that practically every broiler chicken produced in the United States nowadays is a cross between two or more breeds. Some of those parent breeds are new ones developed for that specific purpose. By utilizing the rapid growth resulting from hybrid vigor, the broiler grower gets his birds to market faster, and on less feed, than would be possible in any other way. It is probable that, among the millions of consumers who find chicken the cheapest good meat in North American markets today, very few realize that they owe their bargain in part to hybrid vigor.

TABLE 17-4
Hybrid Vigor from Breed-Crosses in Swine

Item	Pure Breeds	First Crosses	3-Breed Crosses
Sows, number	76	45	24
Pigs, number	715	440	245
Live pigs per litter	8.26	9.22	9.88
Litter size at weaning	5.54	5.95	8.71
Days to 220 lbs.	x	$x - 17$	$x - 17$
Pounds of feed per pig to 220 lbs.	y	$y - 28$	$y - 36$

Source: Data of Winters *et al.*

Crosses between two breeds to produce an F_1 hybrid are sometimes followed by mating that hybrid to still a third breed to produce a three-way cross. Breeders of swine and chickens have found an advantage in such three-way crosses in that the hybrid mothers seem to be more efficient as mothers than their purebred parents. Eggs from F_1 hybrid hens hatch somewhat better than those of purebreds. Hybrid sows tend to have larger litters than purebreds and to lose fewer pigs before weaning. Similarly, hybrid queen bees seem able not only to lay more eggs than others, but also to make the workers under their rule produce more honey than those in colonies not thus stimulated.

Crosses Between Strains Within Breeds

Some years ago, the phenomenal yields of hybrid corn (maize) resulted mostly from the combination of four highly inbred lines. Two such lines, A and B, were crossed to produce an F_1 hybrid, AB. Two others, C and D, were similarly crossed to get another hybrid CD. These "single" crosses had good yields, but their inbred parents were so poor that seed to produce single crosses could not be grown in large quantities. Accordingly, the single crosses were combined, AB × CD, to grow the "double-cross" commercial crop.

Later it was found that, with less inbreeding and some selection, parent lines could be developed that produced ample seed and that, when crossed in suitable combinations, resulted in single crosses with as great an increase from heterosis as had been attained by the double crosses.

Animal breeders seeking to emulate the example set by the breeders of hybrid corn believed at first that they too would have to develop highly inbred lines for crossing. Such a program is very expensive because of the many losses that must be endured to finish up with any highly inbred stock at all. Fortunately some breeders, seeking for hybrid vigor in more economical ways, found that surprising amounts of it come sometimes from crosses between lines or strains that are not specifically inbred for that purpose. The following manifestations of hybrid vigor found at Cornell University in the first crosses between two strains of White Leghorns are fairly typical:

Hatchability of fertile eggs, increase	4.7 per cent
Days to first egg, reduction	5 days
Eggs laid to 500 days, increase per bird	22-25 eggs
Egg size, increase	2 grams
Body size when adult, increase	130 grams

The strains used for these crosses were closed flocks to which no new blood had been introduced for 13 years. Similar results have been obtained by others.

Most of the eggs laid in North America are now produced by hybrids of one kind or another, and extensive testing of strain crosses is under way to find still better ones. To some extent, of course, these are crosses of inbred lines. In the United States, most of the strains of the fowl crossed for egg production are pure White Leghorns. As purebred stock of one breed, they are somewhat more homozygous than mongrels. Moreover, most of the strains tested come from flocks in which, because they are closed flocks, there has been some slight additional degree of inbreeding.

Perhaps it is not surprising that so much hybrid vigor is sometimes found from crossing two strains or more of the same breed. It should be remembered, however, that hybrid vigor does not come automatically in every cross, and that some breeders have to test literally dozens of such crosses to find one good enough for commercial production on a large scale. Just to keep our terminology straight, let us remember that most of the inter-strain hybrids that lay the eggs still qualify as purebred White Leghorns.

Utilization of Hybrid Vigor in the Silkworm

The earliest extensive use of hybrid vigor to raise the productivity of domestic animals was undoubtedly its utilization in Japan to increase yields from the silkworm, *Bombyx mori*. The story is told concisely by Yokoyama (1959). Following establishment of the Sericultural Experiment Station in 1911, and investigations of hybrid vigor there, the first distribution of hybrid eggs was begun in 1914. The hybrids, which were crosses between strains of Chinese and Japanese origin, became so popular with the farmers that by 1919 over 90 per cent of the eggs produced were of hybrid origin, and that figure was soon raised to 100. Average weight of cocoons, which had been brought by slow selection from a little below 20 centigrams in 1804 to a little above that weight in 1910, was increased rapidly as the hybrids took over their production, and by 1932 was well over 40 centigrams.

As with other domestic animals, selection in the silkworm must be directed toward improvement in several different objectives. Weight of the cocoon is of prime importance. Other objectives include better reelability of the filament from the cocoon, quality of that filament (uniformity and low content of sericin), neatness of the silk, and viability of the caterpillars. Strains have been developed to cope with adverse environments, to produce silk for special purposes (including fishing line), and even to grow pupae rich in riboflavin. Breeders of larger domestic animals might well view with envy the remarkable progress made by skilled geneticists in a comparatively short time with this little one.

Breeding for Hybrid Vigor

Many breeders of swine, poultry, beef cattle, and other animals are now trying to find the easiest way to get the most hybrid vigor at the least expense. One of the best solutions is to find a strain that has *general combining ability*, the capacity to cross well with most others and to yield profitable amounts of hybrid vigor. Some stocks may cross well with certain other strains but not with all. Those that "nick" particularly well are said to have *specific combining ability*.

A scheme to enhance specific combining ability was recommended to breeders some years ago. It was known as *reciprocal recurrent selection*. In each of two strains, the breeding stock used to multiply the strain was to be evaluated by performance of its progeny from crosses with the other strain. By thus selecting for ability to cross well with the other strain, both would (presumably) be steadily improved in their ability to nick well with each other.

If this system of breeding had yielded the results hoped for (an increasing degree of hybrid vigor in the crosses), it would have been the answer to many a breeder's prayer. Too many of them have spent their years, youth, and money in developing highly improved strains, only to have those products taken over by newcomers in the field, who capitalize in a year on decades of work by the older breeders. With reciprocal recurrent selection, the breeder could sell the highly productive crossbreds to commercial producers, but never sell to competing breeders the improved parent strains that begot the crossbreds. He could thus retain his superior stocks almost as well as if they were patented.

Unfortunately, experiments set up with fruit flies, chickens, and pigs to demonstrate the merits of reciprocal recurrent selection have failed to do so. Selection to increase hybrid vigor is apparently ineffective. Nevertheless, it can be effectively utilized to attain maximum productivity.

From their experience with strains of White Leghorns bred for more than 30 years at Cornell University for resistance to disease, and for other traits of economic importance, Cole and Hutt (1973, 1974) concluded that when two strains (or breeds) are found on crossing to yield a desirable degree of heterosis in their hybrid progeny, increasing productivity thereafter can be attained by persistent selection (based

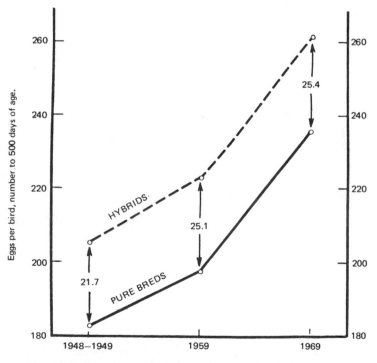

FIG. 17-5 Egg production to 500 days of age, and hybrid vigor over 20 years. The solid line shows the mid-parent average for the two strains that were crossed thrice at intervals of 10 years; the broken line of the hybrids shows a fairly constant increment from hybrid vigor at each crossing. The data are corrected for effects of adverse environments in 1959 and 1969. (From Cole and Hutt in *Animal Breeding Abstracts* as cited.)

on tests of progeny) within each strain and crossing the two to get the full advantage of heterosis in the hybrids.

When their C and K strains were crossed in 1948–1949 and found to nick well, selection within them for several objectives of economic importance was continued for 20 years. They were crossed (reciprocally) 10 years later, and again 20 years later. In those two decades, egg production of survivors to 500 days of age was raised in the C and K lines to 243 and 229 eggs, respectively; both figures are creditable for unculled flocks. The hybrids were still better at 261.4 eggs per bird, a figure higher by 25.4 eggs than the mid-parent mean. In all three crosses the increments from hybrid vigor were remarkably uniform (Fig. 17-5).

Similarly, Osawa and Harada, as reported by Yokoyama (1959), crossed races of silkworms that differed in weights of cocoons, and found that the weights of cocoons by the F_1 hybrids were consistently about 0.3 g heavier than the mid-parent value (Fig. 17-6). Among 12 polygenic traits studied, there were only two exceptions to the general rule that the performance of the hybrids stayed at a fairly constant level above the mid-parent mean.

To sum up, we can say that hybrid vigor provides a means of raising yields in domestic animals considerably above those to be

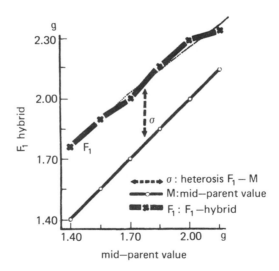

FIG. 17-6 The effect of heterosis on weight of the cocoon of the silkworm. The 122 crosses are grouped into six classes, at intervals of 0.15 g, based on mid-parent values for weight of cocoons. The increment from heterosis is relatively constant over the range of mid-parent values from 1.40 to 2.30. [From Osawa and Harada, 1944 (in Japanese) as reported in English by Yokoyama, 1959.]

expected in pure breeds or pure strains. There is no fixed rule for getting it in the maximum degree. Hybrid vigor is where one finds it. Research to find something about the forces contributing to hybrid vigor should be highly profitable.

Problems

17-1 Just to keep straight on nuances, describe clearly the differences between (a) inbreeding and line breeding, (b) hybrid vigor and heterosis, (c) crossbred and purebred, (d) purebred and Thoroughbred.

17-2 Why are marriages of near relatives not encouraged in most human communities?

17-3 Referring back to Table 17-1, and following the procedure therein, work out the proportion of heterozygotes in the fifth generation.

17-4 Can you devise a general formula with which you can easily determine the proportion of heterozygotes in the nth and tenth generations?

17-5 From what we have learned in this chapter, would you expect the proportion of lethal genes among all genes to be higher in beans than in buffaloes—or *vice versa*—and why?

17-6 If two wild plants are brought under cultivation, one of which proves to be self-pollinating and the other cross-pollinating, which is more likely to respond to selection, and why?

17-7 Make up a pedigree for a marriage of first cousins and, using Wright's formula, calculate the degree to which children from such a marriage are inbred.

17-8 Get a pedigree of any prominent sire, whether a bull, a boar, or a dog, and determine to what extent, if any, he is inbred.

17-9 How does the degree of inbreeding from a brother × sister mating compare with that from parent × offspring?

17-10 Referring back to Table 12-4, compare mean weights for the Polish, Flemish Giant, and F_1 hybrids, and, comparing the mid-parent mean with that for the F_1, determine whether or not the weights at 275 days were influenced by hybrid vigor.

17-11 What has a hinny got in common with a mule, and how does it differ in origin from that hybrid?

Selected References

BENIRSCHKE, K., R. J. LOW, M. M. SULLIVAN, and R. M. CARTER. 1964. Chromosome study of an alleged fertile mare mule. *J. Heredity* **55:** 31–38. (With conclusions about similar cases.)

CALE, G. H., JR., and J. W. GOWEN. 1956. Heterosis in the honey bee (*Apis mellifera* L.). *Genetics* **41**: 292–303. (On the egg production of hybrid queens and yields of colonies headed by them.)

COLE, R. K., and F. B. HUTT. 1974. Combining selection with heterosis for maximum productivity. *1st World Congress on Genetics Applied to Livestock Production*, Madrid, Vol. **1**: 75–83. (Some conclusions from selection and crossing during 20 years with White Leghorns.)

COLE, R. K., and F. B. HUTT 1973. Selection and heterosis in Cornell white Leghorns: A review, with special consideration of interstrain hybrids. *Animal Breed. Abstr.* **41**: 103–18. (Covers 35 years of selection for traits of economic importance.)

DICKERSON, G. E., *et al.* 1954. Evaluation of selection in developing inbred lines of swine. *Missouri Agric. Exper. Sta. Bull.* **551**. (Source of data in Table 17-3 and report of extensive study of inbreeding in swine.)

GAVORA, J. S., A. EMSLEY, and R. K. COLE. 1979. Inbreeding in 35 generations of development of Cornell S strain of Leghorns. *Poultry Sci.* **58**: 1133–36. (Calculated with method of Malecot and a computer.)

JOHANNSON, I., and J. RENDEL. 1968. *Genetics and Animal Breeding*. Edinburgh and London. Oliver and Boyd. (See Chapter 14 for systems of mating, inbreeding, and hybrid vigor.)

LUSH, J. L. 1945. *Animal Breeding Plans*. Ames. The Collegiate Press, Inc. (A standard text in this field; see Chapters 20 to 25 for discussion of inbreeding.)

SHOFFNER, R. N. 1948. The reaction of the fowl to inbreeding. *Poultry Sci.* **27**: 448–52. (Effects of inbreeding on various characters.)

WINTERS, L. M., O. M. KISER, P. S. JORDAN, and W. H. PETERS. 1935. A six-years' study of crossbreeding swine. *Minnesota Agric. Exper. Sta. Bull.* **320**. (Source of data in Table 17-4.)

WRIGHT, S. 1923. Mendelian analysis of the pure breeds of livestock. I. The measurement of inbreeding and relationship. *J. Heredity* **14**: 339–48. (With inbreeding of Favourite slightly higher than the figure in Table 17-2 because the pedigree used by Wright goes farther back. For Comet, see *ibid.*, pp. 408, 414.)

YOKOYAMA, T. 1959. *Silkworm Genetics Illustrated*. Tokyo. Japan Society for the Promotion of Science. (A beautifully illustrated monograph covering genetics and cytology of a domestic animal that has been thoroughly studied genetically.)

CHAPTER
18

Changes in Genes and Chromosomes

In previous chapters we have used the term *mutation* without defining it exactly, but considering it as a general term for hereditary variations that have resulted from changes in genes. If, now, we broaden that conception slightly, and consider mutations simply as changes of any kind affecting the hereditary material, we can include under that heading some that affect whole chromosomes, or segments of them, and others resulting in cells that have more or fewer than the normal number of chromosomes. Most geneticists consider all such changes as mutations of one kind or another.

We shall not delve very deeply into this aspect of genetics, because, thus far, few cases are known in domestic animals in which inheritance is modified by chromosomal aberrations of any kind. However, geneticists and cytologists interested in plants have uncovered many such cases. So have the assiduous devotees of Drosophila, which is particularly good for such studies because of the "giant" chromosomes in its salivary glands. Within recent years geneticists studying heredity in man have found abnormalities in the chromosomes to be responsible for certain afflictions, the cause of which had long been a complete mystery,

It is certain, therefore, that in domestic animals, as in other organisms, chromosomes do go astray, break apart, or get tangled up with each other. The few examples cited in this chapter may encourage further investigations.

Gene Mutations

Let us start with the comparatively simple mutations with which we are familiar—those affecting a single locus. From what we learned in Chapter 6 about the complexity of the gene, and of the structure of DNA, it is not difficult to imagine that any accident in the duplication of that substance, when a new chromosome is formed, could cause a change in the chemical structure, *i.e.*, a mutation. A pair of bases might be lost, or reversed, or placed in a different sequence in the new strand. Other possible kinds of accidents have been suggested, but, in our present state of knowledge, further speculation about them is unwarranted here.

An example of just such a mutation is provided by the abnormal kind of human hemoglobin, called hemoglobin S, about which we read in Chapter 16. A molecule of hemoglobin contains two identical *alpha* chains of amino acids and two identical *beta* chains of amino acids. Hemoglobin S differs from the normal kind only in having in each half-molecule one extra valine, which replaces one glutamic acid at a specific point in one chain. The biochemists, who have been able to pin-point the exact place of that substitution, tell us that it occurs at postion 6 in the *beta* chain.

These two amino acids are much alike. As the triplet RNA codes for glutamic acid and valine are A U G and U U G, respectively, the corresponding bases in the strand of DNA would be a T A C in normal hemoglobin, but A A C in hemoglobin S. One cause for the change from thymine to adenine in that triplet might be the reversal of a single pair of bases from a normal A-T position to one of T-A during reproduction of the gene.

To some of us viewing with wonder the complex models of the DNA molecule, and realizing the infinite number of times it is reproduced, the thing most marvellous is not that mutations occur, but that they do not occur more often. Spontaneous mutations are *recurrent*, that is, the same change (or a similar one) occurs at the same locus over and over again. It does not do so frequently. The rate at which such mutations occur can best be determined in organisms that multiply rapidly, like bacteria, and fruit flies, but it can also be estimated (in various ways) for some mutations that cause recognizable abnormalities in man.

One must remember that the rarity implied by a mutation rate of 2 or 3 per 100,000 gametes is considerably less rare for the species as a whole when one considers the incalculable numbers of gametes produced. It has been estimated that some hereditary abnormalities in man (for example, hemophilia) are such a handicap that they would disappear entirely were it not for recurring mutations at the same gene locus. Mutation rates have not been calculated for domestic animals, but evidence of recurring mutations is provided by the spontaneous appearance of the same condition in entirely different breeds. Recessive achondroplasia has been reported in several breeds of cattle. The dominant type, a breed characteristic of the Dexter, has occurred also in Jerseys. True hemophilia has been found in Greyhounds in the Dutch East Indies, in Irish Setters in the United States, and in other breeds in Canada. Platinum foxes appeared in Norway, but an identical dominant mutation was found also in North America. Other examples of similar or identical independent mutations are familiar to most animal breeders.

As we learned in Chapter 11, mutations of different kinds can occur at one and the same locus, and, altogether, these will comprise a resultant series of multiple alleles. Some of these series contain only a few mutations, so far as is known at present, but the fact that over 500 alleles are known at the *B* locus (which determines various blood antigens) in cattle suggests that certain loci mutate more often than others.

In all the genetic experiments of half a century, only a few cases have been found of "contamination," or change in a gene by its association with an allele. In maize, R. A. Brink found that the gene R^r, determining color of the aleurone layer of the seed, had one effect when the male gamete carrying it came from a homozygote, but a different effect when that same allele came (in a male gamete) from an $R^r R^{st}$ heterozygote. In the latter case, most of the pollen carrying R^r produced effects like those of R^{st}. Presumably R^r had been influenced by R^{st} during their association in the heterozygote, a hitherto unknown phenomenon to which Brink gave the name *paramutation*.

When we think of a mutation as a change, we might well ask: A change from what? The Drosophila geneticist usually thinks of mutations from the wild type, but is familiar with those occurring within lines that are themselves mutations from the wild type. With domestic animals, for some of which the wild type is not well known, it is easier to think of mutations as changes from what is considered normal for the breed, strain, or species. For example, the spontaneous appearance of polled animals among Dorset Horn sheep in North Carolina, and in Australian Merinos, resulted from dominant mutations. If, in polled breeds of sheep like Suffolks, Southdowns, and others, horned animals should appear and their condition be proven hereditary, they too could be charged to a mutation, in that case probably recessive.

Fig. 18-1 In the Silkie, fanciers have preserved mutations from the wild type to white plumage, silky feathers (without hooklets), crest, trifid comb, black skin, feathered shanks, and polydactyly.

The fact that both the polled sheep in the Dorsets and Merinos and the horned ones in the polled breeds might thus result from mutations at one locus is consistent with the evidence from other organisms that mutations are not always in the same direction of change. Changes from the normal to the mutant type are best known, but *reverse mutations*, from the mutant type back to the normal one, also occur. If we consider that the polled breeds of sheep undoubtedly arose by preservation in them of mutations from the horned state of their wild ancestors, then, if in some polled breed the horned condition reappeared by mutation, that could be attributed to reverse mutation, or, as it is sometimes called, to *back mutation*.

One should remember also that any advantageous mutations are likely to be preserved in wild animals, and thus become normal for that species. When domestication occurs, the breeders select further among any mutations observed to preserve those considered desirable. Mutations that would be deleterious in the wild state thus become not only normal for some breeds of domesticated animals, but also indispensable witnesses to the "purity" of their breeding. This is amply attested by the completely albinotic New Zealand White rabbits, polled cattle and sheep, Pekingese dogs, and such bizarre breeds of the fowl as the Frizzle and the Silkie (Fig. 18-1).

Somatic Mutations

Thus far we have been thinking in terms of hereditary mutations, *i.e.*, those that occur in the germ cells. Similar changes must occur in other cells of the body, but comparatively few of these can be detected because they are not transmitted from one generation to the next. Such *somatic mutations* are occasionally recognized in animals when they cause areas (usually small) in which the hair, skin, or feathers are different from the structure or color found elsewhere on the body. Patches of blue fur on black rabbits, or chickens with one leg yellow and the other white, and other similar *mosaics*, fall into this category.

An easy way to see the effects of somatic mutation in a domestic animal is to examine the plumage of almost any Barred Plymouth Rock hen—except one at a poultry show. Feathers of such a hen should all be crossed with alternate bars of black and white from tip to base of each feather. The barring gene, B, mutates to b (non-barred, or solid black), but, as B is sex-linked, and therefore carried by the hen in the hemizygous state (B—), mutations from B to b can often be recognized by the resultant all-black feathers, or parts of feathers. When the mutation occurs in some cell early enough in development to be passed to many daughter-cells, it may even cause a little patch of three or four solid black feathers. Mutations at later stages eliminate barring from a whole feather, from one side only, or from a smaller sector (Fig. 18-2).

There is no use in seeking these somatic mutations in pure barred males. They are *BB*, and a mutation from *B* to *b* in one of their sex chromosomes will not show because the other still carries *B*, which is dominant to *b*; hence the feather is still barred. Presumably such somatic mutations occur just as frequently in males as in females, but to be revealed in males there would have to be simultaneous mutation at the *B* locus in *both* sex chromosomes, an event so unlikely that search for black feathers in a pure Barred Plymouth Rock male is not very rewarding. On the other hand, they do occur in heterozygous males (*Bb*), in which they are just as frequent as in the females. The reason why seekers for somatic mosaicism in Barred Plymouth Rock hens should not hope to find it at a poultry show is simply that any self-respecting exhibitor of Barred Rocks knows enough to pull out the black feathers, so that no judge will ever see them.

Induced Mutations

So far we have been considering only spontaneous mutations, the kind that "just happen" in ways not yet known. They occur in wild animals

FIG. 18-2 Somatic mosaics in feathers of a Barred Plymouth Rock hen. Left to right: a normal barred feather, one completely non-barred, two with smaller non-barred portions.

as well as domestic ones, and are responsible for the frequent reports of albinism, melanism, or other "sports" in the press, and for the privilege, which some of us have enjoyed, of seeing black, grey, and albinotic squirrels—all of one species—disporting themselves in the same trees at the same time.

The rates at which mutations occur can be accelerated by various *mutagenic* forces. Those most extensively used in experimental studies of mutation are probably ionizing radiations of various kinds, particularly X-rays. The demonstration by Muller in 1927 that mutations can be induced by X-rays led to his receiving for the accomplishment (19 years later) a Nobel prize. Other *mutagens* include ultraviolet light, which does not penetrate as much as X-rays, and causes less breakage and other damage to chromosomes. It is used particularly for inducing mutations in bacteria and other simple organisms. Several chemical substances have mutagenic properties. These include mustard gas, formaldehyde, and boric acid. It is evident that X-rays do considerable damage to the chromosomes, and some investigators consider that many of the mutations attributable to X-rays are not changes in single genes, but rather chromosomal aberrations of one kind or another. We shall return to these later in this chapter.

Among other kinds of mutagenic ionizing radiation is the fall-out from explosions of nuclear bombs. There has been much discussion of the danger of such fall-out to future generations that might be the victims of mutations thus induced. We cannot go into any lengthy discussion of that risk here, but it will do no harm to point out that two independent committees of scientists appointed to study that problem came to very similar conclusions. One committee was appointed by the National Academy of Sciences and the National Research Council of the United States, and the other by the British Medical Research Council. Both came to the conclusion that the risk from nuclear bombs is only a tiny fraction of that from the use of X-rays for diagnosis and therapy.

It has been suggested that mutagens could be helpful in agriculture if even a small fraction of any mutations thus artificially induced have any capacity to raise yields of crops or productivity of animals. The possible value of any such application is now being investigated by plant breeders. So far as the larger domestic animals are concerned, the phenomenally high productivity of the best individuals suggests that the variation already existing is adequate to permit further improvement, and that the problem is not to add to that variation, but to utilize it more fully.

For an example of a potentially useful gene mutation artificially induced in a domestic animal, we must go to the one best known genetically—the silkworm. Ordinarily caterpillars of that species eat only leaves of the mulberry, but Tazima induced with X-rays mutants that would eat leaves of beets, beans, and other plants. They were abnormal in the structure of the maxilla, particularly in certain sensory hairs on the lobe, which is believed to guide the caterpillars in selection of food. It is not clear whether the mutants could be used commercially, but it seems probable that, as with many other mutations, further selection (accumulation of modifiers) would be necessary to perfect them.

Polyploidy and Aneuploidy

In Chapter 6 we learned that the normal number of chromosomes in somatic cells is called the diploid, or $2n$ number, and that mature germ cells carry n, or the haploid number of chromosomes. In many groups of plants and some invertebrate animals, related species or genera have numbers of chromosomes that are multiples of some basic number. A familiar example is found in wheat in which the chromosome numbers of somatic cells in different species are 2, 4, or 6, times 7 (Table 18-1).

Species like these are said to form a *polyploid* series. In such a series, the normal number of chromosomes in somatic cells of one species would be the $2n$ number for that species, but could be $4n$ or $6n$ with respect to the basic, lowest common multiple. Durum wheat is thus a *tetraploid* in that series, and common wheat is a *hexaploid*.

Although breeders of domestic animals are not yet concerned with polyploids, it is desirable that we should know some of the polysyllabic terms so freely bandied about by our opposite numbers among the plant breeders. Some of these follow.

Autopolyploids are those in which the extra chromosomes are the same as those normally occurring in the species. Thus an *autotetra ploid* could arise from reduplication of each $2n$ chromosome in the normal way but without the normal division of the cell.

Triploids could arise from fusion of a normal gamete (n) with one from an autotetraploid, which should carry $2n$ chromosomes, to produce a $3n$ plant.

Allopolyploids result from fusion of gametes from two different species, A and B, and are ordinarily sterile because, as there is only a single representative of each A chromosome, and the same for the B set, no chromosome has a homologous mate; hence normal meiosis cannot occur. Such polyploids can be propagated asexually. If any cell in such a plant undergoes doubling of its chromosomes without associated division of the cell, that cell will have two sets of A chromosomes and two of B. At meiosis it could form some viable gametes and so could reproduce. A fertile allopolyploid of this kind is called an *amphidiploid*. The most famous example of such a plant is

TABLE 18-1

Diploid Chromosome Numbers in Different Species of Wheat

Species	Common Name	Chromosomes in Somatic Cells	Cytological Classification
Triticum monococcum	Einkorn	14	Diploid
Triticum durum	Durum	28	Tetraploid
Triticum dicoccum	Emmer	28	Tetraploid
Triticum vulgare	Common	42	Hexaploid

the Raphanobrassica, resulting from a cross between cabbage and the radish (for each of which $2n = 18$), out of which came eventually fertile hybrids carrying 36 chromosomes. From the illustrations, these plants seemed almost capable of pushing the roof off the greenhouse.

Euploidy is a general term including all the kinds of polyploids considered thus far. They have in common the fact that the whole set of chromosomes is multiplied, with none missing and no extra ones.

Aneuploidy is the state of plants (and of some animals, too) that have only one or a few chromosomes duplicated or lacking, but are otherwise $2n$.

Trisomics, the most common aneuploids, are $2n + 1$, but the extra chromosome can be any one of the set, and the appearance of the trisomic will depend on which chromosome is represented thrice. An example is Down's disease (mongolism) in man, in which the affected child is (in most cases) trisomic for Chromosome 21 (one of the two smallest autosomes) and has a karyotype showing 47 chromosomes, *i.e.,* $2n + 1$. It is known to be more frequent in children of mothers over 40.

Most trisomics are believed to have arisen from *non-disjunction*, the failure of two homologous chromosomes (or of two chromatids) to separate at meiosis, so that one gamete gets two members of one pair, instead of one. The other gamete would lack that chromosome entirely. Fusion of two gametes, one normal (n) but the other $n + 1$ would result in a trisomic individual. Gametes lacking one chromosome are not as viable as those that carry an extra one; hence *monosomic* zygotes that are $2n - 1$ are much scarcer than trisomics.

Polysomy is a general term covering any number of extra chromosomes more than one.

Polyploid Plants and Animals

There is ample evidence that polyploid plants have arisen in nature, and many of the improved varieties of tulips, iris, roses, and other flowering plants are either triploids or tetraploids. Most varieties of apples are diploids, with 34 chromosomes, but triploids, having 51 chromosomes, include such old favorites as Rhode Island Greenings, Gravenstein, Baldwin, and King of Tompkins.

Apart from polyploids that occur naturally, plant breeders have been able to make some almost to order with the help of the drug, *colchicine*. This substance, which has been for centuries the standard remedy for gout, has also the unique property of arresting cell division. In proper concentrations it prevents the chromosomes from forming the normal equatorial plate at metaphase of mitosis and, although they do eventually divide, they all remain in one cell, which thus has the tetraploid number.

Polyploid series are known in some invertebrate animals, and polyploid individuals have been found in amphibians, but if any occur in higher animals they are very rare. Polyploid cells have been identified in mouse embryos, but apparently not in any surviving long enough to be born. Some years ago attempts to induce polyploidy with colchicine in rabbits and pigs led to a headline in a farm magazine reading "Super Hogs Ahead!", but not to any big polyploid pigs going to market. Polyploids induced by Japanese investigators in silkworms by centrifugation, high temperature, or colchicine had larger cells than diploids (as do most polyploids) but were considered not likely to be useful for production of silk.

Polyploidy and aneuploidy seem to be fatal at early embryonic stages in mammals and in the fowl. The chick embryo is favorable for such studies because the embryo can be examined at any stage desired, and also because, by candling the incubating eggs, one can detect the dead and dying embryos. In studies of over 4,000 4-day embryos, Bloom (1972) found the frequencies (per cent) of four abnormalities to be:

Haploids 1.4	Tetraploids 0.1
Triploids 0.8	Trisomics 0.2

All of these died during the first week of incubation except a few triploids, none of which hatched. The haploids were consistently underdeveloped at four days, as were also any trisomics still living. Among the embryos that died early, the proportion with abnormal numbers of chromosomes was 10.8 per cent.

Some mature fowls have been reported to be triploid, but it is not clear whether they were triploid in all cells, or were mosaics having both diploid and triploid cells.

Similarly, McFeely (1967) found that, among 10-day blastocysts from normal, healthy sows, 10 per cent had abnormal numbers of chromosomes, and a further 2.3 per cent, which showed no mitotic figures, were already dead or dying.

From the review by Hare and Singh (1979), it seems probable that much of the early embryonic mortality in mammals can be charged to changes in the *numbers* of chromosomes (*i.e.,* polyploids and aneuploids). Structural changes *within* the chromosomes (to which we come shortly) also cause some mortality of embryos, but much less than the changes in numbers.

Monoploids and Drones

Some plants are occasionally found (maize, tomato) which have only the haploid number of chromosomes. Because the term *haploid* is usually reserved for the chromosome numbers of gametes, these plants with only *n* chromosomes are commonly designated as *monoploid*. They are almost always sterile because they cannot undergo normal meiosis.

This is not the case with what is probably the only monoploid domestic animal—the male honey-bee, *Apis mellifera*. The drone's sole function is to fertilize the eggs, but he himself develops from an unfertilized egg. As a monoploid, the drone might be expected to have difficulty in producing viable gametes, but apparently during the course of evolution a way to compensate for his unpaired chromosomes ($n = 16$) has been found. At the first division of the primary spermatocyte a bit of cytoplasm containing no nuclear material is eliminated. At division of the secondary spermatocyte, each chromosome is duplicated, as is normal, and the partners of each pair go separately to two nuclei, as is also normal. One of these, however, gets nearly all the cytoplasm, the other little or none. As a result, only a single functioning spermatozoön is produced from each primary spermatocyte.

Diploidy in Parthenogenetic Turkeys

Following his discovery in 1954 that unfertilized eggs of Beltsville Small White turkeys could occasionally undergo embryonic development, M.W. Olsen of the U.S. Department of Agriculture at Beltsville selected for higher rates of parthenogenesis in that stock, and by 1959 had raised the proportion of eggs showing such development to 41.7 per cent. The proportion of these surviving to later stages of incubation was also greatly increased, and by 1960 Olsen had actually hatched no fewer than 68 parthenogenetic poults. Most of them were weak and had to be helped from the shell by experts in avian midwifery, but a few survived to maturity. One of these (Fig. 18-3) eventually became the proud sire of 122 poults.

All such parthenogenetic embryos and poults were found to be males. Examination of their chromosomes showed that, contrary to what one might expect, all were diploid. Here is a peculiar situation. Unfertilized eggs produce monoploid male bees, but in the turkey, when parthenogenetic development might also be expected to result in monoploid animals, diploidy is the rule. Evidently, the haploid number of chromosomes in the egg is restored in some way to the full diploid complement. This could happen if one of the meiotic divisions were suppressed or followed by fusion of one of the polar bodies with the secondary oöcyte, but exactly what happens in this case is not yet known. If fusion of the second polar body with the nucleus of the secondary oöcyte were responsible, that would be equivalent to self-fertilization, and the resulting zygotes should be homozygous at many loci.

It is easier to understand why all the parthenogenetic embryos

FIG. 18-3 A Beltsville Small White turkey, diploid and male, hatched after parthenogenetic development of an unfertilized egg. He became the sire of 122 normal poults. (Courtesy of M. W. Olsen and U.S. Department of Agriculture.)

are males. As the female turkey is heterogametic, half her secondary oöcytes should carry an X-chromosome, but the other half would not. Any doubling of chromosomes in oöcytes carrying an X would yield a viable cell with the normal XX complement of males. On the other hand, doubling of chromosomes in an oöcyte lacking X would result in a cell with only two sets of autosomes and no X. From analogy with the situation in Drosophila, in which flies having two Y-chromosomes (but no X) are not viable, it seems probable that such cells could not survive. Accordingly, the only viable cells resulting from restoration of the diploid condition to these unfertilized ova of the turkey would be genetically males.

Aberrations Within Chromosomes

Under this heading come various kinds of breaks in chromosomes and the consequences thereof. These abnormalities can be caused by harsh mutagens, like X-rays, but there is ample evidence that they also occur spontaneously without any such adverse environmental cause. As before, many of them have been studied in plants, and in Drosophila and other insects, but until recently very few had been identified in animals with backbones. This does not mean that such conditions do not occur in domestic animals, but only that they had been little sought.

With the newer techniques (for banding) by which individual chromosomes and parts thereof can now be identified, many of these aberrations have been found, and new ones are being reported every year. In this chapter, we can describe only a few examples, but all those known to date in domestic animals are reviewed in the monograph by Hare and Singh (1979).

Deletions and Duplications

Also known as *deficiencies*, deletions of segments in a chromosome may involve only a few loci or many. It is believed that many lethal mutations, particularly those induced by X-rays, are deletions. In suitable material for cytological study, they can be identified in cells carrying the deletion in the heterozygous state (*i.e.*, one chromosome normal, its mate deficient in a segment), because, while the homologous sections of the two chromosomes pair at meiosis, the part opposite the deletion cannot do so. It forms a loop.

Segments of chromosomes that occur thrice in the cell instead of the normal two times are called *duplications*. The duplicated part may be attached to one of the chromosomes to which that segment belongs, or to some other chromosome, or it may even be inserted in one of these. With any genes in such a duplication occurring thrice instead of twice, normal Mendelian ratios for those genes are not to be expected. Aneuploid cells could result from faulty meiosis.

Inversions

An inversion is a segment of a chromosome turned end-for-end within the chromosome. If the normal arrangement of genes in a chromosome were *a b c d e f g h,* an inversion might place them *a b e d c f g h.* Presumably, inversions result from breaks at two places followed by recovery of the fragment thus cut out, but with that fragment reversed from its normal position. In pairs of chromosomes heterozygous for such an inversion, crossing-over may result in fragments with two centromeres or none, and most gametes are then nonviable, but, when (through recombination) both members of the pair carry the same inversion, then crossing-over is normal, but the linkage relations of genes in the inverted segment will, of course, be different from those in chromosomes not carrying the inversion.

Studying spermatogenesis in healthy bulls with low fertility at the Royal Veterinary College in Stockholm, Knudsen found four that were heterozygous for an inversion. These were identified by the presence of typical inversion loops at meiotic divisions of primary spermatocytes and by characteristic *inversion bridges* (Figs. 18-4 and 18-5). The inversion bridge is a consequence of crossing-over within the loop when the centromere of the chromosome lies outside it. One result of such a cross-over is the formation of a *dicentric* chromatid with two centromeres instead of one. As the other chromosomes go to opposite poles of the spindle, the two centromeres of the dicentric chromatid pull in opposite directions to form an inversion bridge—and a gamete that is non-viable.

Clearly, further studies like Knudsen's would be helpful, not only in elucidating causes of low fertility, but also, perhaps, in early detection of chromosomal aberrations responsible therefor.

Translocations

Earlier we considered briefly the inversions resulting from breaks in a chromosome followed by restoration (albeit the wrong way around) of the fragment thus cut out. Sometimes such a fragment fails to get back where it belongs, and joins up with an end of some other, non-homologous chromosome to form a *translocation.* In some cases

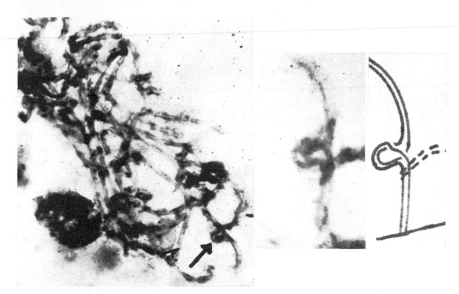

FIG. 18-4 Abnormal behavior of chromosomes at meiosis of primary spermatocytes of an infertile bull that proved to be heterozygous for an inversion. Left, pachytene stage with arrow pointing to an inversion loop, and (right) the same loop, greatly enlarged, with diagrammatic representation. Photographs through electron microscope. (From O. Knudsen in *Internat. J. of Fertility*.)

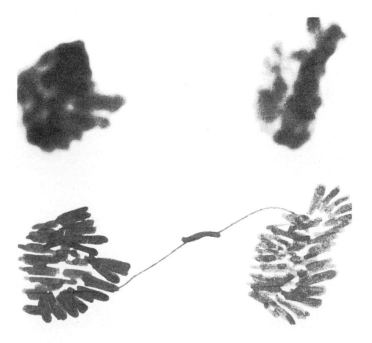

FIG. 18-5 Chromosomes (at late anaphase of meiosis) of bull heterozygous for an inversion, showing inversion bridge resulting from dicentric chromatid. Photograph through electron microscope, with diagram below for greater detail. (From O. Knudsen in *Internat. J. of Fertility*.)

non-homologous chromosomes actually exchange end segments in *reciprocal translocations,* sometimes called *segmental interchanges.* These can be recognized (in microscopic preparations of cells in meiosis) by characteristic configurations like crosses, or (at later stages) rings formed around them.

Organisms heterozygous for reciprocal translocations produce gametes some of which are unbalanced because they lack a piece of one chromosome, but carry a duplication of another. Because such gametes would form zygotes incapable of normal development, fertility (as measured by normal, viable offspring) of plants or animals heterozygous for translocations is subnormal.

A Swedish Yorkshire boar that was investigated because he sired small litters was found by Hageltorn *et al.* (1976) to be heterozygous for a reciprocal translocation. As a result, the boar's karyotype showed the expected 19 pairs of chromosomes, but one normal Chromosome 13 (in descending order of length) had to be paired with a very short partner, and one normal Chromosome 14 with an unusually long one (Fig. 18-6). By special techniques for staining the transverse bands, and photoelectric recordings of the varying densities along the chromosomes, it was recognized that almost all of Chromosome 13 had become attached to almost all of Chromosome 14. The two small remnants of numbers 13 and 14 had joined together (Fig. 18-7). According to recommendations of a conference of cytogeneticists at

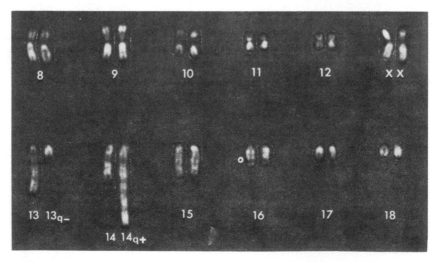

FIG. 18-6 The 12 smaller pairs of chromosomes of the pig, with G-banding showing that the translocation involves Chromosomes 13 and 14, as described in context. (From Hageltorn *et al.* in *Hereditas* as cited.)

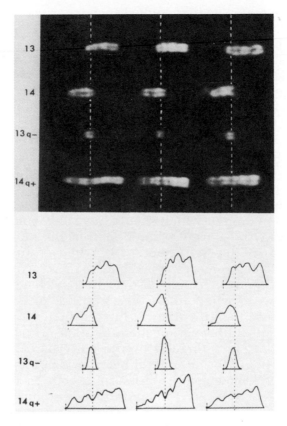

FIG. 18-7 Above: Q-banding (fluorescence), and below: Photoelectric recordings of density along the lengths of Chromosomes 13 and 14, and translocations in the pig, with presumptive points of breakage and fusion as indicated by the vertical broken lines in the three karyotypes. Centromeres of the chromosomes are at the left. (From Hageltorn *et al.* in *Hereditas* as cited.)

Reading (England) in 1976, this translocation should be labeled 38,XY t(13q-; 14q+). In controlled progeny tests it was found that about half of the boar's offspring had received the translocation from their sire and that the rest had normal karyotypes.

Robertsonian Translocations

A type of translocation that has been found in several species of domestic mammals results from fusion of the centromeres of two different chromosomes. It is commonly called a Robertsonian translocation after the man who first studied such translocations in insects of the order Orthoptera.

A chromosome is said to be *acrocentric* (or *telocentric*) when its centromere is at one end, and *metacentric* when the centromere is near the middle as in many X-chromosomes. Fusion of the centromeres of two acrocentric chromosomes results in the formulation of a new chromosome, which is metacentric if its founding chromosomes were of equal length, or *sub-metacentric* if one was longer than the other. In a species having all the autosomes normally acrocentric (*e.g.*, cattle, dogs), such a new chromosome (translocation) is easily recognized by its extra length and two arms. It thus resembles the X-chromosome.

When the cytologist finds two such metacentric chromosomes in cells of a male mammal (which should have only one), he knows that one of them may be a translocation or that his material may have come from a Klinefelter (see Chapter 19). When a translocation caused by fusion of two centromeres is present, the karyotype will show one unpaired autosome, and usually also (by its extra length and metacentric centromere) to what other autosome the missing partner has become attached. Identification is further facilitated by the newer staining techniques that permit identification of individual chromosomes (and even parts thereof) by the patterns of their transverse bands. Furthermore, for an animal heterozygous for a translocation, the karyotype will show not 2n chromosomes but 2n-1. In the homozygotes, which are much rarer, the number of chromosomes is 2n-2.

There can be different kinds of Robertsonian translocations in one and the same species, and they can occur in different breeds. Four of these known in the dog are listed in Table 18-2.

The normal diploid number of chromosomes in the dog is 78, but a translocation of the Robertsonian type reduces the number in heterozygotes to 77, and in homozygotes to 76. As taxonomists are likely to classify as separate species (or as sub-species) animals that differ in chromosome numbers, there has been some speculation about the possible role of translocations in the differentiation of species. In the four cases of Table 18-2, no abnormal effects could be attributed specifically to the translocations, but, as we shall see shortly, when breeding tests are made, there is ample evidence that a translocation can reduce the fertility of animals carrying it.

For translocations known in cattle, sheep, goats, pigs, and dogs, one should consult the useful monograph of Hare and Singh (1979). Apparently, none has yet been reported in horses or in cats, but that probably means only that they have been less sought in those two species.

TABLE 18-2
Robertsonian translocations in dogs

Breed	Chromosomes Involved	Authority
Miniature Poodle	1/19	Hare & Bovee (1974)
Golden Retriever	13/17	Larsen *et al.* (1978)
Crossbred	15/35	Ma & Gilmore (1971)
Crossbred Terrier	2/10	Shive *et al.* (1965)

The 1/29 Translocation in Cattle

No chromosomal abnormality in domestic animals has been more thoroughly investigated than the 1/29 translocation studied by Gustavsson and his associates in cattle (Fig. 18.8, 18.9). It was first detected in 1964 (see references) in the Swedish Red and White breed, but it has now been reported in about 30 different breeds in 20 countries. A full description, along with studies of its effects on fertility, was given by Gustavsson (1969), and his 1979 review summarizes what was learned in 15 years about its distribution and effects.

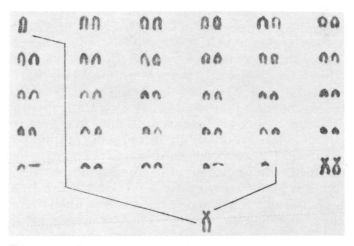

FIG. 18-8 Karotype of a cow heterozygous for the 1/29 translocation showing its origin. (From Gustavsson and Rockborn in *Nature* **203**:990. Reprinted by permission. Copyright © 1964 MacMillan Journals Limited).

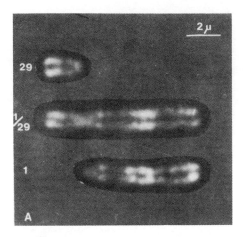

FIG. 18-9 R-banding patterns of the 1/29 translocation and of the two chromosomes that fuse to make it. (From Gustavsson and Hageltorn in *J. Heredity* 67:175-78. Copyright © 1976 by the Amer. Genetic Association).

The currently wide distribution of the translocation is believed to have resulted mostly from the international trade in cattle rather than from recurrent mutation, but, because it has been found in some non-European breeds that do not ordinarily travel afar, the possibility of recurrent mutation can not be ruled out. The translocation has no visible effects on animals carrying it, and, until its adverse effects on fertility became known, there had been no selection in the Swedish cattle either for it or against it. It is more common in beef breeds than in dairy cattle. In two extensive cytogenetic surveys of the Swedish Red and White breed, the frequencies of animals carrying the translocation were found to be 12.4 and 14.3 per cent.

Because a translocation could cause formation of faulty gametes, with resultant imbalance of chromosomes and early death of embryos, some reduction in fertility of bulls heterozygous for the 1/29 translocation was possible. It was found. In these days of artificial insemination, fertility in cattle is measured by the proportion of cows served that does not have to be returned for a second service. In a controlled test in which 12 bulls that carried the translocation were compared with 45 others that did not, fertility was determined by non-returns at 28 days and at 56 days. For the 12 carriers, fertility was lower than that of the normal bulls by 3.0 to 3.5 per cent, when measured by non-returns at 28 days and by 4.5 per cent at 56 days. Among daughters of bulls heterozygous for the translocation, fertility was reduced by 3 to 6 per cent, presumably as a result of early embryonic mortality. Similar results were reported for daughters of five bulls carrying the 1/29 translocation in the Norwegian Red breed. Because of such findings, it is now routine procedure in several countries to screen cytogenetically the bulls used in stations for artificial insemination.

It should be made clear that the comparatively slight reduction in fertility just described applies only to the 1/29 translocation in cattle. Other translocations, whether in cattle or in other species, could be less serious or worse.

Useful Translocations

Translocations are usually disadvantageous to an organism because they lower viability or the efficiency of reproduction, but in species amenable to experimental treatment, translocations can sometimes be utilized to advantage in animal husbandry. Yokoyama lists eight translocations that had been induced in the silkworm up to 1959, and two of these had been turned to remarkable economic advantage. Male caterpillars are preferred over females because the former produce more silk. Tazima found a strain carrying a spontaneous translocation—a fragment of chromosome II attached to the Y-chromosome. (Female silkworms are XY; males XX.) The translocated fragment carried a gene (sable) which causes spotting of the larval skin, hence the spotted females could be distinguished easily from the males, which were non-spotted. Because of the unbalanced condition caused by the extra genes in the fragment, these females were not hardy enough to be useful in practice. However, after repeated irradiation with X-rays, Tazima succeeded in breaking off the superfluous part of the translocation, but leaving the valuable sable gene still on the Y-chromosome. The resulting stock, thus freed of its handicap, was hardy enough for commercial use, and very valuable therefor because it permitted early and easy identification of the males that the industry preferred.

A translocation experimentally induced by Tazima proved even more valuable because it permitted identification of sex at an age still earlier—in the egg. In this case the Y-chromosome acquired a small translocation carrying a gene that causes the eggs to be black. As the X-chromosomes lacked that gene, white eggs produced males, and black ones females (Fig. 18-10). A means of sorting the eggs by photoelectric cells was devised, and the growers were thus provided with eggs that could yield only males. Tazima's accounts of this interesting work are cited in the monograph by Yokoyama, from which this report is taken.

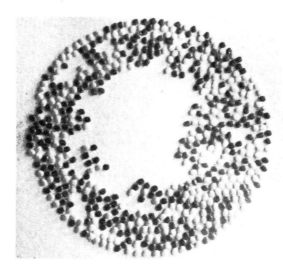

FIG. 18-10 Sex difference in eggs of the silkworm produced artificially by inducing a translocation to the Y-chromosome. Black eggs produce females, white ones males. (From Y. Tazima, by courtesy of T. Yokoyama and the Sericultural Experiment Station, Japan.)

In Brief

Mutations occur, and recur, spontaneously. Over unknown thousands of years those most useful to animals in the wild have been retained. Others have been preserved since wild animals were domesticated and selection to improve them was begun. The rate of mutation can be raised by various mutagens. Most of those induced by X-rays are lethal. Polyploid series are common in plants but rare in animals. While many polyploid plants are economically very valuable, no polyploid animals of value have yet been produced.

Parthenogenetic development of eggs produces male honey-bees and poults, the former being monoploid, the latter diploid. Chromosomal aberrations commonly cause poor reproduction, but translocations of great potential economic value have been experimentally induced in silkworms. The role of chromosomal aberrations in reducing fertility of domestic animals merits further study.

Problems

18-1 Back in Problem 2-24, we have considered the aniridia transmitted by the Belgian stallion, Godvän. What evidence was there to show that this condition had arisen by mutation?

18-2 Rank the following kinds of mutations in order according to the rapidity with which they would become evident in animals after the change in the gene: (a) a recessive mutation in an autosome, (b) a dominant mutation in an autosome, (c) a recessive mutation in the X-chromosome.

18-3 What dominant mutations in domestic animals can you recall that we have considered earlier, and which have been of great economic value? If you can't think of any, look again in Chapters 4 (cattle), 7 (fowl), and 11 (fox).

18-4 Newspapers delight in reporting abnormal animals of all kinds, from pink-eyed, albinotic ground-hogs, frogs, or deer, to chickens with three legs and calves with two heads. Some of these result from gene mutations, others from accidents in development. By what kinds of criteria might you distinguish between the genetic and non-genetic variations (a) if the animals were available for breeding tests, or (b) if they weren't?

18-5 Compare the Silkie in Fig. 18-1 with the jungle fowl in Fig. 8-1 with respect to ability to survive in the wild. How would each of the mutations characteristic of the Silkie affect its chances of survival?

18-6 Would experimental induction of mutations that might improve yields or productivity be equally valuable in (a) animals, (b) cross-fertilizing plants, and (c) self-fertilizing plants? If not, why not?

18-7 Why are black feathers more common in males heterozygous for barring than in pure Barred Plymouth Rock males?

18-8 Singer and MacIlraith noticed that Barred Rock females carrying numerous black feathers also showed a greenish-grey iris more often then hens of the same flock that lacked black feathers. If the greenish-grey color results from extra melanin in the iris, can you account for the association?

18-9 Do you know the "wild type" for any domestic animal bigger than a fox? What breed of dogs comes closest to it, in your opinion?

18-10 A few years ago, a dealer in lobsters in Kennebunkport, Maine, found one that was half crimson (*i.e.*, one whole side) and half black. If you were asked to account for it, what would you say?

18-11 If you keep an eye out for them, you will occasionally find an apple in which a whole sector, from stem to calyx, is colored differently from the rest, or a Bartlett pear with a sector of russet skin. Such a fruit is a *sectorial chimera*, having genetically different tissues in the two parts. In what ways could such interesting curiosities be caused?

18-12 On eugenic grounds, would you advise a queen bee to mate with a drone from her own hive, or to seek out another, as some people believe they do? Why?

18-13 Most of us animals carry around a load of undesirable mutations which Dobzhansky has labelled collectively as "genetic junk." Would you expect *Apis mellifera* to have more of this sort of stuff than *Homo sapiens*, or less, and in either case, why?

18-14 Most trisomic individuals result from non-disjunction of chromosomes. Mongolism in man is a consequence of trisomy, but its frequency rises as does age of the mother above 35 years. What does this suggest about factors contributing to non-disjunction?

18-15 After reviewing processes by which all of Olsen's parthenogenetic turkeys might become diploid, can you suggest any reason why most of them had to be helped from their shells?

18-16 Would you expect offspring of the male turkey shown in Fig. 18-3 to be similarly weak, or not? Why?

18-17 From what indications would you eliminate other possible causes and suspect that low fertility in some strain of domestic animals could be caused by an inversion or a translocation?

Selected Referencs

BLOOM, S. E. 1972. Chromosome abnormalities in chicken (*Gallus domesticus*) embryos: Types, frequencies, and phenotypic effects. *Chromosoma* **37**:309–26. (In 4,182 embryos examined at 4 days of incubation.)

GUSTAVSSON, I. 1969. Cytogenetics, distribution, and phenotypic effects of a translocation in Swedish cattle. *Hereditas* **63**:68–169. (Including extensive studies of its effects on fertility.)

GUSTAVSSON, I. 1979. Distribution and effects of the 1/29 Robertsonian translocation in cattle. *J. Dairy Sci.* **62**:825–35. (A review covering 15 years of study.)

GUSTAVSSON, I., and G. ROCKBORN. 1964. Chromosome abnormality in three cases of lymphatic leukaemia in cattle. *Nature* **203**:990. (It was not related to the leukaemia.)

HAGELTORN, M., I. GUSTAVSSON, and L. ZECH. 1976. Detailed analysis of a reciprocal translocation (13q-;14q+) in the domestic pig by G- and Q-staining techniques. *Hereditas* **83**:268–72. (Source of Figs. 18-6 and 18-7.)

HARE, W. C. D., and K. BOVEE. 1974. A chromosomal translocation in Miniature Poodles? *Vet. Rec.* **95**:217–18. (As listed in Table 18-2).

HARE, W. C. D., and E. L. SINGH. 1979. *Cytogenetics in Animal Reproduction.* Slough (England) Commonwealth Agric. Bureaux. (Chromosomal abnormalities in domestic mammals and their effects. Methods of preparation of cells for study and of five banding techniques are described. Extensive bibliography. 96 pp., plus 42 figures.)

KNUDSEN, O. 1958. Studies on spermiocytogenesis in the bull. *Internat. J. Fertil.* **3**:389–403. (Details of his cytological studies of infertile bulls, and source of Figs. 18-4 and 18-5.)

LARSEN, R. E., E. DIAS, and J. CERVENKA. 1978. Centric fusion of autosomal chromosomes in a bitch and offspring. *Amer. J. Vet. Res.* **39**:861–64. (She was heterozygous for a translocation of Chromosomes 13 and 17; so were about half of her pups.)

MA, N. S. F., and C. E. GILMORE. 1971. Chromosomal abnormality in a phenotypically and clinically normal dog. *Cytogenet.* **10**:254–59. (Centric fusion of Chromosomes 15 and 35.)

McFEELY, R. A. 1967. Chromosome abnormalities in early embryos of the pig. *J. Reprod. Fertil.* **13:**579–81. (In 10-day blastocysts.)

OLSEN, M. W. 1960. Nine year summary of parthenogenesis in turkeys. *Proc. Soc. Exper. Biol.* **105:**279–81. (A concise review; 926 virgins produced 68 poults.)

SHIVE, R. J., W. C. D. HARE, and D. F. PATTERSON. 1965. Chromosome studies in dogs with congenital cardiac defects. *Cytogenet.* **4:**340–48. (As cited in Table 18-2. The translocation was not responsible for the heart defect.)

YOKOYAMA, T. 1959. *Silkworm Genetics Illustrated.* Tokyo. Japan Society for the Promotion of Science. (With lists of induced mutations and details of the translocations mentioned in this chapter. In English.)

CHAPTER

19

Intersexes, Sex Determination, and Sex Ratios

Back in Chapter 7, where we considered sex determination and sex linkage, the reader was warned that the situation is not quite so simple as it was there presented, and that complications lay ahead. Here they come. They are very interesting, not just because sex is hardly a dull subject (insofar as can be judged by the unusually rapt attention of students during lectures about it), but also because after centuries of speculation about sex determination, including fifty years of genetics as a science, something new about sex determination in mammals was learned as recently as the decade of the 1970's.

That discovery was so long delayed for two reasons. In the first place, geneticists accepted somewhat too readily the concept that what the intersexes in Bridges' fruit flies had shown him about sex determination in Drosophila applied equally well to other species (including all the mammals) in which females are XX and males XY. We know now that it doesn't. In the second place, it was not until the newer techniques of cytology were applied to man and other mammals after 1950 that it became possible to learn much about the chromosomes of intersexes in such species.

If, in species having heterogametic males, $XY = \male$, and $XX = \female$, how, then, do intersexes arise? Similarly, if normal gametogenesis in the heterogametic sex results in the production of $\male$-determining and $\female$-determining gametes in equal numbers, why then is the sex ratio at birth not always $1:1$, instead of being, as in some mammals, $105\male\male : 100\female\female$? This chapter is written to answer these questions and others related to them.

Sex Determined by the Environment

To facilitate recognition of the fact that sex is not always determined by the presence of one sex chromosome or two, let us consider a few cases in lower animals in which sex is apparently determined more by the environment than anything else. The classical example in this class is provided by the marine annelid, Bonellia, a species remarkable not only for its casual treatment of sex, but also for its extreme sex dimorphism. The female has a bulky body about two inches long, and shaped something like a bean, with a proboscis five times as long, and forked at the distal end. The male is a minute creature which lives as a parasite in the uterus of the female. Baltzer found that in this species sex is determined almost by pure chance. The free-swimming larvae are sexually neutral. If one settles on the bottom of the sea, it becomes a female, but if by chance, or possibly even by attraction, a larva settles anywhere on the long proboscis of the female, then it develops into a male. When larvae that had settled on a female proboscis were removed therefrom and forced to continue their development away from females, Baltzer found that they became intersexes, and that the degree of intersexuality depended on the length of time during which they had rested on the female.

A somewhat parallel case is found in the slipper limpet, Crepidula, a well-known pest in oyster beds. The free-swimming larva settles on an older limpet and goes in succession through a neutral stage, then a male phase in which it produces spermatozoa, next a period of hermaphroditism during which it produces both eggs and spermatozoa, and finally becomes a female, producing eggs only. If the young larva is prevented from settling on a limpet that is larger and older (hence a female), the male phase is eliminated and the larva goes directly from its indifferent state to that of a functional female.

Sex Determination in the Honey-Bee

If we consider that full sexual development is not attained unless the individual is fertile, a case of environmental modification of sex is provided by a familiar domestic animal, the honey-bee. A single mating, which occurs during a nuptial flight by the queen and is followed immediately by the death of the drone, provides the queen with enough spermatozoa to last for her lifetime, which might be four or five years. A few of her eggs in each brood are not fertilized, and these give rise to drones (Fig. 19-1).

The question whether a fertilized egg becomes a normal, fertile female (a queen) or a sterile one (a worker) depends entirely on an environmental influence—royal jelly. This whitish substance, produced in certain glands in the heads of worker bees, is a normal part of the diet for all bees. Ordinary female larvae, whose destiny holds for them no hope of posterity but the prospect only of life as busy little bees improving the shining hours, are fed royal jelly for two to three days only, and become adults in about 21 days from the egg. For a prospective queen, matters are different. The cell in which she is to develop is enlarged by the workers, and in her larval stage she is supplied with royal jelly for five days. The result of this preferential treatment (*i.e.*, environment) is that she emerges only about 16 days from the egg, bigger and better than any of her less-favored sisters, and able to reproduce.

Our apicultural friends tell us that the queen is capable of laying her own weight in eggs in one day, and that this prodigious output is possible only because she is again supplied with royal jelly during the process. It is scarcely surprising that this magical compound, which can apparently be collected fairly easily, should be extolled in another

FIG. 19-1 Honey-bees, showing differences among queen (below), worker (upper left), and drone (right). The queen has been marked with white paint on her thorax. (Courtesy of Maurice Smith.)

species as a potent elixir, able to convert to queens the wrinkling drudges of other hives, to delay departure from this vale of tears, and to restore the lost virility of tired old men.

Modification of Sex by Autosomal Genes

In Drosophila and in other species, sex is influenced, as we shall see shortly, by an unknown number of genes in the autosomes. All of these together tend to induce maleness in Drosophila. A single recessive gene with remarkable effects on sex in that species was discovered by Sturtevant, who called it "transformer" because (when homozygous) it converted diploid females to a phenotype little different from that of normal males. The transformed flies showed normal male genitalia and male color. They even had testes, but these were smaller than normal, and the flies were sterile.

In certain Cyprinidont fishes, autosomal genes play an important role in determination of sex. Winge developed a strain of *Lebistes reticulatus* in which all fishes were XX and sex was controlled by a pair of autosomal genes. From his studies, and those of Gordon with diffcrent races of *Xiphophorus helleri*, it would appear that the sex chromosomes in these fishes differ little from autosomes, and that under some circumstances, such as the accumulation by selection of genes influencing sex, one of the autosomes could take over the role of a sex chromosome.

In larger animals, the gene causing goats to be polled provides a good example of an autosomal gene affecting sex (Chapter 13), because females homozygous for that gene are intersexes.

Balance of Sex Chromosomes and Autosomes in Drosophila

After discovery of a number of intersexes in the progeny from a single mating, and of abnormal segregation of certain genes in the same stock, Bridges concluded that the mother of the intersexes was a triploid and that the intersexes were aneuploids. Subsequent experiments, along with flies from other sources, led to recognition of various types of flies ranging from "superfemales" to "supermales."

If all the chromosomes in one set of autosomes be designated as A, then, when the ratio of X to A is 1:1, the fly is a female, whether it is a normal fly (2:2), a triploid (3:3), or a tetraploid (4:4). Similarly, while the ratio for normal males is 1X:2A, those having 2X:4A are also normal. Flies in which the two kinds of chromosomes deviate from these ratios are abnormal. With an extra X-chromosome they become superfemales; with one fewer than they should have, they are intersexes; and with a ratio of 1X:3A they are supermales.

Triploid females are fertile, but the intersexes and the so-called superfemales and supermales are not. Actually, these "super" flies are not Drosophiline Amazons or Tarzans, but are weak and subnormal in viability. Their name indicates only that they exceed the normal X/A ratios.

The H-Y Antigen

The general conclusion from these studies, that a balance between genes in the sex chromosomes and others in the autosomes is the decisive factor in determining sex, was considered for many years to apply not only to Drosophila, but also to most other forms having the simple XX-XY or XX-XO distribution of sex chromosomes. A corollary, therefore, was that the Y-chromosome plays no role in sex determination in Drosophila, and for many years this was believed to apply also to other species, including mammals. It was known, however, that fruit flies lacking Y are sterile, and also that that chromosome did carry some genes, so it was not considered entirely inert.

That fairly simple concept of the Y-chromosome had to be abandoned when new discoveries showed it to be far more important in the determination of sex than had hitherto been supposed. It was found to carry a histocompatibility locus, so called because a gene (or genes) there causes females of certain inbred strains of mice to reject skin grafts from males of the same strain. It does so by producing an antigen (found on the surface of male cells) which is called the histocompatibility-Y antigen or H-Y. To this, lymphoid cells of the rejecting inbred strains are strongly antagonistic, but in some strains the female lymphoid cells are not rejectors, and in others they are only weakly antagonistic.

Apart entirely from the rejection of grafts, the H-Y antigen plays an all-important role in the determination of sex. When it is present, the undifferentiated gonad of the mammalian embryo develops into a testis; when it is not present, that gonad becomes an ovary. H-Y is associated with the heterogametic sex rather than with males only. In birds, which are heterogametic, the H-Y antigen is present on the cells of females, is associated with the W-chromosome, and is involved in primary differentiation of the ovary—not of the testis. As usual in biology, there are still a few non-conforming exceptions to be explained.

Details of the discovery of the H-Y antigen and of its effects are given by Ohno (1979). According to Wachtel (1977), its full effectiveness may require supplementary action by other genes.

Sex Determination in *Bombyx mori*

Studies by Japanese geneticists with the silkworm, *Bombyx mori*, have shown that in the moth the ♂-determining forces in the Z-chromosome (=X) are comparatively weak, and that sex is determined entirely by the presence or absence of the W-chromosome (=Y). Moths that are ZZW, or even ZZZW, are females, *not* intersexes. (Table 19-1).

From studies of strains that had deletions from different regions of the Y-chromosome, Tazima found that the ♀-determining force is not scattered along the whole length of that chromosome, but is localized in one particular region of it. This, again, is quite different from the situation in Drosophila, in which the ♂-determining genes resident in the autosomes are apparently distributed through chromosomes II and III, but are absent from IV.

TABLE 19-1

Sexes in Various Combinations of Sex Chromosomes and Autosomes in the Silkworm

Ploidy	Autosomal Sets (A) (number)	Z-Chromosomes[a] (number)	W-Chromosomes[a] (number)	Sex
2n	AA	ZZ		Male
		Z	W	Female
		Z	II·W·Z^L	Female[b]
3n	AAA	ZZ		Male
		ZZZ		Male
		ZZ	W	Female
		ZZ	WW	Female
4n	AAAA	ZZZ	W	Female
		ZZ	WW	Female

[a]Z and W = X and Y.
[b]These females carried in their W-chromosome translocations from Autosome II and from Z.
Source: From Yokoyama, p. 47.

From the few examples considered thus far, and from many other variations that are known, it seems clear that caution is necessary before assuming that what is known about sex determination in one species is applicable to others, especially in groups that are phylogenetically far apart.

Genetic Intersexes in Mice and Men

In the previous chapter we learned that most trisomic individuals arise from failure of two homologous chromosomes to separate at meiosis—a process known as non-disjunction. When the two that fail to separate are the X-chromosomes of a (female) mammal, the resulting female gametes, instead of carrying each an X-chromosome and being all equipotential in sex discrimination, have either two X's or none. These unusual gametes give rise to unusual individuals. Such abnormal gametes are designated XX and O. Their fusion with normal male gametes (half carrying X and the remainder Y) yields aneuploid zygotes which are shown in Fig. 19-2. The names and chromosome numbers there assigned to the four types of zygotes are those for the corresponding conditions in man. All have normal autosomes (*i.e.,* 2 A = 44).

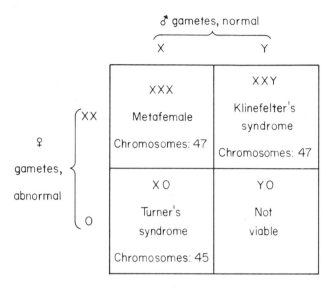

FIG. 19-2. Abnormal offspring resulting from non-disjunction of X-chromosomes in a woman mated with a normal male.

Turner's syndrome, XO, is found in females who are usually short in stature, show infantilism, lack normal secondary sex characters of a female, and are mostly sterile. The ovaries are usually absent or rudimentary, but at least one case is known to have had offspring. This XO type has also been found in the mouse, in which species it is fertile and apparently indistinguishable from normal females unless it happens to carry some sex-linked genes that should show only in males.

Klinefelter's syndrome, XXY, is found in males that are normal in appearance and behavior, but have small testes in which spermatogenesis does not occur. Gynecomastia (enlargement of the breasts) is usual. Reports indicate that, contrary to earlier belief, the frequency of Klinefelter's syndrome is remarkably high—even as common as 1 in 400 or 500. The XXY condition has also been found in the mouse. As we shall see shortly, there is evidence that it occurs also in cats.

YO zygotes have not been found in man or mouse and are presumably non-viable as in lower forms. Even if any mammals of this type survived to be born, it seems possible that they would soon perish because the X-chromosome, which they lack entirely, has been shown (in man and in the dog) to carry at least two genes necessary for normal blood clotting.

Metafemales, XXX, or triple-X's, are reported to show little abnormality and some, at least, are fertile. This is the type long designated in Drosophila as superfemales. Now that the same condition has been found in man, Stern's suggestion that it be called metafemale will be welcomed by every teacher who has had to explain to hopeful male students that the term *superfemale*, as used in human genetics, does not mean 40–24–39.

According to most reports, persons with any of the three viable syndromes just considered are subnormal in mental ability, but further evidence on this point is desirable.

In addition to the abnormalities in man briefly considered in this section, many baffling cases of intersexuality have been described, some of which were designated as hermaphrodites, and others as *gynandromorphs* (to be considered shortly). After the comparatively simple variations shown in Fig. 19-2 had been identified, cytological studies of intersexual individuals revealed more complex combinations of X- and Y-chromosomes. Those in man, which include females having up to five X-chromosomes, and males with three or four Y-chromosomes, were reviewed by Stern (1973). In domestic animals, fewer cytological studies have been made of intersexes, but it is reasonable to expect that abnormalities of the sex chromosomes found in man will also occur in other mammals. Those already described in domestic animals are thoroughly reviewed by Benirschke (1981). The XXY male (Klinefelter's syndrome) has been identified in the cat, dog, pig, cattle, and sheep. Chromosomes of an XXY Hereford bull studied by Dunn *et al.* (1980) are shown in Fig. 19-3.

In some intersexual animals, all cells of the body have the same chromosomal abnormality. In others, most cells are normal, but some have an abnormal number of sex chromosomes. Animals with such a mixture are cellular mosaics, or *chimeras*. The proportion of such exceptional cells could vary, depending on whether the abnormality of sex chromosomes was present (in a gamete) at fertilization, or occurred later by aberrant behavior of a chromosome during mitosis.

Finally, one significant finding from all this study of intersexes is that in mammals, unlike the situation in Drosophila, but much like that in the silkworm, it is the Y-chromosome that makes a male a male.

Sex Chromatin, or Barr Bodies

One difficulty besetting the study of intersexes and hermaphrodites in man in earlier years was that of knowing the genetic sex of the abnormal individuals. Without knowing whether the original zygote was male or female, deviations from the normal phenotype for either were difficult to interpret. To a very large degree, if not entirely, this

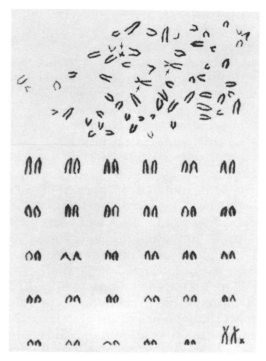

FIG. 19-3 Metaphase chromosomes of a Klinefelter bull. The arrows point to the big X-chromosomes and the small Y. (From Dunn *et al.* in *Cornell Vet.* as cited.)

problem was solved by studies beginning in 1949 when M. L. Barr discovered that somatic cells of normal men and women differ by having or lacking a small body of chromatin which can be detected in suitably stained cells (Fig. 19-4). This material has been called sex chromatin, "Barr body," and nuclear chromatin. It is usually adherent to the inner surface of the nuclear membrane, but, as it is only about one micron in diameter (about 25,000 microns make an inch), good preparations (thin sections, properly stained) and experience are necessary for its identification. Cells from the mucous membrane of the mouth are commonly examined in smear preparations. In sections of other tissues not over 5 microns in thickness, sex chromatin can be recognized in 60 to 80 per cent of cells from females, and in particularly good preparations it can be seen in almost every female nucleus. Conversely, it is rarely found in cells from males, but up to 10 per cent of nuclei from male somatic cells contain a tiny body of chromatin smaller than the sex chromatin of females, the exact significance of which has yet to be determined.

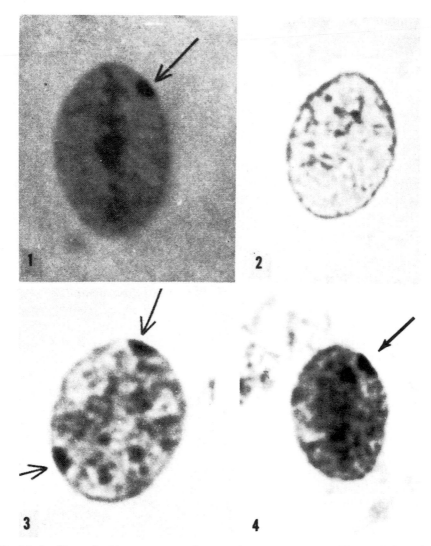

FIG. 19-4 Barr bodies, or sex chromatin, in man. 1. Normal female. 2. Normal male, with no Barr body. 3. Two bodies, as in XXX females and XXXY males. 4. Enlarged Barr body in female having one X-chromosome, and one large "isochromosome" believed to have been formed from the long arm of the X-chromosome. (Courtesy of M. A. Ferguson-Smith, V. A. McKusick, and *Quart. Rev. Biol.*)

Search for sex chromatin in human intersexes has revealed that Turners (XO), like males, have none. Klinefelters (XXY), like females, have one Barr body, and among the rare individuals who have three or four X-chromosomes the Barr bodies number one fewer than those chromosomes. These facts and others have led to recognition of a situation that is practically a rule, namely, that the number of Barr bodies is always one less than the number of X-chromosomes. Accordingly, normal males should have no Barr bodies, and females should have one. With any deviations from these numbers, some abnormality in the numbers of X-chromosomes is indicated, and the number of Barr bodies should give a clue to the nature of the deviation from normal. In other words, genetic intersexes arising from too many X-chromosomes, or too few, can be distinguished from any not thus caused.

The earlier opinions of some investigators that sex chromatin is not related to the X-chromosome are at variance with the general rule just given, but that rule, in turn, became evident only when cases with extra X-chromosomes were found. The hypothesis now considered closest to reality is that advanced in 1961 by Mary Lyon and others which postulates that all X-chromosomes more than one are inactivated early in development, and that each such inactivated X becomes a chromatin body.

The Tortoise-shell Tom-cat

As we learned in Chapter 7, the tortoise-shell (calico) cat, with its attractive mixture of white, orange, and black (or tabby) is heterozygous for a pair of sex-linked genes. As the male cat, though heterogametic, can be hemizygous, but not heterozygous, there should be no tortoise-shell males—only females. The proportions in which these last are expected in different matings were given in Table 7-1, where no provision was made for tortoise-shell males. Against the fact that such males should not happen stands the inconvertible fact that they do happen.

The genetic basis for the tortoise-shell male stood for over half a century as one of the great unsolved mysteries of animal genetics. The problem was complicated by the fact that most such males are sterile. They have small testes in which spermatogenesis does not normally occur, but a few fertile tortoise-shell males have been reported. Many ingenious hypotheses were elaborated over the years to account for the tortoise-shell males.

Eventually, Thuline and Norby showed that the tortoise-shell male is (in some cases) the feline counterpart of Klinefelter's syndrome in man—the XXY male (Fig. 19-5). Among 12 males examined

FIG. 19-5 A tortoise-shell male cat, the feline counterpart of Klinefelter's syndrome. (Courtesy of H. C. Thuline.)

which showed three colors, only one was a typical calico with black and orange patches. When screened by search for nuclear chromatin in smears from the buccal mucosa, 10 showed none, but two had Barr bodies, as do normal females. White blood cells from these two animals were grown in tissue-culture for cytological study, and proved to carry 39 chromosomes instead of the 38 that are normal in diploid cells of the cat.

According to Lyon (1962), the mosaic color pattern of the tortoise-shell cat (whether male or female) would result from early inactivation of one X-chromosome (which would then become a Barr body), but with one X being inactivated in some cells, the other one in others. If the X-chromosome inactivated in one cell were to carry the gene for orange (O), all cells descended from that one would carry o, and hence be black. If the chromosome thus inactivated were to carry o, all descendant cells would be orange. The Lyon hypothesis holds that *both* kinds of inactivation could occur in the same animal thus giving rise to the mosaic color pattern. Her theory was not devised to account solely for tortoise-shell cats, but rather to account for Barr bodies and for mammals showing mosaic patterns with respect to any sex-linked genes.

It should be clear that XXY tom-cats do not have to be tortoise-shells. Only when sex-linked genes are involved, with the male showing the heterozygous condition normally found in females, can such intersexes by detected by some "marker" such as color. Presumably the counterpart of Klinefelter's syndrome occurs in other mammals, but has not yet been recognized for the very reason that few sex-linked genes are known in mammals bigger than mice. The XO females studied in mice by the Russells were fertile and normal, and would not have been detected if they had not shown sex-linked characters which were expected in males. As such useful markers are not yet available in the larger domestic animals, it would seem that routine screening for Barr bodies might be useful in any studies seeking to account for subnormal reproduction in the larger domestic animals.

Subsequently, after some tortoise-shell males had been found to have anomalies of the X- and Y-chromosomes more complex than the XXY of Klinefelter's syndrome, Centerwall and Benirschke (1973) examined six more of them and reviewed all the known cases in which the chromosomes had been studied. The normal diploid number in the cat is 38. The deviations from it found in 25 tortoise-shell males illustrate some of the variations that can occur in combinations of X- and Y-chromosomes in mammals (Table 19-2).

Only four cats were simple Klinefelters in all cells. The rest were mosaics carrying two, three, or four "cell lines" (kinds of cells), each with a different combination of X- and Y-chromosomes. Six cats had both normal diploid cells and other cells carrying the triploid number (57). Three others were XY in all cells, as are normal males, but must have had X-chromosomes of two different kinds—one bearing the gene O, the other carrying its allele o. Otherwise, the cat could not have shown the mixture of black and orange that is tortoise-shell.

Centerwall and Benirschke discuss the ways in which these assorted anomalies could have arisen. They cite an estimate by T. C. Jones, based on a population of 17,000 cats, that among tortoise-shell cats only about one in 3,000 is a male.[1] The occasional fertility of tortoise-shell males was attributed to the presence of XY cells in the germinal tissue.

Gynandromorphism

A true hermaphrodite is an individual that has gonads of both sexes and produces both eggs and sperms. An intersex is an individual of one sex in which the accessory ducts, the genitalia, or the secondary sex characters are modified in the direction of the other sex. A *gynandromorph* is quite different from either of these; it is a *sex-mosaic, i.e.,* having part or parts of its body genetically male, and part genetically female.

Gynandromorphs are most easily recognized in animals in which sex dimorphism is very marked, and in which differences between genetically different tissues are not obscured, as they are to a consid-

[1]No wonder that a rumor persisted in New York State for decades that Cornell University would pay $50.00 for a genuine tortoise-shell male cat! It won't.

TABLE 19-2
Chromosomes Found in 25 Tortoise-Shell (or Calico) Cats

One Cell Line	Second Cell Line	Third Cell Line	Fourth Cell Line	Cats (number)
39, XXY	—	—	—	4
38, XX	38, XY	—	—	6
38, XY	38, XY	—	—	3
38, XY	39, XXY	—	—	4
38, XX	57, XXY	—	—	4
38, XY	57, XXY	—	—	2
38, XY	39, XXY	40, XXYY	—	1
38, XX	38, XY	39, XXY	40, XXYY	1

Source: From Centerwall and Benirschke.

erable degree in mammals and birds, by sex hormones that circulate to all parts of the body. Many have been described in insects and other arthropods, and a number in birds. From studies of gynandromorphism in Drosophila, it was concluded that most cases (in that species, at least) result from the loss, at some early stage during development, of one X-chromosome, which lags behind during the anaphase of a mitotic division and is excluded from the daughter-cell to which it should have gone. As a result, only one of the two new cells carries the XX combination, as it should, and the other is XO. Cells and tissues derived from the first type are female (in species having homogametic females), those from the second type are male. It follows that, if the original zygote were heterozygous with respect to sex-linked genes, there would be corresponding differences in the sex-linked characters manifested by the male and female parts. It was this situation, and the fact that many sex-linked mutations are known in Drosophila, which gave the clue to the origin of gynandromophism in that species.

The proportion of the body that is male in such a fly will depend upon the stage to which development had progressed before the X-chromosome was lost. Some gynandromorphs are bilateral, *i.e.*, female on one side and male on the other, thus indicating that the X-chromosome went astray very early, probably at the first equation division of the fertilized egg. In others, only small segments show male tissue.

A few true gynandromorphs have been reported in the fowl, but apparently none has yet been recognized in domestic mammals. Interpretation of the origin of these cases is often very difficult, but one such case, at least, was evidently caused in the same way as are most of those in Drosophila. This bird, a pure Barred Plymouth Rock, had plumage and shanks lighter in color on the right side than on the left. It was clearly homozygous for barring (*BB*) on the right side, but hemizygous (*B—*) on the left. The right side exceeded the left in size by about the normal difference between males and females (Fig. 19-6). On autopsy, a testis and *vas deferens* were found on the larger side, an ovary, oviduct, and rudimentary *vas* on the left one. It seemed likely that the bird had originally been a male, but that very early loss of one sex chromosome (and with it one barring gene) had given rise to cells genetically female, and hence to a bilateral gynandromorph.

Not all gynandromorphs are caused by accidents during mitosis. In the silkworm, Goldschmidt and Katsuki found a hereditary type of gynandromorphism induced by a gene, which, when homozygous, causes retention in the ovum of chromosomes that would normally go into a polar body. As a result, the egg has two nuclei, one of which is ♂-determining, the other ♀-determining. On fertilization of both, the resulting zygote has both male and female parts in equal proportions.

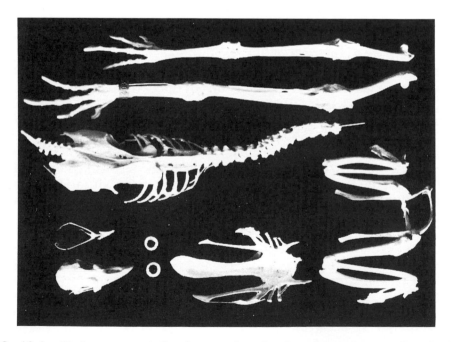

FIG. 19-6 Skeleton, partially disarticulated, of a gynandromorphic fowl, showing larger size on right (male) side than on the left and resultant curvature of every bone in which the two sides are joined together. (From F. B. Hutt in *Genetics of the Fowl*, McGraw-Hill Book Co., Inc.)

Hormonal Intersexes and the Freemartin

The vertebrate intersexes considered hitherto in this chapter were all genetic intersexes, *i.e.*, they resulted from abnormalities of the sex chromosomes. Hormonal intersexes occur when the genotype is normal with respect to sex, but the *sex hormones* (endocrine secretions of the gonads, which circulate in the blood) are abnormal. The freemartin in cattle provides a good example of a naturally occurring intersex caused by such hormones.

The freemartin is an intersex, genetically female, born co-twin to a bull calf, and usually sterile. Although the external genitalia are usually little affected, the uterus is not normally developed and the ovaries may show both ovarian and testicular tissue. Johansson estimated from all available data that about 92 per cent of the heifers born co-twin to bull calves are freemartins. Independent concurrent studies of Keller and Tandler in Austria, and of Lillie in the U.S., led to the same conclusion about the cause of these intersexes. At an early stage of development there is a fusion of fetal membranes, followed by anastomosis of the blood vessels in those membranes, so that any hormone produced in one sex can circulate through the other. It was concluded that male hormones are produced by the testes of the bull calf before the ovary of the heifer calf is fully developed. These male hormones circulating through the female fetus prevent normal development of the ovary and of the female reproductive tract.

Obviously, such vascular anastomosis in fetal membranes and resultant common circulation of blood could cause no complications when both fetuses are of the same sex. It should be remembered also, that from Johansson's estimate, there is no such fusion in about 8 per cent of twins of unlike sexes. Freemartins have also been reported in sheep, but apparently anastomosis of fetal blood vessels in that species is not common. Stormont estimated the frequency of freemartins to be considerably less than one per cent among ewes.

The condition also occurs very rarely in pigs. Intersexes are not rare in swine, but the causes of them have yet to be discovered. A hereditary type has been reported to occur in the New Hebrides, where intersexual pigs are apparently prized for certain religious ceremonies.

Recently, it has been suggested that what causes faulty development of the freemartin's reproductive system is not male hormone from her co-twin's testes, but H-Y antigen from his Y-chromosomes. This theory, which implies that the antigen acts somewhat like a hormone, has not yet been proven correct. One mystery, related and unsolved, is why anastomosis of the fetal membranes, which occurs also in primates, does not cause freemartins in that group. After reviewing all the evidence, Benirschke (1981) concluded that a satisfactory explanation for the occurrence of freemartins is still awaited.

Hormonal Intersexes in the Fowl

The remarkable difference between male and female fowls in the secondary sex characters, particularly the size of the comb and the structure of the plumage in the neck and saddle regions, is caused by differences in the hormones produced by their gonads. Any abnormalities, such as a lack of adequate sex hormone, or an excess in one sex of the hormones appropriate for the other, are quickly reflected in the growth or subsidence of the comb and (after a moult) in the plumage. For this reason the fowl has been a favorite species for study by endocrinologists. The extent to which its secondary sex characters can be modified by gonadectomy, or by transplantation of gonads, is shown in Fig. 19-7.

In the absence of male or female hormones, the neutral fowl has a small, pale, and somewhat shrivelled comb, and long, lacy plumage in the neck or saddle. The counterpart of the capon, the ovariectomized female, or *poularde*, has the same appearance as the capon but is smaller, as is normal. The sex hormones apparently do not affect size; hence the hope of some turkey raisers, that, by removal of the ovary, females of that species could be made to grow to the size of males, though a fertile idea, is also a futile one. Plumage of capons and poulardes is of the same type as in males, but the feathers of the dimorphic areas are not as long in the latter as in the neutral birds. If the poularde is given a graft of testicular tissue, she assumes the upright comb of the male, but the plumage is little affected. On the other hand, when a castrated male gets an ovarian graft, his plumage becomes (after a moult) like that of normal females. A female carrying both ovarian and testicular tissues usually develops the upright comb of the male, but retains the female type of plumage (Fig. 19-7).

Birds in which the gonads have not developed, or do not function normally, are called *developmental capons*, or *developmental poulardes*. Occasionally, the ovary of a female is so inactive at the time of the moult that the new feathers are not influenced by female hormones, as they should be, and hence are like those of the poularde, *i.e.*, much like those of the cock. If in such a bird the ovary subsequently resumes its normal activity, the hen could lay eggs. As her plumage could not be changed until her next moult, such a bird becomes conspicuous as a bird that looks like a cock but lays eggs. In the twentieth century such a hen causes only titillating bewonder-

FIG. 19-7 Modification of secondary sex characters (comb and plumage) by hormones in Brown Leghorns.
1. Male, normal
2. Male, castrated
3. Male, castrated, with engrafted ovary
4. Male with both testis and ovarian graft
5. Female, ovariectomized and with testes, engrafted
6. Female, ovariectomized
7. Female, normal
8. Female, with ovary and engrafted testis
(From G. F. Finlay in *Brit. J. Exper. Biol.*)

ment and speculation, but in days when witchcraft was suspected at every turn, unorthodox behavior of this kind was a serious matter. In 1474, at Basle, one such "cock" was burned at the stake for the heinous and unnatural crime of laying an egg.

The intersex most commonly seen in the fowl is the female that assumes secondary sex characters of the male. Such birds grow combs, and, if they live long enough to go through a moult, acquire plumage of the male type. Some crow vigorously, like males, and even tread other females. In practically every such case that has been examined, the ovary was found to show some pathological condition. Most commonly it is invaded by a tumor. As has been shown experimentally, when the ovary (which normally occurs in the fowl only on the left side) is removed or destroyed, a gonad is subsequently developed on the right side, and that gonad is either a testis or an ovotestis. The male hormone produced by that ovotestis is responsible for the assumption of male characteristics in such birds.

Sex Reversal in the Fowl

Many such masculinized hens have been reported in the literature. For a delightful account of the excitement they have caused over the centuries the record compiled by Forbes, which begins with Aristotle's era, is well worth reading. An extreme case of sex reversal was reported by Crew, who studied a Buff Orpington, which, after serving dutifully as a layer of eggs and a mother, assumed at 3½ years the characteristics of a male, and, a year later, became the sire of two chicks. On its death, the bird was found to have two large testes, with *vasa deferentia*, an ovary almost destroyed by disease, and a small oviduct on the left side. One other such case has since been reported. (See also Problems 19-6 to 19-8.)

Reversal in the opposite direction, *i.e.*, cocks becoming like hens, does not occur spontaneously, but it can be induced experimentally. A 15-milligram pellet of diethylstilbestrol implanted under the skin of male fowls effectively converts them to a female state of metabolism, so that their combs shrink to resemble those of females. Bellicose cocks can thus be tamed, and, if the treatment coincides with the moult, they develop the female type of plumage. A similar effect can be induced in young cockerels by feeding them a synthetic estrogen—dienestrol diacetate.

The facility with which plumage of the fowl can be altered from the type characteristic of one sex to that normal for the other is shown by the pseudo-gynandromorph illustrated in Fig. 19-8. This bird, an F_1 hybrid from the cross: Barred Plymouth Rock ♂ × Silver-spangled Hamburg ♀ , was a male. Lamoreux plucked the feathers on the right side and administered a female hormone while the new ones were growing in. As a result, so far as the plumage could indicate sex, the bird eventually appeared to be male on one side and female on the other.

Sex Ratios in Domestic Animals

Before proceeding to actual sex ratios that have been observed in domestic animals, it may be helpful to list the conditions that would have to prevail to yield a perfect ratio of 1 ♂ : 1 ♀ at birth or hatching.

FIG. 19-8 A pseudo-gynandromorph produced artificially as described in the text. Plumage on the right side of the bird is that of a normal female (left photo), while that on its left side is typical of the male (right photo). (From W. F. Lamoreux in *J. Heredity*.)

These are:

1. The heterogametic sex must produce ♀ -determining and ♂ -determining gametes in equal numbers.
2. Both kinds must be equally viable under all conditions.
3. Neither kind could have any advantage over the other in capacity to fertilize, or to be fertilized.
4. Male and female zygotes must be equally viable up to birth.
5. The number of animals considered must be large enough to be affected only negligibly by chance variations in small numbers. (Any student wondering what small numbers are might toss a coin 1,000 times to see how close he comes to the expected 500 heads : 500 tails.)

Thus far no case has been demonstrated of differential production or viability of gametes in domestic animals, but there are indications that one kind of gamete may be more effective than the other in some species, and there is also evidence of differential viability of zygotes after fertilization.

Sex ratios in some species, if not all, vary according to the age to which they apply. Three kinds of ratios are commonly considered:

Primary: at fertilization of the egg

Secondary: at birth or hatching, in living animals

Tertiary: at any specified later age, as for example, at weaning time or, in man, at age 70

In practice, the secondary sex ratio is the one cited most commonly for domestic mammals. The primary sex ratio can be determined in domestic birds but not in mammals. The tertiary sex ratio can be determined for any age in man, and the same could be done with domestic animals if all of both sexes were kept.

The sex ratio is sometimes expressed as (a) the proportion of males among all the animals considered, and sometimes as (b) the number of males per hundred females. A condensed summary of sex ratios for some domestic animals, as compiled in the *Handbook of Biological Data*, is given in Table 19-3.

It should be clearly understood that the single figures given in this table would apply only to very large numbers, and that in smaller groups considerable deviations are to be expected. For example, while the figure given for cattle is 52, the range within which 95 per cent of observed sex ratios might be expected to fall is 43 to 58. Johansson's figure for 124,166 calves was 51.5 per cent males. In most species of mammals listed in Table 19-3 the proportion of males is slightly over 50 per cent, but a few cases call for interpretation and comment.

In the dog there is a conspicuous excess of males, but a reason for that distinction is not yet known. A summary by Burns and Fraser of sex ratios reported for 18 breeds shows the number of males per 100 females to exceed 100 in all but two reports (Border Collies and Setters), and it seems probable that those two exceptions may be attributable more to chance variations in small numbers than to

TABLE 19-3

Sex Ratios at Birth[a] (Males per 100 Individuals) in Some Domestic Animals

Man, white, U.S.	51.4	Swine	52
Man, not white, U.S.	50.3		
		Horse	52
Cattle, single births	52	Dog, German Shepherds	55
Cattle, twins	49	Dog, Terriers	56
Cattle, various breeds	44–52	Dog, various breeds	51–54
Zebu	51	Mink	50–51
		Rabbit	50–57
Sheep, single births	50		
Sheep, twins and triplets	49	Fowl	49–50
Sheep, Merino	47	Turkey	50

[a]See also list of sex ratios and literature in *Growth Including Reproduction and Development.* Washington. Federation of Amer. Soc. Exper. Biol., 1962. Tables 118–20, pp. 433–37.
Source: From extensive list in *Handbook Biol. Data*, p. 519.

anything else. For 13 of their breeds the figure is above 110, and for 5 of these (Bloodhounds, Retrievers, Airedales, Bulldogs, and Toy breeds) it is over 120. Among 1,440 puppies of breeds not stated, but born under uniform conditions, the ratio observed by Whitney was 124.3 ♂♂ : 100 ♀♀. To what extent the variation in this respect among the 18 breeds listed by Burns and Fraser may be caused by real genetic differences, or by chance fluctuations in numbers not too large to eliminate them, remains to be seen. There are similar indications in other species of significant differences among breeds in sex ratios. The proportion of males among Rhode Island Reds is evidently higher than in other breeds of fowls.

For the horse, six earlier reports consistently showed a slight excess of females at birth, and it is not clear how the figure in the *Handbook* got up to 52. For Thoroughbreds only it is given as 50. The secondary sex ratio in man varies in different races and countries, but it is generally found to be 103 to 107♂♂ : 100♀♀, and most other mammals show a similar slight excess of males.

Differential Viability of the Sexes

The fact that males outnumber females at birth in most mammals is all the more interesting because there is evidence that in some mammals, at least, mortality of fetuses during gestation falls more heavily on males than on females.

Pigs are particularly favorable animals for such studies, partly because they usually have big litters, but also because fetal litters can be readily obtained at almost any abattoir. Both Parkes and Crew found the proportion of males to be very high in the smallest (youngest) fetuses, but to decline at successive stages thereafter. In spite of that differential mortality, there is still an excess of males among pigs born alive (Table 19-3). Crew and Parkes both concluded that the proportion of males at fertilization (the primary sex ratio) must be as high as 60 per cent. That is equivalent to 150♂♂ per 100♀♀.

Somewhat similarly, it has been found in man that among aborted and still-born fetuses there is an excess of males, and this has led to estimates that the primary sex ratio may be upwards of 125♂♂ per 100♀♀. In some species (including sheep and the mouse), search for differential mortality at fetal stages has revealed none.

There has been much speculation about ways in which a great excess of males might be conceived. A favorite theory is that the ♂-determining gamete would have an advantage in any race to the egg because the Y-chromosome is usually smaller than the X; hence the Y-bearing sperm might be lighter and more agile than those encumbered with the X. This is doubtful because the movement of the spermatozoa is not entirely dependent on their own efforts. They are helped along by muscular contractions and by the beating of cilia in the oviducts. It is also possible that the two kinds of gametes may not be equally viable in the reproductive tract. Until more is known, little is to be gained by further speculation.

All of this applies to mammals. In the fowl (which, as has been delicately hinted throughout this book, has many advantages over mammals for some biological investigations) it is comparatively simple to determine the primary sex ratio. During every breeding season, a very few hens will have every egg fertile, and every egg will hatch. With any differential embryonic mortality thus eliminated, Hays found the sex ratio in chicks hatched from such hens to be 432♂♂ : 438♀♀, about as close to equality as anyone could hope to find. In this case the secondary sex ratio was also the primary one.

Apart from pre-natal mortality, it is well known that in most mammals, including man, the females outlive the males. In our own species this has varied consequences, all of them interesting to biologists. Some figures indicate that at ages 65 to 75, the number of males left per 100 females does not exceed 80, while after age 85, the corresponding figure drops to about 55. It is not difficult to see why life-insurance companies pay smaller annuities to women than to men, why a financial matriarchy owns more than half of some of the largest industrial companies in the country, or why wealthy widows abound on cruise ships.

Biologists have long been curious about reasons for the higher mortality of males. A favorite theory, that male mammals should be affected more than females by any adverse effects of sex-linked genes, prevailed for some time. However, if deleterious sex-linked genes were responsible, then in species having heterogametic females, the female sex would experience higher mortality than males. It was eventually shown by MacArthur and Bailie that this does not happen. Both in birds and in Lepidoptera, just the same as in mammals, mortality is higher in males than in females. It has been suggested that the differential mortality may be related to the fact that metabolism is generally higher in males than in females.

Sex Ratios in Hybrids

Among interspecific hybrids, sex ratios are often markedly different from those in the parent species. In mammalian hybrids, in contrast to the prevailing excess of males shown in Table 19-3, there is usually a deficiency of that sex. Among 1,416 mule foals, Craft found the proportion of males to be only 44.3 per cent, and there was a similar deficiency of males among the other rarer hybrids for which he could get records. This does not apply to species that cross with good fertility and normal reproduction. Craft's records for calves from the cross *Bos taurus* × *Bos indicus* showed the normal excess of males, the figures being 52.8 and 50.9 per cent in two different groups.

In hybrid birds, this situation is reversed, and females are scarcer than males. Crosses of the ring-neck pheasant, *Phasianus colchicus*, with the domestic fowl yield a normal sex ratio when the cross is Gallus ♂ × Phasianus ♀, but in the reciprocal cross there is a marked shortage of females. In crosses between pheasants and turkeys, Asmundson and Lorenz found a deficiency of female hybrids. From crosses between pigeons and ring-doves, Cole and Painter found male hybrids to outnumber females by more than 30 to 1.

The contrasting effects of hybridization on sex ratios in mammals and birds are not as opposite as they seem. Both groups conform to Haldane's rule, which holds that, whenever, in the offspring of an interspecific cross, one sex is missing, rare, or sterile, that sex is the heterogametic one.

Conditions Affecting Sex Ratios

It is not possible in this text to review all the evidence showing how sex ratios appear to be influenced by various conditions. Too often what seems to be conclusive evidence from one investigation is refuted by equally conclusive but opposing figures from the next one. Perhaps much of the confusion in the field comes from the fact that conditions affecting sex ratios in one species, or one breed, do not necessarily influence others in the same way.

Among dog breeders, it is a common belief that the proportion of males depends to some extent upon the season of the year when the litter is conceived. There are extensive figures showing that for litters of warm and cold months the sex ratios are 126 and 122 ♂ ♂ per 100 ♀ ♀, respectively, but in Whitney's German Shepherds the corresponding figures were 116 and 143.

In rats and mice the feasibility has been demonstrated of breeding strains that differ in sex ratio. In her famous inbred rats at the Wistar Institute, Helen Dean King selected (after six generations of brother × sister matings) two distinct lines, one (A) bred from litters having an excess of males, the other (B) from those with a majority of females. From the eighth to the twenty-fifth generations, the average number of males per 100 females was 122.3 in line A but only 81.8 in line B. Most of the differentiation was effected during the first two generations of selection. While it is understandable that sex-linked genes affecting viability of male fetuses could reduce that sex in line B, a satisfactory explanation for the excess of males in line A has not yet been adduced.

In the mouse Weir found that selection to differentiate two lines differing in the pH (alkalinity) of the blood was effective, and that the two strains differed consistently in sex ratio. The one with the higher pH (7.48) showed the higher proportion of males, the other (with pH 7.43) a remarkably low one. In reciprocal crosses between the two strains and in matings of males of both to females of other stock, sex ratios in the progeny were always like that of the sire's strain (Table 19-4).

The degree to which males of these two lines differ in sex ratio of their progeny is more conspicuous if the averages in Table 19-4 are converted to the other measure of sex ratio, *i.e.*, males per 100 females. On that scale, the figure for the "high" strain is 114, against only 67 for the "low" one. This is very striking, because, as the data show, the consistent deficiency of males in the low strain was not matched by any corresponding decrease in litter size. One cannot attribute the difference, therefore, to any sex-linked lethal genes

TABLE 19-4
Consistently Different Sex Ratios in Progeny from Males of Two Strains of Mice

| Strain of Mother | Age at Classification | Progeny by Males of: | | | |
| | | High[a] Strain | | Low[a] Strain | |
		Males (per cent)	Litter Size (average)	Males (per cent)	Litter size (average)
High[a]	Weaning	52.8	6.4	38.8	6.9
Low	Weaning	55.5	6.4	41.8	5.7
Four others	Weaning	55.5	6.4	39.1	7.4
K	Birth	52.6	7.7	39.0	8.3
Crossbred	Birth	52.8	9.5	36.7	8.0
Total mice		—	4,839	—	5,125
Average sex ratio		53.2		40.4	

[a]High means here both higher pH of blood and higher proportion of males; conversely for low.
Source: Data of Weir.

killing males in the low strain. The difference between the high and low strains is clearly a genetic one, but the physiological basis for the effect has yet to be determined. The difference between the strains in pH appears slight, but is consistent.

Other investigators have been similarly able to breed lines of mice differing in pH of the blood, but have not found therein corresponding differences in sex ratios, so the striking differences in that respect shown in Table 19-4 may need some other explanation. Furthermore, other attempts to change the sex ratio in the mouse and in other species by selection have not been successful. Animal breeders should not be too hopeful about any prospects for breeding Holsteins to produce more heifers, or Beagles to produce fewer bitches.

Problems

19-1 Work out (as in Fig. 19-2) the consequences of non-disjunction of the X- and Y-chromosomes in a male mammal when the resulting gametes fertilize eggs of a normal female.

19-2 If the male of the previous problem were mated with a female in whose gametes the XX chromosomes had failed to disjoin, what kinds of viable zygotes could result?

19-3 Parthenogenetic development of eggs of the honey-bee gives rise to haploids that are males. If the same thing were to occur in Drosophila, would you expect the resultant haploid flies to be (on Bridges' theory) males or females, and why?

19-4 How many Barr bodies would you expect to find in nuclei from somatic cells of (a) a normal, black tom-cat, (b) a tortoise-shell female, (c) a tortoise-shell male?

19-5 If Barr bodies occur in birds and have the same origin as in mammals, how many should there be in nuclei of (a) a normal cock, (b) a normal hen, (c) a hen that has assumed male characteristics?

19-6 By experimental treatment, Humphrey was able to convert females of the Mexican axolotl (a salamander) to males. Females of this species are XY. Reversed females mated with normal ones yielded 1,588♀♀ : 509♂♂. How do you account for that ratio?

19-7 In what respect are Humphrey's data at variance with the results of matings in man and in Drosophila when one parent is aneuploid, as shown in Fig. 19-2?

19-8 If Crew's fowl that underwent sex reversal had sired 100 chicks instead of two, what sex ratio would have been expected among them if the same viability of zygotes lacking X-chromosomes occurs in the fowl as (a) in Humphrey's axolotls? (b) in Drosophila?

19-9 There is evidence (from man and the pig) that in the primary sex ratio for some mammals there are many more males than females. In the fowl, however, the primary sex ratio is 1:1. Can you think of reasons why such a difference might be expected?

19-10 What seem the most likely ways in which the secondary sex ratios of domestic mammals could have more males than females?

19-11 Considering the record in Table 19-4, how would you explain the differing sex ratios in Weir's high and low strains?

Selected References

BENIRSCHKE, K. 1981. Hermaphrodites, freemartins, mosaics and chimeras in animals. In *Mechanism of Sex Differentiation in Animals and Man. Chapter 11*, ed. by C. R. Austin and R. G. Edwards. London and New York. Academic Press, pp. 421–63. (A thorough review with bibliography.)

BURNS, M., and M. N. FRASER *The Genetics of the Dog*, 2nd ed. Edinburgh: Oliver and Boyd. Philadelphia: Lippincott. (Sex ratios and litter sizes for 18 breeds on p. 17, with references cited in bibliography.)

CENTERWALL, W. R., and K. BENIRSCHKE 1973. Male tortoise-shell and calico (T-C) cats. *J. Heredity* **64:** 272–78. (With cats in color and their chromosomes in black).

DUNN, H. O., D. H. LEIN, and K. McENTEE 1980. Testicular hypoplasia in a Hereford bull with 61, XXY karyotype: The bovine counterpart of human Klinefelter's syndrome. *Cornell Vet.* **70:** 137–46. (Source of Fig. 19-3.)

FORBES, T. R. 1947. The crowing hen: Early observations on spontaneous sex reversal in birds. *Yale J. Biol. Med.* **19:** 957–70. (Folklore and explanations from Aristotle to the present century; extensive literature citations.)

Handbook of Biological Data, ed. by W. S. Spector. 1956. Philadelphia, W. B. Saunders Co. (For sex ratios in many species, see Table 440, on p. 519.)

HUTT, F. B. 1949. *Genetics of the Fowl*. New York. McGraw-Hill Book Co., Inc. (Sex dimorphism, intersexes, and sex ratios in Chapter 13, with references to pertinent literature.)

LYON, M. 1962. Sex chromatin and gene action in the mammalian X-chromosome. *Amer. J. Human Genet.* **14:** 135–48. (Details of her theory that one X is inactivated early in development and that mosaics with respect to sex-linked genes can result therefrom.)

McKUSICK, V. A. 1962. On the X chromosome of man. *Quart. Rev. Biol.* **37:** 69–175. (Reviews all known sex-linked characters, evidence for linkage, and facts about intersexes, with extensive bibliograpy.)

OHNO, S. 1967. *Sex Chromosomes and Sex-Linked Genes*. Berlin, Heidelberg, New York. Springer-Verlag. (Status of knowledge about how they determine sex, and variations in different species, before recognition of the role of the H-Y antigen.)

OHNO, S. 1979. *Major Sex-Determining Genes*. Monographs on Endocrinology, Vol. 11. Berlin, Heidelberg, New York. Springer-Verlag. (About the H-Y antigen and hormones that differentiate the sexes.)

STERN, C. 1973. *Principles of Human Genetics*. 3rd ed. San Francisco. W. H. Freeman & Co. (See Chapter 20 for anomalies of X- and Y-chromosomes.)

THULINE, H. C. and D. E NORBY. 1961. Spontaneous occurrence of chromosome abnormality in cats. *Science* **134:** 554–55. (With evidence that the tortoise-shell male may be XXY.)

WACHTEL, S. S. 1977. H-Y antigen and the genetics of sex determination. *Science* **198:** 797–99. (With the conclusion that factors other than the H-Y antigen may be necessary for its action.)

WEIR, J. A. 1962. Hereditary and environmental influences on the sex ratio of PHH and PHL mice. *Genetics* **47:** 881–97. (Source of data in Table 19-4, with reports of other studies of sex ratios in the two strains.)

YOKOYAMA, T. 1959. *Silkworm Genetics Illustrated*. Tokyo. Japan Society for the Promotion of Science. (For sex determination, see pp. 44–51.) In English.

Genetically Aberrant Physiology

Effects of genes that influence form and color in higher organisms are much easier to recognize than genetic variations in physiology, and many genetic variations in cellular chemistry can be identified only by special kinds of chemical analyses. It is these last that are often classified as the special province of "biochemical genetics." Studies in this field have been facilitated in recent years by improved techniques for biochemical analyses, such as microbiological assays, the differentiation of proteins by their rates of flow in electrophoretic currents, and identification of chains of polypeptides.

Garrod and Genetic History

It is of interest to note, however, that the first investigations in this important area actually antedate the re-discovery of Mendel's laws. It is commonly written that the father of biochemical genetics was A. E. Garrod, whose book, *Inborn Errors of Metabolism*, published in 1909, was based on his Croonian lectures to the Royal Society of London in the preceding year. His studies of alkaptonuria began at least a decade earlier. Alkaptonuria is a rare "error of metabolism" in man recognizable by brown color of the urine, which becomes black on exposure to air.

Bateson's report of December 17, 1901, to the Royal Society of London on his experiments with combs and colors in the fowl carried (when published in 1902) this footnote:

> *Recently, however, Garrod has noticed that no fewer than five families containing alkaptonuric members, more than a quarter of the recorded cases, are the offspring of* first cousins. *In only two other families is the parentage known, one of these being the case in which the father was alkaptonuric. In the other case the parents were not related. Now there may be other accounts possible, but we note that the mating of first cousins gives exactly the conditions most likely to enable a rare and usually recessive character to show itself. If the bearer of such a gamete mate with individuals not bearing it, the character would hardly ever be seen; but first cousins will frequently be bearers of similar gametes, which may in such unions meet each other, and thus lead to the manifestation of the peculiar recessive character in the zygote. See A. E. Garrod, "Trans. Med. Chir. Soc," 1899, p. 367 and "Lancet" November 30, 1901.*

Here was biochemical genetics waiting in the wings when, with the re-discovery of Mendel's laws, the new science, later to be known as genetics, came to the forefront of the biological stage.

The Role of Neurospora

Genetic variations in biochemical processes determine the degree of efficiency with which organisms utilize available foodstuffs, *i.e.*, the nutritional requirements of plants and animals. Genetic variations in this respect were first extensively studied in the pink bread mould, Neurospora. Although we cannot delve far into that field in this text, it is appropriate that we pause a moment, before going on to the animals which are our chief concern, to recognize just what Neurospora has contributed, not only to biochemical genetics, but also to our knowledge of biochemical processes. Those contributions resulted from studies begun at Stanford University in 1941 by G. W. Beadle and E. L. Tatum, and continued later in other laboratories by them and by their disciples.

Normal strains of Neurospora are able to synthesize their nutritional needs from very simple substances. These last include (a) a source of carbon, such as a sugar or starch, (b) inorganic salts (to provide calcium, potassium, sulphur, and phosphorus), and (c) a single vitamin—biotin. From these materials all amino acids, vitamins, and other substances needed are synthesized to permit normal growth of the plant. By irradiation with X-rays, Beadle and Tatum induced mutations which prevented the synthesis of essential substances, and hence prevented also normal growth. In its growing (asexual) stage, the mould is haploid, so any mutations induced would not be obscured by their normal alleles, as are single recessive mutations in diploid cells. To recognize such mutations, the irradiated stocks were first grown on complete diets, so that they did not have to synthesize any essential nutrients, and then, after their ability to thrive under those conditions had been demonstrated, conidia (spores formed asexually) were transferred to what we might call the diet of raw materials. Many of the transfers grew equally well there, indicating that they could synthesize what was needed, but some could not do so. By adding one at a time various purified amino acids, vitamins, or purines, it was possible thus to identify which one a mutant strain needed for growth, or, in other words, which one it could not synthesize.

As the result of such studies, mutations were eventually found which prevented the synthesis of almost every known amino acid and vitamin. A most significant outcome of the work with Neurospora was the discovery that, in the synthesis of some one essential nutrient, the process might be blocked at different stages by entirely different mutations. Complex organic compounds are not built from simpler substances by a single big change, but by a whole series of little ones.

These little ones have been compared to steps on a staircase, leading from simple beginnings at the bottom to the complex substance at the top. Expressed otherwise, *anabolism*, or the synthesis of complex compounds from simpler ones, is sometimes referred to as "stepwise metabolism" because the process proceeds from each stage, or step, to the next. As we shall see shortly, there are corresponding steps in *catabolism*, the breaking-down of complex compounds to simpler ones.

While Neurospora may be viewed by bakers and handlers of bread only as an unmitigated nuisance, geneticists and biochemists should both be grateful to it, not only for its contributions to their two sciences, but also for bringing those two closer together. Let us not forget, either, that the contribution of Neurospora to science was made possible because its special potentialities for revealing mutations affecting biochemical processes were recognized and utilized by Beadle and Tatum.

Blocks in the Catabolism of Phenylalanine

Any synthesis of essential nutrients by animals is on a small scale compared with that in plants. In general, animals use complex substances that have been built up by plants, and do so by breaking them down to simpler substances. Before proceeding to consideration of genetic blocks in catabolism in domestic animals, we might consider briefly, as perhaps the best known examples of such defects, some of the conditions caused in man by genetic blocks in the utilization of phenylalanine. That amino acid is used to some extent for formation of proteins, but much of it is oxidized to tyrosine, and to derivatives therefrom such as melanin, adrenaline, and thyroxine. Four blocks in the catabolism of phenylalanine are shown in Fig. 20-1.

Phenylketonuria. Individuals homozygous for this condition lack the enzyme *phenylalanine hydroxylase* which normally converts phenylalanine to tyrosine (A). Because the phenylalanine cannot be oxidized to tyrosine, it accumulates in the blood, and a derivative of it, *phenylpyruvic acid*, is excreted in abnormally large amounts in the urine (from which it is absent in normal persons). This metabolic disorder, which was first recognized by Følling in Norway, causes a form of idiocy which was originally called *phenylpyruvic oligophrenia*, and it has been variously estimated to be responsible for 0.6 to 0.8 per cent of the mentally defective people in institutions. A simple test has been devised which, when applied to wet diapers, will reveal at about six weeks of age whether or not a newborn infant is

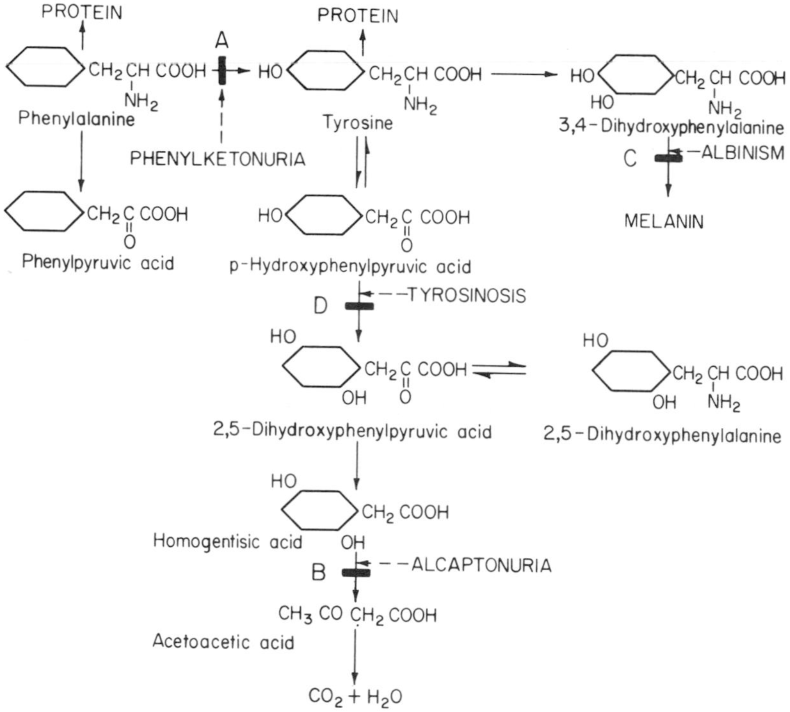

FIG. 20-1 Metabolism of phenylalanine. Arrows show reactions important in normal metabolism. Blocks in metabolism, indicated by solid bars, cause: A, phenylketonuria; B, alkaptonuria; C, albinism; D, tyrosinosis. (After G. W. Beadle in *Chem. Reviews* after J.B.S. Haldane.)

excreting phenylalanine. For such cases, diets low in phenylalanine have been formulated, and there is evidence that such diets may be helpful in forestalling trouble. A test has also been devised (feeding phenylalanine and assaying the amount excreted) whereby it can be determined with a high degree of accuracy whether or not a normal adult person might carry in the heterozygous condition the recessive autosomal gene that causes phenylketonuria.

Alkaptonuria. This condition, the name of which comes from excretion in the urine of *alkapton,* or *homogentisic acid,* is caused by a gene which, in the homozygous state, eliminates the enzyme *homogentisic acid oxidase* that is necessary to break homogentisic acid down to simpler substances (B). The condition is very rare, but should be easily recognized because the homogentisic acid in the urine is oxidized when exposed to air and turns black. Contrary to the situation with phenylketonuria, alkaptonuria is not a serious affliction, although it causes blackening of the bones and cartilage (and bed sheets), along with a tendency to arthritis.

Albinism. Another block in the catabolism of phenylalanine results in albinism. It is more common than the two considered just previously, and the only one of the four which has been found thus far in animals other than man. A compound with a polysyllabic name which workers in the field have conveniently shortened to *dopa* is prevented (C) from being converted to melanin. There are different degrees of albinism, all recessive, some caused by autosomal genes, but two (in the fowl and the turkey) are sex-linked.

Tyrosinemia. This rare condition is attributed to a block as shown at D in Fig. 20-1, but only a few affected individuals have been found thus far. It is apparently caused by a simple recessive mutation.

One Gene—One Enzyme

It will have become evident that, as is shown in Fig. 20-1, each block in the catabolism of phenylalanine is caused by a deficiency or inactivation of some one specific enzyme that is necessary for the change from one chemical compound to another. Similarly, in the synthesis of complex substances from simple ones in Neurospora and other organisms, the genes that block the synthesis are considered to do so by preventing the formation of the enzymes necessary for the chemical changes. As each such change apparently requires the catalytic action of a specific enzyme, and blocks at separate steps of the synthesis are each caused by separate mutations, it is considered that one gene controls in some way the action of one specific enzyme. This postulated control mechanism is sometimes designated the "one gene—one enzyme" hypothesis. This does not imply that the effect of the recessive mutant gene must be to prevent formation of the enzyme. The enzyme might be present, in almost complete form, but need some final contribution from one gene to permit its activation. As enzymes are themselves complicated proteins, conceivably their synthesis can be blocked in much the same way as synthesis of amino acids.

Purine Metabolism in Dalmatians

In rodents, carnivores, marsupials, and ungulates, purines are broken down to uric acid, and most of that in turn, by action of the enzyme *uricase*, to allantoin, which is excreted. In man and the chimpanzee, that last step is omitted and the chief end-product of purine catabolism is uric acid. Unlike all other breeds of dogs tested thus far, Dalmatians (Fig. 20-2) follow the same pattern as in man and the chimpanzee, and excrete uric acid at what is an inordinately high level for their species. Furthermore, that level is not just at the extreme end of the range for other dogs, but is far above it. In genetic studies of this condition, Trimble and Keeler were able to classify their animals as either low or high in uric acid, without any overlapping of the two classes. The amounts of that substance excreted in 24 hours per kilogram of body weight by their two kinds of dogs were, for dogs with low uric acid: 4 to 10 milligrams, and for those with high uric acid: over 28 mg.

The earlier conclusion of Onslow that the difference between these two types of dogs is genetic in origin was confirmed in F_2 and backcrosses by Trimble and Keeler, who found the high excretion of uric acid to be a simple recessive autosomal character.

Their further discovery that it is entirely independent of the spotting which is a breed character in Dalmatians raises the interesting question: How did the Dalmatians get their peculiar physiological distinction? Obviously, while the spotting pattern was deliberately incorporated by the breeders as an indispensable characteristic of Dalmatians, the same could not be done with excretion of uric acid, which can be measured only by chemical analyses. It has no known

FIG. 20-2 Dalmatian. Champion Pennydale Pal Joey. (Courtesy of Mr. and Mrs. Arthur W. Higgins, of Pennydale Kennels, Syosset, New York.)

effects, either harmful or beneficial. Dalmatians were bred to run behind coaches, not to excrete uric acid in abnormal quantities. It is easy to ascribe the distinction to the inclusion of the trait by chance in a comparatively small group of ancestors which gave rise to the whole breed, but that tells us little. In any case, the Dalmatians, apart entirely from commanding our respect as splendid dogs, deserve a special mark of merit for preventing us from jumping too hastily to arbitrary definitions of normality. If what is normal for the breed is abnormal for the species, what, then, does the word "normal" mean?

Other investigators have found that the Dalmatians do not lack the enzyme, uricase, and the incomplete metabolism of purines in that breed is attributed to failure of the renal tubules to resorb uric acid, as do those in other breeds.

DEFICIENT ENZYMES

Many diseases are caused by enzymes that are genetically unable to function normally. The enzyme may be entirely lacking, or its activity may be subnormal. In some instances, as a result of mutation it may form some abnormal product that is useless but cannot be eliminated. Diseases in which such products accumulate are now commonly called storage diseases.

McKusick (1978) lists 170 disorders in man that are now known to result from inadequacy of 170 specific, named enzymes. In this book we give only a few examples to show the diversity of ailments in domestic animals that are caused by malfunctioning of enzymes.

Porphyria in Cattle and Swine

Porphyria results from an abnormality of metabolism which causes the red pigment, porphyrin, to be produced in excessive amounts and accumulated in the blood, bones, teeth, and other parts of the body. Affected animals can be recognized by the facts that their urine turns reddish brown, and their teeth are pink. They are extremely sensitive to sunlight, and exposure to it results in the formation of blisters which eventually become deep ulcers and leave bad scars (Fig. 20-3). These lesions are found around the eyes and nostrils, and often on the middle of the back where parting of the hair leaves little protection. Spotted animals have more lesions on white parts than on dark ones.

This condition is sometimes called congenital porphyria, or erythropoietic porphyria, to distinguish it from abnormalities (in man)

FIG. 20-3 A cow affected with porphyria and showing on the white areas of the back and upper side extensive lesions resulting from sunburn. The black areas are not affected. (Courtesy of D. W. Baker and New York State Veterinary College, Cornell University.)

which resemble it but appear later in life. To avoid entanglement with the complex chemical reactions involved, it may suffice here to say that a form of porphyrin is a normal constituent of hemoglobin and that the excessive accumulation and excretion of porphyrins are attributed to an enzymatic block in the formation of the *heme* part of hemoglobin from a precursor, *porphobilinogen*.

Cattle with porphyria apparently are healthy enough if kept from sunlight. One such animal, which, after the age of 8 years, was kept indoors during the day and allowed out of its paddock at night, lived to 18 years, when it had to be destroyed because of bad teeth (Fourie). The condition adds one more to the series of cases in which the welfare of the individual is not determined solely by its genes, but is also dependent on the environment.

Porphyria is rare in man, but apparently less so in cattle. It has been found in Shorthorns in South Africa, England, and Denmark; in Holstein-Friesians in the United States; and in Jamaica Reds and

Jamaica Blacks. It is transmitted in cattle as an autosomal recessive trait.

In pigs, porphyria has been found in New Zealand and in Denmark. The condition may not be biochemically the same as in cattle, for its onset is delayed somewhat, and the affected pigs are apparently not photosensitive, as are the cattle. All the Danish cases were descended from a single boar. Porphyria in swine is transmitted as a dominant trait, but it is not yet clear that a single gene is responsible. The condition is also known in cats.

Excessive Phylloerythrin in Sheep

An enzymatic block differing from that causing porphyria is undoubtedly responsible for the photosensitivity studied in Southdown sheep by Clare and Hancock in New Zealand.

In afflicted lambs, the liver fails to excrete *phylloerythrin,* which is a product of the digestion of chlorophyll. All of the lambs are normal at birth and have no difficulty until they begin to graze at five to seven weeks. The severity of their injuries after that age varies according to the intensity of light and the proportion of their diet that comes as milk from the ewe (with the remainder from grass). If the lambs are left outdoors, they develop eczema over the face and ears (Fig. 20-4), and death ensues within two to three weeks after the appearance of the first symptoms. On the other hand, if the lambs are kept indoors during the day, and allowed out only at night, they learn to do their grazing at night, to keep out of the sun during the day, and so to survive despite their abnormal physiology.

In genetic studies with these sheep, Hancock found both sexes to be equally affected, and ratios in the different matings were as follows:

| | Offspring | |
Parents	Normal	Affected
Homozygous normal × affected	74	0
Heterozygotes × affected	18	16
Affected × affected	0	14

From these unusually good numbers for animals as big as sheep, it is clear that this abnormality of metabolism is a simple, autosomal, recessive character. Although it has not yet been found in other

FIG. 20-4 Sheep photosensitized by excessive phylloerythrin, showing lesions on drooping ear and on upper eyelid, also swollen lips, and "Roman nose" in puffy face. (Courtesy of N. T. Clare, Ruakura Animal Health Research Station, New Zealand, and of Cornell University Press; from *Genetic Resistance to Disease in Domestic Animals*, by F. B. Hutt.)

animals, the possibility should not be overlooked that some of the photosensitization commonly caused by the eating of certain plants might be aggravated in some animals by genetic variations in ability to break down the sensitizing pigments that those plants contain.

Neuronal Ceroid-Lipofuscinosis in Dogs

In Koppang's remarkable study of this disease in English setters, during a period of eight years (in which there were 250 dogs in his breeding trials), he had 70 cases from his own genetic analyses and over 30 more from other breeders. The disease is hereditary and lethal, caused by an enzymatic block in the catabolism of fats. In earlier years, the very similar disease in man was called juvenile amaurotic idiocy. That too is lethal.

According to Koppang (1970), the dogs affected showed clinical symptoms at 12 to 15 months of age. These included failing vision (amaurosis) and increasing dullness. The disease was detected earlier by other dogs, which disliked those afflicted and sometimes even killed them. At 15 to 18 months, the amaurotic dogs showed difficulty in walking and in ability to localize sounds. They clicked their teeth. Later on, some had muscular spasms, and few survived to two years of age. At post-mortem examination, they were found to have extensive deposits of lipid granules in the nerve cells.

As some of the affected dogs could be used in breeding trials before the onset of severe disease, it was possible to make the usual experimental matings. The large numbers of progeny from these, unusual for animals as big as English setters, showed clearly that the disease is caused by a simple recessive gene (Table 20-1). It was proven (by reciprocal matings) to be autosomal.

Subsequently, it was determined that the disease was caused by reduced activity of the enzyme *p-phenylenediamine-mediated peroxidase*. On a scale for measuring that activity, the average was 12.5 in normal controls, 6.0 in heterozygotes, and only 0.88 in afflicted dogs.

Mannosidosis in Cattle

A lethal storage disease in Aberdeen-Angus cattle in New Zealand was found by Hocking *et al.* (1972) to be caused by an almost complete lack of the enzyme *α-mannosidase*. The disease became evident at several weeks or months of age. Affected animals were unthrifty but

TABLE 20-1
Inheritance of Neuronal Ceroid-Lipofuscinosis

Mating	Progeny	
	Normal	Diseased
Carrier × non-carrier (Reciprocal crosses)	18	0
Carrier × carrier	104	25
Affected × carrier (Reciprocal crosses)	43	44
Affected × affected	0	1
Affected × non-carrier	15	0

Source: Summarized from data of Koppang, 1970.

tended to be aggressive. They showed head tremors and ataxia, and usually died within their first year.

The storage product in this case was found to be an oligosaccharide containing mannose and glucosamine. Genetic evidence showed that the animals affected were homozygous for a recessive gene. In heterozygotes, there was a partial deficiency of *α-mannosidase* but viability and reproduction were apparently normal. (See also Jolly, 1978.)

Other Mutant Enzymes

Many other hereditary storage diseases are known. We saw one of them in Chapter 14 in the "obese" mice, and another (in the rat) is shown in Fig. 20-5. Both are probably caused by malfunction of enzymes. In the rats, the level of fatty acids in the blood serum was ten times the normal. Even after an 18-hour fast, the blood lipids re-

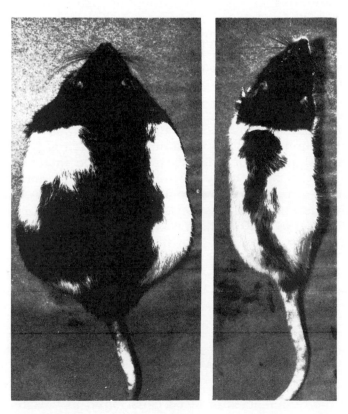

FIG. 20-5 Genetically fat rat and brother at ten months; weights: 1,035 and 447 grams, respectively. (Courtesy of L. M. and T. F. Zucker in *J. Heredity.*)

mained high, but blood sugar was normal whether the rats were fed or fasted.

Not all mutations of enzymes are pathogenic, and some might even be beneficial. Sawin and Glick (1943) found that more than half of 181 rabbits tested carried an enzyme, *atropinesterase*, with which they could destroy the toxic alkaloid, atropine. The remainder could not. The differences proved to be caused by a recessive gene in the rabbit's sixth linkage group. As atropine is common among plants of the family *Solanaceae*, the ability to destroy it could be advantageous to rabbits obliged to forage in the wild for their food.

Lysozyme is an enzyme common to most animals, but a marked deficiency of it in rabbits, found by Prieur and Cámara (1979) to be a simple recessive trait, apparently caused no ill effects whatever. The deficient animals were healthy enough to maintain the reputation of their species for efficient reproduction. Similarly, dogs genetically deficient in the enzyme *catalase* were found by Allison to show no obvious anemia or susceptibility to oral gangrene such as have been reported in humans lacking that enzyme, but their blood was notably subject to hemolysis.

Another mutation prevents some rabbits from forming the enzyme *xanthophyllase* which, in most rabbits, destroys the yellow pigment xanthophyll and thus causes the fat to be white. A lack of that same enzyme may be responsible for the yellow fat that is hereditary in sheep.

The fact that the examples given in this section of genetic defects in enzymes are drawn mostly from laboratory animals and dogs does not mean that the larger domestic animals are exempt from them. It means mostly that genetic analysis is easier and faster with small animals than with large ones. We must remember, moreover, that most domestic animals are mammals, and so is man. A mutation occurring in one mammalian species might also occur in any other. Let us not forget that hemophilia A is known in man, dogs, and horses, or that hereditary epilepsy is known, not only in man, but also in species as far apart as dogs and chickens. Some day we may recognize in swine, sheep, cattle, or horses most of those 170 enzymatic disorders now listed by McKusick as hereditary in man. By that time the list for man will probably far exceed 170.

ABNORMAL BLOOD CELLS

Apart from the deficiencies of blood cells found in various types of anemia, there are some genetic disorders associated with abnormalities in the cells themselves. A few examples follow.

The Chediak-Higashi Syndrome

This disease, so named after the first two investigators who described it in man, is now known in cattle, minks, and mice. It is characterized by abnormal leukocytes which carry large granules in their cytoplasm. In cattle, it causes partial albinism, resultant photophobia, and marked susceptibility to infections.

In minks there is a color variety called Aleutian, gunmetal, or blue, which was popular until such minks were found to be unusually susceptible to disease when other color varieties in the same environment were not. The affliction thus became known as "Aleutian disease." Subsequently, studies by Padgett *et al.* (1967) showed that the disease was caused by virus and that all the Aleutian minks had the Chediak-Higashi disorder.

In both cattle and minks, the syndrome is a simple recessive autosomal character. The causative gene (*a*) was tried out by the ever-hopeful mink breeders in color combinations other than black, but the homozygotes (*aa*) were all found to have the Chediak-Higashi syndrome just as in Aleutians. Heterozygotes (*Aa*) did not.

Stomatocytosis in Alaskan Malamute Dogs

This abnormality was first studied genetically as a kind of chondrodystrophy because it left the afflicted animals dwarfed and bow-legged (Fig. 20-6). It proved to be a simple recessive trait. In subsequent studies, the visible abnormalities were found to be associated with a mild hemolytic anemia. Eventually, that, in turn, proved to be associated with abnormal red blood cells of a type called stomatocytes (Pinkerton *et al.*, 1974). Such cells show a short streak that does not stain (Fig. 20-7), and, at high magnification, are found to be somewhat cupped.

Cyclic Neutropenia

In most Collie dogs of a certain shade of grey, infections causing enteritis or pneumonia result in death before six months of age. Cheville (1968) found that the underlying reason is a genetically determined periodic decline in the numbers of certain polymorphonuclear leukocytes known as neutrophiles. They contain stainable granules and hence are also known as granulocytes. As these cells are phagocytic, they destroy invading bacteria and thus, ordinarily, help the animal to resist infection.

In the grey Collies, these important cells decrease in number about every 10 to 12 days, at which time the animal shows typical

FIG. 20-6 A: A dwarf Alaskan Malamute showing bowed front legs and thick carpal joints that were first ascribed to chondrodysplasia, but later to an anemia associated with stomatocytes (below) in the blood; B: Normal Malamute, with straight legs. (From Subden *et al.* in *J. Heredity,* copyright © 1972 by the Amer. Genetic Association).

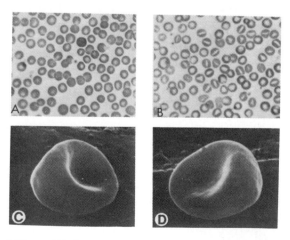

FIG. 20-7 Blood films (A) from a normal Malamute, and (B) from an anemic one showing stomatocytes among the normal red cells. C and D: Stomatocytes highly magnified. (From Pinkerton *et al.* in *Blood,* as cited, by permission).

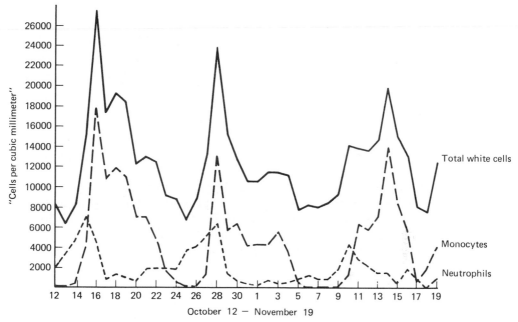

FIG. 20-8 Cyclic neutropenia in a grey Collie. During the 38 days shown, there were three episodes of neutropenia in which the neutrophiles decreased markedly. (After Lund in Animal Models for Biomedical Research III. Washington, D.C., Nat. Acad. Sci. 1970, p. 75).

signs of distress. This lasts for two or three days, but, if the attack is not fatal, the neutrophiles then rise to abnormally high levels before starting the next decline (Fig. 20-8).

Cyclic neutropenia is apparently caused by periodic failure of the bone marrow to produce the neutrophiles in normal amounts. The defect is a simple recessive character. The grey color of the coat is a pleiotropic effect of the genotype responsible for this remarkable lethal defect. The condition occurs also in man, but with the period between episodes of neutropenia about twice as long as in the dog.

Hemolytic Anemia in Basenji Dogs

This condition, now known to be a simple autosomal recessive disease, is caused by deficiency of the enzyme *pyruvate kinase* (PK) in the erythrocytes. Along with chronic anemia and some debility, there is an enlargement of the spleen. The disease is not always recognizable from clinical symptoms but is easily detected by examination of the blood.

The degree of activity of pyruvate kinase can be measured quantitatively in units. The procedures for making such assays are too complex to warrant description here, but they yield results that are expressed as units of PK activity per standard sample of erythrocytes. In samples of blood from about a thousand Basenjis, Brown and Teng (1975) found a bimodal distribution of PK activity, with one group ranging from 0.36 to 0.82 units, and a larger class varying from 0.84 to 2.00 units. From this and their genetic tests, they concluded that dogs in the former group were probably heterozygous, *Pk pk*, for the gene causing anemia, but that those above 0.84 were probably *Pk Pk*, homozygous for the normal allele.

This was confirmed by Andresen (1977), who used Basenjis with genotypes known (from breeding tests) or presumed (from pedigrees). His method of assay differed from that of Brown and Teng. For 12 dogs known not to carry *pk*, the average unit of pyruvate-kinase activity was 1.32, but the corresponding figure for 30 known carriers (*Pk pk*) was barely half as great at 0.63.

The ability to identify the two genotypes (one wanted and the other not) by a laboratory test makes breeding tests unnecessary, and should thus greatly facilitate the elimination of the undesirable gene *pk* from the breed. Andresen determined that in his tests a value of 0.99 units would distinguish between the two genotypes with errors fewer than 5 per cent. As he pointed out, the figure for discrimination between the two genotypes could vary depending on the method of assay of PK activity and the temperature at which it is used.

ENDOCRINE DISORDERS

There are thick books describing abnormalities of physiology that have been shown to be caused by malfunction of endocrine glands, and, as some of those diseases are genetically determined, it is unnecessary to belabor that point here. Two examples of endocrine disorders—one of the thyroid gland and the other of the pituitary—are described below because they are not as well known as some others.

Autoimmune Thyroiditis

Autoimmunity is a condition in which the body forms antibodies against tissue components of one or more of its own organs. After noting signs of hypothyroidism in some of his White Leghorns, Cole (1966) raised the frequency of such birds to 83 per cent in females of his "obese" strain by selection for six generations. By 1979, it was 99 per cent. Contrary to many other genetic defects, this one is polygenic.

The condition becomes recognizable by 6 to 8 weeks of age. In addition to obesity it is characterized by elongated, lacy feathers, slow growth, and often delayed sexual maturity (or none) in the females (Fig. 20-9). However, when supplementary thyroxine is supplied, the obese birds perk up, develop red combs, lay eggs, and reproduce well.

Pathological changes begin in the thyroid gland at about two weeks of age with infiltration of lymphoid cells and others. Eventually, much of the gland is destroyed. Cole's obese birds are now serving in several laboratories around the world as animal models useful for the studies of autoimmune disease in man.

A very similar hereditary condition was found in Beagle dogs, but it is more difficult to recognize. It has been reported in other breeds, but, because it is more likely to be noted in large colonies of animals than in small kennels, it has been most studied in Beagles. Fritz *et al.* (1976) found it to be associated with lymphocytic orchitis. Among his Beagles all descended from common ancestors, the frequency of thyroiditis and of orchitis was 85 and 65 per cent, respectively. Those

FIG. 20-9   A white Leghorn female affected with auto-immune thyroiditis. The lacy plumage and lack of development of the comb are characteristic. (Courtesy of R. K. Cole, Cornell University.)

with orchitis had smaller testes than the others and were less fertile.

The lesions found in fowls and dogs affected with autoimmune thyroiditis are similar to those in Hashimoto's disease in man, which is known to be hereditary.

Pituitary Dwarfism

A genetically determined inability of the anterior lobe of the pituitary gland to produce the hormone necessary for growth occurs in the mouse, as well as in dogs. In both species, it is a simple recessive autosomal character. It was first studied in German Shepherds by Andresen *et al.* (1974), but was later found also in Carelian Bear-Dogs.

In affected pups, growth was retarded in the first two months and ceased a few weeks later. At 16 months, one weighed only 5½ kg. and her brother dwarf only 14 kg. (Fig. 20-10). Proportions of different parts of the body were normal, but, because of their small size, these dwarfs seemed "fox-like."

The growth hormone of the pituitary is believed to induce growth through elements in the blood called somatomedins. Assays of these

FIG. 20-10   Three German Shepherd dogs from the same litter at 17 months of age. Two have pituitary dwarfism; the other is normal. (From Andresen *et al.* in *Nord. Vet.-Med.* **26:** 692–701.)

were found by the Danish investigators to be high in unrelated dogs of normal size, extremely low in dwarfs, and intermediate in eight normal relatives of the dwarfs. Some of these last must have carried the causative gene.

NUTRITION

Genetic differences among domestic animals in rates of growth and in conversion of feedstuffs to meat, milk, eggs, and other desirable products must depend in large measure on the efficiency with which the animals can digest, assimilate, and utilize the nutrients given them. From what we have read in this chapter, genetic differences in those important processes are to be expected as in other aspects of physiology. Apart from the drastic abnormalities which cause recognizable pathological conditions like porphyria and storage diseases, there must be lesser ones with effects not evident except in subnormal productivity or in the amount of feed required per pound, gallon, or dozen of finished products.

These lesser variations are more easily detected in small laboratory animals than in the large domestic ones. The white rat, that favorite of the nutritionists, has revealed that (within the species) different strains vary greatly in their requirements of riboflavin, choline, thiamine, and vitamin D.[1] Similar differences and others are to be expected in any domesticated species.

In the domestic fowl, which, in addition to its many other virtues, is particularly suitable for studies of nutritional requirements, breeds differ significantly in ability to utilize vitamins D, E, and thiamine, in requirement of manganese for normal development of bones, and in requirement of the amino acids methionine and arginine for growth. These are all genetic differences, though less visible than those in form and feathers which differentiate the breeds. The block in the utilization of riboflavin, whereby certain hens are genetically unable to pass that vitamin from the feed to their eggs, was considered in Chapter 14. Hens on one and the same diet differ also in the amounts of thiamine they deposit in the egg. Presumably, that thiamine comes from a surplus in the blood above the hen's own requirement, and hence differences in the levels of thiamine in the eggs are indirect measures of the efficiency with which the bird utilizes that vitamin.

---

[1]Such differences suggest that, when we are warned that saccharine, or ABC, or XYZ causes cancer in rats, we should ask not only: At what dosage?, but also: In what rats?

That White Leghorns are significantly better in that respect than are heavy breeds is shown by the following figures (from Howes and Hutt) for thiamine per 100 grams of yolk in eggs from hens all on the same feed:

	Strains (number)	Thiamine (micrograms)
White Leghorns	13	198
Rhode Island Reds	5	138
Barred Plymouth Rocks	3	134
New Hampshires	4	146

Within the White Leghorns, significant differences have been demonstrated among families (*i.e.*, genetic differences) in the thiamine content of their eggs.

Utilization of Arginine

With the exception of that block in the utilization of riboflavin, all of the differences in nutritional physiology mentioned above differ in one way from most of the genetic variations considered in this chapter. They are not caused by mutation of single genes. They are polygenic. They result from the combined action of an unknown number of genes.

Similar genetic differences in the ability of White Leghorn chicks to utilize the amino acid, arginine, were studied at Cornell University. When one strain was found to have an abnormally high requirement of arginine for growth to four weeks of age, selection was begun in two other stocks (both normal) to see if two strains could be differentiated, one with a high requirement of arginine (HA) and the other with a low requirement (LA).

Ability to utilize arginine was measured by weight at four weeks of age on a diet containing only 0.9 per cent arginine. In the fifth selected generation, chicks of the LA strain attained 84 per cent of the weight of controls given ample arginine, but the corresponding figure for the HA chicks was only 28 per cent. The selection practised had differentiated two distinct strains.

To determine more clearly the genetic basis for that differentiation, the two strains were crossed (using reciprocal crosses to ensure adequate sampling) and the F_1 generation was then backcrossed (in reciprocal crosses) to each of the parental pure strains. The F_1

generation was found to be intermediate between the two parental strains, and reciprocal backcrosses were intermediate between the F_1 and the parental strain to which they were backcrossed (Fig. 20-11).

Clearly, genetic differences in the utilization of arginine are polygenic. Activity of the enzyme, *arginase*, in the kidneys of chicks was inversely proportional to the rate of growth and was about five times as high in the HA strain as in the LA strain. Chicks of the former line also showed higher levels of free lysine, which are believed to be responsible for the excessive activity of arginase in the HA chicks.

One problem still not answered is: How did that original strain that prompted this investigation get its high requirement of arginine?

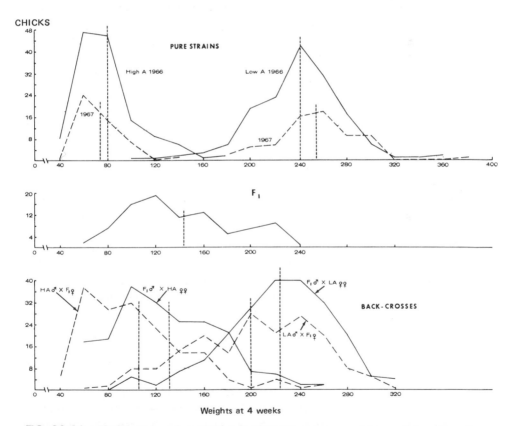

Weights at 4 weeks

FIG. 20-11 Utilization of arginine by chicks, measured by weight after four weeks on a diet deficient in arginine. After breeding two strains, one with high requirement (HA), the other with low requirement (LA), these were crossed, and the F_1 was backcrossed (reciprocally) to the two parent strains. The distributions show typical polygenic inheritance. (From Hutt and Nesheim in *Canad. J. Genet. Cytol.* as cited.)

The birds had been intermingled for many years with those of the two strains that had no such abnormal requirement. All had had the same diet, and there had been no selection for nutritional requirements. That significant differences among strains in requirement of one amino acid could arise without deliberate selection therefor, and remain undetected until revealed by specific tests, suggests that similar unsuspected genetic differences may account for some of the diverse results sometimes found in experiments with animals.

Applications of Biochemical Genetics

It seems desirable, perhaps as a stimulus to further work in this field, to mention briefly some useful applications that could result from extension of our knowledge about genetic variations affecting biochemical processes in domestic animals.

One of these would be the identification of genetic abnormalities of those processes that interfere with normal viability or productivity. Not all of these are likely to show such drastic effects as those in hemophilia, mannosidosis, porphyria, and the excess of phylloerythrin. Some cases of diabetes are recognized only by special search. A glance through the lengthening list of pathological conditions which are now known to be genetic "errors of metabolism" in man suggests that their counterparts in domestic animals may be equally numerous, and that some of them, although not causing conspicuous disease, could be responsible for many of the animals that the stockman conveniently classifies as "unthrifty." The smooth-tongue syndrome in cattle (Chapter 13), which is really a hypochromic anemia, is a good example.

Once such genetic deviations from efficiency are identified, and the mode of inheritance is determined, the genes responsible therefor should be quickly eliminated. The fact that in man every effort is made to preserve the biologically unfit and to give them the opportunity to pass their bad genes to succeeding generations does not mean that we must do the same with domestic animals. We can eliminate genetic junk from pigs, if not from people.

Secondly, further studies of genetic variation in requirement of specific nutrients could lead, not only to the breeding of stock more efficient in converting raw materials to edible foodstuffs, but perhaps even to the development of stock better able to cope with shortages of essential nutrients that are sometimes difficult to obtain. The amino acid, lysine, is one of these. Can plants generally considered unpalatable, inedible, or even toxic, become useful if animals are bred that can handle them? Remembering that some of us are homozygous for a recessive gene that removes all aversion to the unpalatable PTC, and that the rabbits having atropinesterase can cope with the toxic alkaloid, atropine, of Solanaceous plants, only little imagination is needed to prompt the hope that some day we may breed sheep not only willing to browse on the bracken but also genetically able to thrive on it.

One attractive application of biochemical genetics lies in the possibility of discovering some relationship between, on the one hand, productivity, fertility, viability, or adaptation to environment, and, on the other, variations in simply inherited traits such as red-cell antigens, hemoglobins, serum proteins, or enzymes. Such an association might result from pleiotropy in which genes with specific biochemical effects influence other traits as well. We have already considered many examples of such pleiotropy, most of them undesirable (*e.g.*, phenylketonuria and mental defect). In Chapter 23, we consider a better one—the helpful action of the red-blood cell allele B^{21} in combatting Marek's disease in the fowl.

A relationship of the kind sought could come from linkage of two genes, one of which affects some economically desirable trait and the other some biochemical process. If discovery of such desirable relationships proceeds only slowly, we at least know some that are undesirable, such as the hemolytic disease of foals which results from incompatibility of maternal and fetal genotypes.

Pharmacogenetics

One aspect of biochemical genetics with important practical applications to the health of man and animals is *pharmacogenetics*, the study of genetic differences in sensitivity to certain drugs. For example, there are sex-linked mutations which make some people deficient in the enzyme, glucose-6-phosphate dehyrogenase, and such people develop hemolytic anemia when given primaquine, a drug used in combatting malaria. Sawin's rabbits that can or cannot cope with atropine illustrate a similar variation.

In sheep, variation in arylesterases has been shown to be important in sensitivity to Haloxon, an anthelmintic drug which kills such worms as *Haemonchus contortus* by inhibiting the action of cholinesterases. In a study of plasma esterases of sheep, Lee showed that the plasma of some sheep, which he called Halon-high, hydrolyzed Haloxon more rapidly than others (Halon-low). He found that Halon-high is inherited as a simple dominant trait, and that when Halon-high sheep were given large doses of Haloxon (much higher than the anthelmintic dose) they did not show neurotoxic symptoms as the Halon-low sheep did. Tucker and her co-workers showed that arylesterase types could be identified by starch-gel electrophoresis. These investigators also developed a quick tube test to identify esterase-positive (EsA+, or Lee's Halon-high) and esterase-negative (EsA−, or Halon-low).

In order to test the sensitivity of the two types to Haloxon, Baker and his associates gave large doses of the drug to eight EsA+ and eight EsA− sheep. The EsA+ sheep were unaffected, but seven out of eight of the EsA− sheep developed such a severe ataxia or paralysis of the hind legs that they died or had to be destroyed. The EsA+ sheep were apparently unaffected because of their ability to hydrolyze Haloxon. Although the affected sheep suffered considerable loss of motor function, sensory function of the nerves was retained. They did not show symptoms until 21 days after they were treated with Haloxon, and the paralysis was not progressive. The sheep just died from injury or from inability to eat. Although the toxic doses of Haloxon are much higher than the anthelmintic doses ordinarily used, this experiment showed that genetic differences in sensitivity should be considered in drug usage.

Problems

20-1 What ways can you think of by which the high excretion of uric acid could have become a characteristic of the Dalmatian dogs?

20-2 Would the previous question be easier to answer if Trimble and Keeler had not tested for linkage of high uric acid with spotting? If so, why?

20-3 In studying the inherited inability to excrete phylloerythrin, Hancock mated affected sheep to others from the cross, normal × carrier, and found among 22 offspring 5 lambs with the defect. What number was to have been expected?

20-4 There are scores of genetic aberrations of physiology yet to be discovered in domestic animals. When you think that you have found one, by what criteria would you suspect that it might be (a) genetic, or (b) non-genetic?

20-5 If it proves to be hereditary, how would you determine if it is (a) dominant or recessive, (b) autosomal or sex-linked, or (c) polygenic?

20-6 Would the evidence needed to answer 20-5 come most easily from cattle, horses, sheep, swine, or Irish setters? Why?

20-7 How would you find out whether or not a similar abnormality is already known in man?

20-8 If you were appointed as animal geneticist in some agricultural or veterinary experiment station, (a) what problems would you tackle in the field of biochemical genetics of domestic animals, (b) how would you go about it, and (c) in what science or sciences would you seek associates to collaborate in the studies?

Selected References

ANDRESEN, E. 1977. Haemolytic anaemia in Basenji dogs. II. Partial deficiency of erythrocyte pyruvate kinase (PK; EC 2.7.1.40) in heterozygous carriers. *Animal Blood Groups Biochem. Genet.* **8:** 149–56. (Details of his genetic and biochemical studies.)

ANDRESEN, E., P. WILLEBERG, and P. G. RASMUSEN. 1974. Pituitary dwarfism in German Shepherd dogs: Genetic investigations. *Nord. Vet.-Med.* **26:** 692–701. (About the dwarf dogs described in this chapter.)

BAKER, N. F., E. M. TUCKER, C. STORMONT, and R. A. FISK. 1970. Neurotoxicity of haloxon and its relationship to blood esterases of sheep. *Am. J. Vet. Res.* **31:**865–71. (Details of experiment demonstrating differences in sensitivity of sheep to Haloxon.)

BROWN, R. V., and Y. TENG. 1975. Studies of inherited pyruvate kinase deficiency in the Basenji. *J. Amer. Animal Hosp. Assoc.* **11:** 362–65. (Reports extensive tests to detect carriers.)

CHEVILLE, N. F. 1968. The gray Collie syndrome. *J. Amer. Vet. Med. Assoc.* **152,** Part I: 620–30. (Study of cyclic neutropenia.)

CLARE, N. T. 1945. Photosensitivity diseases in New Zealand. IV. The photosensitizing agent in Southdown photosensitivity. *New Zealand J. Sci. and Tech.* **27,** Sect. A: 23–31. (About the excess of phylloerythrin.)

COLE, R. K. 1966. Hereditary hypothyroidism in the domestic fowl. *Genetics* **53:** 1021–33. (Details about the hypothyroid Leghorns and the selection that increased their numbers.)

FRITZ, T. E., L. S. LOMBARD, S. A. TYLER, and W. P. NORRIS. 1976. Pathology and familial incidence of orchitis and its relation to thyroiditis in a closed Beagle colony. *Exper. Molec. Pathol.* **24:** 142–58. (With descriptions of the affected dogs.)

HANCOCK, J. 1950. Congenital photosensitivity in Southdown sheep, a new sublethal factor in sheep. *New Zealand J. Sci. and Tech.* **32,** Sect. A: 16–24. (Description, with evidence of a genetic basis.)

HOCKING, J. D., R. D. JOLLY, and R. D. BATT. 1972. Deficiency of α-mannosidase in Angus cattle. *Biochem. J.* **128:** 69–78. (About this lethal storage disease.)

HUTT, F. B., and M. C. NESHEIM. 1968. Polygenic variation in the utilization of arginine and lysine by the chick. *Canad. J. Genet. Cytol.* **10:** 564–74. (Source of the data shown in Fig. 20-10.)

JOLLY, R. D. 1978. Lysosomal storage diseases of animals with particular reference to bovine mannosidosis. *Proc. Amer. Coll. Vet. Pathologists,* 29th Ann. Meeting, San Antonio, Texas, Nov. 14–18, 1978. (A review listing 11 storage diseases in domestic animals; 50 references.)

KOPPANG, N. 1970. Neuronal ceroid-lipofuscinosis in English setters. *J. Small Anim. Pract.* **10:** 639–44. (Report of eight years of study.)

LUSH, I. E. 1966. *The Biochemical Genetics of Vertebrates Except Man.*

Amsterdam: North Holland Publishing Co. Philadelphia: W. B. Saunders. VIII + 118 pp. (Reviews those genetic aberrations then known.)

McKUSICK, V. A. 1978. *Mendelian Inheritances in Man: Catalogs of Autosomal Dominant, Autosomal Recessive, and X-Linked Phenotypes*. 5th ed. Baltimore. Johns Hopkins University Press. XCI + 975 pp. (Covers all genetic disorders known in man, with extensive references, and much about man's chromosomes.)

PADGETT, G. A., C. W. REIQUAM, J. R. GORHAM, J. B. HENSON, and C. C. O'MARY. 1967. Comparative studies of the Chediak-Higashi syndrome. *Amer. J. Pathol.* **51:** 533–71. (In man, minks, and cattle.)

PINKERTON, P. H., S. M. FLETCH, P. J. BRUECKNER, and D. R. MILLER. 1974. Hereditary stomatocytosis with hemolytic anemia in the dog. *Blood* **44:** 557–67. (Trouble associated with peculiar erythrocytes in Alaskan Malamutes.)

PRIEUR, D. J., and V. CÁMARA. 1979. Inheritance of lysozyme deficiency in rabbits. *J. Heredity* **70:** 181–84. (A genetic abnormality that apparently does no harm.)

SAWIN, P. B., and D. GLICK. 1943. Atropinesterase, a genetically determined enzyme in the rabbit. *Proc. Natl. Acad. Sci.* (Washington) **29:** 55–59. (Only in the rabbits with the gene *As*.)

TRIMBLE, H. C., and C. E. KEELER. 1938. The inheritance of "high uric acid excretion" in dogs. *J. Heredity* **29:** 280–89. (About the distinction of the Dalmatians.)

Blood Groups and Protein Polymorphisms

In Chapter 3 we were introduced to MN blood groups in humans as examples of codominance. Codominant inheritance of blood groups is the rule in all species, but we know that there are exceptions. We considered one of these, the inheritance of O in the ABO system, in Chapter 11, where ABO blood groups were considered as an example of multiple alleles. The polymorphisms for MN and ABO blood groups in man can be detected only by special techniques; a person's appearance gives us no clue to his blood type. Similarly, in other animal species there are polymorphisms which are composed of a variety of normal types, so that within strains or breeds which superficially appear alike we can find a wealth of variability. In this chapter we will look at blood groups of animals in some detail, and we will also consider some other polymorphisms, including those of proteins of the blood plasma and milk, which contribute to normal variation in our domestic animals.

RED BLOOD CELL ANTIGENS

An *antigen* is a substance which, when introduced into an animal, stimulates the production of *antibodies* which react with the antigen in question. Research in *immunogenetics,* a combination of immunology and genetics, has made use of antigen-antibody reactions to identify blood groups, such as human ABO, MN, and Rh groups. The antigens of the red blood cells are specific chemical constituents on the surface of the red cells, and the antibodies are modified globulins in the blood plasma which are ordinarily produced in response to stimulation by immunization with red blood cells. An antigen may stimulate the production of more than one kind of antibody, and the reaction of the component of the antigen with antibody specific for it identifies a *blood factor* or *antigenic factor.* An antigen may therefore have a number of different blood factors.

The blood factors can be identified only if the antigen-antibody reactions can be seen. There are a number of ways to make them visible. In a blood-typing test a drop of serum which contains antibodies is mixed with a drop of red blood cells suspended in physiological salt solution. If there is an antigen-antibody reaction, single antibody molecules may combine with antigens on more than one red cell, so that the red cells are drawn together into clumps which are easy to see. This clumping of red cells is called *agglutination,* and such a saline agglutination test is used for many blood-typing tests. However, for some tests, such as for red-cell antigens of cattle, agglutination tests are unsatisfactory. For cattle blood-typing tests, hemolysis is used instead of agglutination. *Hemolysis* is the release of the red pigment of the blood from the red cells, and in the hemolytic test, a drop of complement (present in fresh rabbit or guinea pig serum) is added to the mixture of red cells and antibody. If the test is positive, hemoglobin is released into the supernatant fluid surrounding the ghosts (colorless membranes) of the red cells, so that the supernatant turns the bright cherry red color of hemoglobin.

Antibodies against human A and B antigens are ordinarily present in the blood plasma of persons lacking the respective antigen, but this is not true for most blood-typing antibodies. Antibodies against Rh factors may be produced by a pregnant woman negative for the factor in response to immunization by a fetus positive for the factor. Antibodies for human blood typing may also be obtained by deliberate injection of volunteer recipients lacking a specific blood factor with blood from a donor having the factor. Such recipients are carefully monitored to avoid reactions to the transfused blood. Antibodies for animal blood typing are practically always produced by deliberate injection to immunize one animal with blood of another. Fortunately, adverse reactions to such immunization are rare. *Isoimmunizations,* where the donor and the recipient are of the same species, are usual, but occasionally *heteroimmunization* (donor and recipient of different species) is the method of choice. If an antiserum produced by immunization contains antibodies for only one antigenic factor, it is referred to as a monospecific antiserum and, appropriately diluted, may be used as a *blood-typing reagent.* If an antiserum contains antibodies for more than one blood factor, the antibodies may be separated by absorption techniques, whereby red cells with specific factors are added to the serum, the antibodies for the factors are absorbed onto them, and the red cells coated with antibodies are removed from the antiserum to separate them from those remaining.

Blood factors which identify antigens controlled by alleles at a single locus are part of the same *blood-group system.* Some blood-group systems have large numbers of multiple alleles, each responsible for an antigen characterized by one or more blood factors. Table 21-1 summarizes (in greatly condensed form) blood-group systems in domestic animals as known in 1975. New alleles are still being identified.

Blood Groups in Cattle

There are 12 different genetic systems of blood groups in cattle: A, B, C, F, J, L, M, N, S, Z, R′, and T′ (Table 21-2). All of the factors in these systems are identified by hemolytic tests, and each factor is

TABLE 21-1

Blood-Group Systems in Some Domestic Animals

Species	Systems (loci)	Minimum Number of Alleles at Most Complex Locus
Cattle	12	500
Sheep	8	60
Swine	15	13
Horses	8	6
Fowl	12	35

Source: Data summarized from Rasmusen, 1975.

TABLE 21-2

Blood Groups in Cattle

Locus	Blood Factors	Minimum Number of Alleles
A	A_1, A_2, D_1, D_2, H, Z'	10
B	B_1, B_2, G_1, G_2, I_1, I_2, O_1, O_2, O_3, O_x, P_1, P_2, Q, T_1, T_2, Y_1, Y_2, A', A'_2, B', D', E'_1, E'_2, E'_3, E'_x, F', G', I', J', K', O', Y'	500
C	C_1, C_2, E_1, E_2, R_1, R_2, W, X_1, X_2, L'	70
F	F_1, F_2, V_1, V_2, V_3	5
J	J	4
L	L	2
M	M_1, M_2, M'	3
N	N	2
S	S_1, S_2, U_1, U_2, U', U'', H'	8
Z	Z_1, Z_2	3
R'	R'_1, R'_2, S'	3
T'	T'	2

Source: Abridged from Rasmusen, 1975.

inherited as a codominant. When cattle blood typing was initiated at the University of Wisconsin in the early 1940s, the factors were arbitrarily named A, B, C, etc., in order of isolation of monospecific reagents for the factor, so that A and A', B and B', bear no particular relationship to each other, and the factors within each system are not grouped by their position in the alphabet. Factors with the same base letter but different subscripts, such as A_1 and A_2 are linear *subgroups*, so that if A_1 is present, A_2 is always present also. A_2, on the other hand, can be present without A_1. Within each blood-group system, when an antigen has more than one blood factor, these are grouped together into *phenogroups* (products of individual alleles).

Table 21-3 gives six examples of different groupings into phenogroups of the more than 40 different blood factors in the B system. These six phenogroups are only a small sample of the more than 500 phenogroups possible in the B system. Each was identified by studies of inheritance of the blood factors, and each has been assigned a standard code number for easy identification. (It is easier to say and remember B28 than $BGKO_2Y_1A'B'E'G'K'O'Y'$!) As you can see from Table 21-3, B phenogroups vary considerably in their distribution among different breeds. Some are characteristic of only one of the five breeds listed, whereas others are found in more than one. B89 (I') is a widely distributed, though infrequent, phenogroup. The other phenogroups in Table 21-3 have been chosen to illustrate phenogroups of highest frequencies in the breeds listed. There are, of course, a number of other B phenogroups in each breed. In Holstein-Friesians in the United States there are more than 100.

Since there is so much variation in B phenogroups from breed to breed, the B phenogroup formula of a cow may reveal the breed (or

TABLE 21-3
Examples of B Red-Blood Cell Antigens in Cattle

Code Number	Phenogroup	Guernsey	Jersey	Holstein-Friesian	Milking Shorthorn	Hereford
B89	I'	<.01	.03	.03	.02	.10
B43	$IO_xE'_1$	.23	0	<.01	0	0
B28	$BGKO_2Y_1A'B'E'G'K'O'Y'$	.11	.24	0	0	0
B39	$GY_2E'_1$	0	0	.21	.01	0
B22	$BO_3Y_1A'E'_3G'$	0	0	.03	.42	0
B81	$Y_1D'I'$	0	0	0	0	.38

Allele Frequencies in: (spanning Guernsey, Jersey, Holstein-Friesian, Milking Shorthorn, Hereford)

Source: Condensed from Stormont, 1958.

breeds) of her ancestors. Variation within a breed makes it unlikely that two cows will have identical blood types (unless they are monozygotic twins), so that B phenogroups are especially useful to verify a cow's identity if she has been previously typed. The B system is also frequently useful in cases of disputed parentage. Barring mutation, a calf can not have any blood factors which it did not inherit from its parents. In a common kind of parentage case a cow has been inseminated with semen from two different bulls. If either bull lacks a B phenogroup (or any blood factor in any blood-group system) which the calf must have inherited from its sire, that bull is excluded as a parent and the other bull is presumed to be the sire of the calf. (This assumes he has the appropriate blood type.) This then is a case of *parentage exclusion*.

Exclusions may be possible in incomplete parentage cases, where, for example, the dam has not been typed, if the genotype of the calf can be determined directly from the blood type. This is frequently possible for types in the B, F, S, and R' systems.

Blood typing can also be used to show that cattle twins are not monozygotic. If cattle twins have different blood types, they, of course, must be *dizygotic* (from two eggs) rather than *monozygotic* (from one egg). However, even though on the average only about 10 per cent of like-sexed twin pairs are monozygotic, 90 per cent or more of all twin pairs, including male-female pairs, have identical blood types. Owen showed that this is due to each twin's having a mixture of two types of red cells, referred to as *erythrocyte mosaicism*. It is known that fusion of the embryonic circulation occurs in about 90 per cent of bovine twins; if the twins are a male-female pair, this leads to the development of the *freemartin*, a sterile individual which is a modified female. Owen's studies showed that with the fused circulation there is an interchange of primordial blood cells, which become established and persist in the blood-forming tissues of their new hosts. As a result, both twins have the same blood type.

Identical blood types can extend to higher multiple births as well. Owen and co-workers found identical blood types in a set of quintuplet calves which must have originated from at least four eggs, since one was a solid-colored female (freemartin), one was a solid-colored male, one was a spotted male, and two were white-faced males (Fig. 21-1). By the special techniques of *differential hemolysis*, the blood of each calf could be shown to be a mixture of two types of blood, with all calves having 85 per cent of their red cells positive for the A factor and 15 per cent negative. This showed that all five calves had erythrocyte mosaicism. Typically, in newborn twin calves with erythrocyte mosaicism, the same type of blood will predominate in both calves, and the genotype of each calf for red-cell antigens which are mosaic can be determined only by breeding tests (which are impossible for freemar-

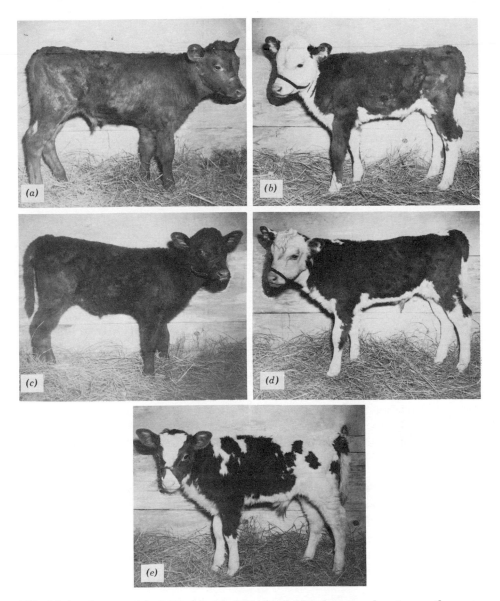

FIG. 21-1 Quintuplet calves with identical blood types due to erythrocyte mosaicism. (From R. D. Owen, H. P. Davis, and R. F. Morgan in *J. Heredity.*)

tins!). Since all five of the quintuplet calves had identical blood types and the same type of blood predominated in each, it seems that the mosaicism involved all five of them.

Differential hemolysis is also useful to identify erythrocyte mosaicism for diagnosis of freemartinism. It is useful to type both twins for red-cell antigens before making this test, since if a heifer has a different blood type for her bull co-twin (except for J), freemartinism may be excluded. J is an exception because the J substance is acquired by the red cells from the plasma rather than being an intrinsic character of the red cells. If the twins have identical types, or differ only in J, and mosaicism can be identified by differential hemolysis, the heifer can be diagnosed a freemartin.

Owen's work on erythrocyte mosaicism was of great importance in the development of the concept of immunological tolerance. If skin is grafted from one mosaic co-twin to another, the grafts succeed between the twins because they are tolerant of each other's tissues; they are *histocompatible*. The interchange of primordial blood cells provides a natural graft which takes place early in embryonic development and makes the twins tolerant of skin and other tissues as well as blood cells. The time factor is important in development of immunological tolerance; after birth, if tissues are transplanted between calves, the transplants will not succeed. The species of animal and the age at which it was first exposed to the foreign antigen are important in determining the degree of tolerance, as well as the specific histocompatibility antigens involved. As a rule, histocompatibility antigens, like red-cell antigens, are inherited as codominants. Owen's pioneering work helped lay the groundwork for research for which Sir Peter Medawar later became a Nobel Prize winner in physiology and medicine.

In linkage studies with cattle blood groups, the locus for A red-cell antigens has been shown to be approximately two map units from the locus for hemoglobin variants, and the J locus is approximately four map units from the locus for β-lactoglobulin (milk protein) types.

There have been a number of studies of possible effects of red-cell antigen type on economic traits such as milk and butterfat production. There is some evidence that alleles at some of the blood-group systems do have an effect on performance, but the effects are not large enough to use blood groups as a basis to select superior cows.

Blood Groups in Sheep

Eight loci have been shown to be involved in blood groups in sheep: A, B, C, D, I, M, R, and X. The B and C blood-group systems were named after the B and C systems in cattle, with which they are homologous, and the M system is homologous to the S system in cattle. In some cases, the same reagent may be used to identify blood factors for both cattle and sheep, and, originally, the blood factors were named basically in the same way as cattle blood factors, with in addition some regard being given to homologies in the two species. In 1973, the sheep factors were renamed, with symbols for the loci in capital letters followed by a small letter for the blood factor within the system, for example, Aa, Ab. (A similar scheme for nomenclature is commonly used for blood groups in horses and pigs as well.)

There is one exception to the new system of nomenclature: In the R system, factors are designated R and O, and and factor O is recessive to R. (In all the other systems, blood factors are inherited as codominants.) R is related to A of humans, and O is related to O of humans, but R and O are not intrinsic characteristics of the red cells but are plasma substances, like J of cattle, acquired by the red cells. Some sera from nonimmunized cows used as sources of reagents for J of cattle may also be used to test for sheep R. Since O is recessive to R, O may be present on the red cells of lambs which have two type-R parents, an exception to the rule that a blood factor in an individual must be present in at least one parent. In addition, R may also be present in a lamb and absent in both parents: some sheep lack both R and O, because they are homozygous, *ii,* for a gene at another locus, which prevents expression of R and O. Since i is epistatic to R and O, a mating of O × i may give an R lamb. The gene interaction involved in R, O, and i phenotypes may be diagrammed as follows:

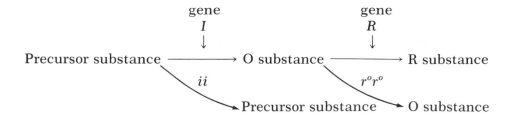

According to this scheme, gene *I* is necessary to convert a precursor substance to O substance, and gene *R* is necessary to convert O to R substance. If *I* is lacking, sheep are type i, because O substance is not produced. Even if they have gene *R*, they are i, because O substance is necessary to produce R substance.

Type-O sheep have a type of alkaline phosphatase in their plasma, called the B band, which is lacking in R and i sheep.

The M system is of special interest because of the relationship of M antigens to levels of potassium ions in red blood cells of sheep. Sheep can be divided into two types with regard to red-cell potassium: high potassium (HK) and low potassium (LK), with red cells of HK sheep having a threefold or higher concentration of potassium ions than LK sheep. In 1966 Rasmusen and Hall showed that alleles for M factors were also responsible for determining whether sheep are of the HK or LK type. The relationships among genotypes, blood factors, and red-cell potassium are:

	Blood Factors		Red-Cell
Genotype	Ma	Mb	Potassium
M^aM^a	+	−	high (HK)
M^aM^b	+	+	low (LK)
M^bM^b	−	+	low (LK)

Although the blood factors Ma and Mb are codominant, the HK type is recessive to the LK type. It is not known how the genes exert their two seemingly different effects on M antigens and potassium levels, but the effect on potassium levels appears to be a property of the red-cell membrane which makes it behave as a pump that provides active transport of potassium ions in and out of the red cell.

The frequencies of HK and LK sheep within different breeds vary considerably in different parts of the world, and it seems likely that potassium levels in their red cells might be important in adaptation to their environment. Rasmusen and Lewis did find an extremely high mortality rate of 48.5 per cent in Suffolk lambs in Illinois from mating of $M^aM^a \times M^aM^a$ compared to mortality rates of from 0 to 40.9 percent in lambs from other types of matings (Table 21-4). Regardless of the sires of the lambs, mortality rates in lambs from homozygous M^aM^a ewes were high. Clearly, in this flock it would have been economical to have used only LK (M^aM^b or M^bM^b) ewes for breeding. It has been suggested that LK types might do better in hot, dry climates and that HK types might do better in cool, wet climates. There is evidence that LK sheep have finer wool than HK sheep.

Blood Groups in Horses

Eight systems of blood groups in horses have been described (A, C, D, K, P, Q, T, and U), and both agglutination and hemolytic tests, as well as special techniques, are used to identify the blood factors. Equine blood groups have been used, as in cattle, for animal identification and for exclusion of parentage. Blood factors in two systems, A and Q, appear to be important in hemolytic icterus of newborn foals. This disease was considered in Chapter 14 as an example of one kind of maternal influence.

TABLE 21-4
M Types of Parents and Mortality of Suffolk Lambs

Mating Type	Lambs (number)	Mortality (per cent)
$\male\male \times \female\female$		
$M^aM^a \times M^aM^a$	21	48.5
$M^aM^b \times M^aM^a$	13	31.8
$M^bM^b \times M^aM^a$	14	40.9
$M^aM^a \times M^aM^b$	22	28.6
$M^aM^b \times M^aM^b$	55	15.7
$M^bM^b \times M^aM^b$	34	23.9
$M^aM^a \times M^bM^b$	14	0
$M^aM^b \times M^bM^b$	27	4.7
$M^bM^b \times M^bM^b$	24	38.7

Source: Data after Rasmusen and Lewis, 1973.

Blood Groups in Pigs

Fifteen loci (A, B, C, D, E, F, G, H, I, J, K, L, M, N, and O) for blood groups in pigs have been identified. In the A system, A is related to A of humans, J of cattle, and R of sheep, and there is an O factor in pigs which is recessive to A. There is also a type negative for A and O, comparable to the i type in sheep. Genotypes at the H locus have been shown to be associated with the lack of expression of A and O.

Genes at the C and J blood-group loci are linked, with about 6 map units between them, and the J alleles have also been shown to be linked with genes of the main histocompatibility complex in pigs, with 9.8 per cent recombination. The K blood-group locus is 4 map units from the locus for variants of heme-binding globulins, and the I blood-group locus is 2.5 map units from the locus for serum amylase variants.

The H system is of particular interest for several reasons. In addition to their association with effects on expression of A and O, alleles at the H locus have been shown to be linked to genes at at least two other loci. The H locus has been reported to be 2.6 map units from the locus for phosphohexose isomerase (PHI) variants of red-cell isoenzymes and 3.4 map units from the locus for 6-phosphogluconate dehydrogenase (6-PGD) variants. The locus for sensitivity to halothane anesthesia is also close to the H locus. Halothane sensitivity is a recessively inherited trait with about 95 per cent penetrance, which is important as an indicator of the porcine stress syndrome (PSS). PSS is a cause of sudden death in pigs which are stressed, and it can cause considerable economic loss to pork producers, especially if the deaths occur in pigs on the way to slaughter. The details of relationships between H, expression of A and O, PHI, 6-PGD, PSS, and halothane sensitivity remain to be worked out, but the genes for all of these traits are located close together on the same chromosome.

H types are also associated with effects on growth and reproduction. In some breeds, such as Durocs, Ha-positive pigs have been found to be faster growing and to have smaller litters than Ha-negative types.

Blood Groups in Fowls

Agglutination tests have been used to identify factors in twelve different blood-group systems (A, B, C, D, E, H, I, J, K, L, P, and R) in the domestic fowl. The A and E loci are linked, with approximately 0.5 map units between them.

The B system is of particular interest for several reasons. B genotypes have been shown to be important in hatchability of eggs and in vigor and egg production of hens, although not important enough that birds are selected for breeding or production on the basis of their B types. The B locus is a major histocompatibility locus as well as a blood-group locus, so that B genotypes are important in transplantation of skin and organs.

B genotypes are also important in susceptibility and resistance to Marek's disease, a virus disease which is a frequent cause of fatal lymphoid tumors. For details, see Chapter 23.

Protein Polymorphisms

We have mentioned a number of protein polymorphisms determined by genes linked to genes for red-cell antigens: hemoglobin and beta-lactoglobulin types in cattle, and heme-binding globulin, amylase, PHI, and 6-PGD types in pigs. All of these protein polymorphisms are identified by electrophoresis, a procedure based upon the differential migration of molecules in an electric field.

Genetically Different Hemoglobins

The first genetic variation in proteins which was demonstrated by electrophoresis was found in human hemoglobin by Pauling and co-workers. Hemoglobin, the red pigment of the blood, has a heme part which is the iron-containing part, and a globin part, which consists of two identical alpha and two identical beta polypeptide chains. Using paper electrophoresis, Pauling and co-workers demonstrated a difference between normal hemoglobin A and hemoglobin S, the hemoglobin responsible for sickle-cell anemia. Since then, many other variants of human hemoglobin have been found. Most are beta-chain variants, and almost all of them can be shown to differ from hemoglobin A by a change in a single amino acid of the beta chain. The vast majority of people are homozygous for hemoglobin A, and other hemoglobins are considered abnormal. In sheep, cattle, and horses, on the other hand, there are two or more kinds of normal hemoglobin.

Hemoglobins in Sheep

In sheep, there are two normal hemoglobins, A and B, the products of one pair of alleles. Under electrophoretic conditions ordinarily used (starch-gel electrophoresis, pH 8.5 to 9.0), hemoglobin A migrates faster than hemoglobin B. Both A and B can be identified in blood of heterozygotes; they are codominants, as in most protein polymorphisms identified as electrophoretic variants. Hemoglobins A and B of sheep differ, surprisingly, in eight amino acids of their beta chains. Both hemoglobins are widely distributed among breeds in different parts of the world, but the "fast" hemoglobin A is predominant among breeds at latitudes higher than 40°. Evans and co-workers, who compared A and AB sheep, found that type-A sheep harbor fewer intestinal worms (*Haemonchus contortus*) than AB. It appears likely that hemoglobin A is of some advantage in harsh, cold climates and in environments where intestinal parasites are common. In spite of the apparent advantage of hemoglobin A over B in those environments, Evans and Turner reported that in Australia, Merino ewes of type B weaned more lambs than those of type AB, and the AB ewes in turn weaned more lambs than did ewes of type A.

Sheep with the gene for hemoglobin A are unique in that when they are made severely anemic by infestation with large numbers of worms (or experimentally, by bleeding), they produce a second hemoglobin, called hemoglobin C, and production of hemoglobin A stops. Hemoglobin B in sheep of type AB or B is not affected. It is not known why this change from production of hemoglobin A to C takes place, or whether this is any particular advantage to the sheep. It does provide a useful model for researchers interested in mechanisms involved in gene action and the change from production of one protein to another.

Hemoglobin in Cattle

Two types of hemoglobin, A and B, which differ in their beta chains, have also been identified in cattle, but if one uses the same technique for typing as for sheep, B is the faster moving of the pair. Most cattle breeds which originated in northwestern Europe have only hemoglobin A; as we move south, this changes so that in southwestern England there is some hemoglobin B in South Devon cattle. B is also found in Channel Island breeds (Jerseys and Guernseys). Further south in Europe hemoglobin B becomes even more common, and it is frequent in zebu cattle. The dividing line between breeds with and without hemoglobin B is roughly that of the boundaries of the old Roman Empire; whether or not this is coincidental is anybody's guess.

Hemoglobin in Horses

In horses there are also two kinds of hemoglobin: fast, or A_1, and slow, or A_2. The vast majority of horses have both A_1 and A_2, so they do not appear to be allelic alternatives, even though the difference in hemoglobin A_1 and A_2 has been shown to be due to a single amino acid. A_1 has glutamine, and A_2 has lysine at position 60 in the alpha chain. The relative proportion of the two kinds of hemoglobins is under genetic control; it appears that a modifying gene modulates the production of hemoglobin, and the relationships between the genotypes for the modifying gene and the percentage of A_2 have been suggested to be as follows:

Genotype	Percentage of Hemoglobin Which Is A_2
$Hb^{m+}Hb^{m+}$	34–41
$Hb^{m+}Hb^{m-}$	15–23
$Hb^{m-}Hb^{m-}$	0

This appears to be an example of a gene that modifies the amount of a specific protein (hemoglobin chain A) which is produced.

At least one other gene is responsible for biochemical differences in horse hemoglobins, but the two variants are not distinguishable by electrophoresis. These variants differ by a single amino acid at position 24 in the alpha chain.

Serum Transferrins

In the middle 1950's, Oliver Smithies, working at the University of Toronto, showed that starch gel was an excellent medium to use to study certain protein polymorphisms. With starch-gel electrophoresis, the proteins are separated because of the size of their molecules as well as because of their electrical charge. Many refinements and variations in technique have been made with gel electrophoresis in order to separate different proteins and to demonstrate protein variants. Figure 21-2 shows the results of one method of separation of proteins of cattle sera using polyacrylamide gel electrophoresis. Among the

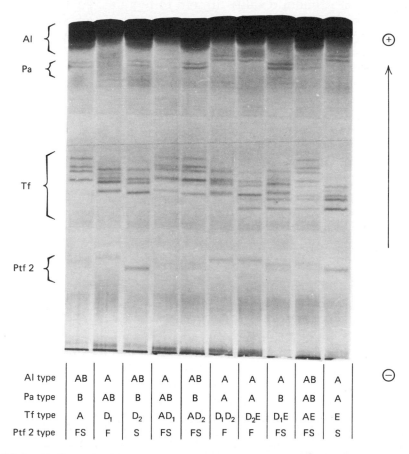

Al type	AB	A	AB	A	AB	A	A	A	AB	A
Pa type	B	AB	B	AB	B	A	A	B	AB	A
Tf type	A	D_1	D_2	AD_1	AD_2	D_1D_2	D_2E	D_1E	AE	E
Ptf 2 type	FS	F	S	FS	FS	F	F	FS	FS	S

FIG 21-2 Different phenotypes of albumin (A_1), post-albumin (Pa), transferrin (Tf), and post-transferrin (Ptf_2) of cattle plasma. (From B. Gahne, R. K. Juneja, and J. Grolmus in *Anim. Blood Groups and Biochem. Genet.*)

protein polymorphisms illustrated are serum transferrins. Transferrins, as the name suggests, are iron-binding proteins. They are in the beta-globulin fraction of the serum and have various functions in the animal body which are not well understood. Four alleles, Tf^A, Tf^{D_1}, Tf^{D_2}, and $Tf^{E,}$ are commonly found in European breeds of cattle, and Figure 21-2 illustrates transferrin types resulting from various combinations of these four alleles. There are at least three more alleles in zebu cattle. The transferrins have also been found to be polymorphic in domestic animals other than cattle. They are especially useful for parentage exclusion in many breeds of cattle, horses, and sheep.

The fact that males as well as females can be typed for transferrins has facilitated study of any possible relationship they might have to traits of economic value limited to females, such as yields of milk and butterfat. In an early study of genotypes of 141 bulls in which he did not differentiate between transferrin D_1 and D_2, Ashton found that the milk yield of daughters of bulls of transferrin type D was higher than that of daughters of bulls of type A. More recent studies have confirmed Ashton's results in part: for example, Kiddy and co-workers found that transferrin D increased milk production in American Holstein-Friesians when compared to transferrin A, but seemed to decrease it in Jerseys and Guernseys.

In other studies, Ashton found that transferrin type had an influence on fertility in cattle. Among other effects he found that fertility was significantly higher when both partners in a mating were homozygous at the transferrin locus than when one or both were heterozygous at that locus. Studies of effects of transferrin types on other species, including sheep and pigs, also indicate that transferrins do make a difference in fertility, but effects have not been entirely consistent among different breeds.

Variations in Proteins in Milk

In addition to polymorphisms of serum proteins, a number of polymorphisms have been described for proteins of other body fluids. Considerable progress has been made in determining different genetic types of milk proteins. The mammary glands of ruminants, such as cattle, synthesize large amounts of six proteins: four kinds of caseins which make up the largest proportion of milk protein (about 80 per cent of the total), α-lactalbumin, and β-lactoglobulin.

Polymorphism of milk proteins was first reported by Aschaffenburg and Drewry (1957), who used paper electrophoresis to identify two different β-lactoglobulins, A and B, produced by two alleles, Lg^A

and Lg^B. As in most protein polymorphisms, the alleles are codominant, so that heterozygotes have both kinds of β-lactoglobulin. Since the discovery of β-lactoglobulins A and B, two additional, rather rare types have been identified. The function of β-lactoglobulins is not understood, but when yields of milk and casein are fairly constant, cows of the genotype $Lg^A Lg^A$ produce nearly twice as much β-lactoglobulin as do cows homozygous $Lg^B Lg^B$.

Casein in milk can be divided into two principal components referred to as αs_1-casein and β-casein, and two minor components, αs_2-casein and κ-casein. Genetic variants have been identified for all four of these components, and they have provided material for detailed genetic and biochemical analyses. Two or more genetic variants have been identified for each of the casein components, and the different components have been shown to be inherited together. Since the exact amino-acid sequences have been determined for the various casein types, it is possible to postulate the exact relationship between gene and gene product, and it has been proposed that the different casein components are the products of genes which have resulted from gene duplications (see Chapter 18, pp. 415).

Blood Groups, Protein Polymorphisms, and Productivity

The most attractive application of research in blood groups and protein polymorphisms lies in the possibility of discovering some relationship between, on the one hand, productivity, fertility, viability, or adaptation to environment, and on the other hand, variations in simply inherited traits such as red-cell antigens, hemoglobin, serum proteins, or enzymes. There are at least four ways in which relationships between these traits and productivity might come about:

1. Pleiotropy, where a gene responsible for a specific qualitative trait affects productivity as well. We have already considered many examples of pleiotropy, such as the alleles in sheep which influence red-cell potassium levels and M antigens.

2. Linkage between two loci, one of which determines a simply inherited trait and the other, an effect on productivity. Unless the linkage were complete (and thus not distinguishable from pleiotropy), the simply inherited trait would have a positive relationship to productivity in some families and a negative one in others. Such relationships are not likely to be identifiable and would be difficult to utilize unless the linkage were very close.

3. Heterozygosity, where the heterozygous individual is superior in productivity to either homozygote. Such overdominance, if identified, would provide a means of making matings to utilize hybrid vigor resulting from genotypes at specific loci.

4. Maternal-fetal interactions. A dramatic example of a trait which results from a difference between the genotypes of mother and fetus is hemolytic disease of newborn foals. There are probably less striking examples of undesirable consequences of maternal-fetal interaction which, if understood, could be avoided in breeding programs. Interactions of this sort may complicate utilization of heterosis to increase productivity.

Problems

21-1 A type-i ram and a type-O ewe had twin lambs; one was type R and one was type i. What are possible genotypes of the ram and the ewe?

21-2 When a Duroc boar negative for A and for O was mated to a type-O Yorkshire sow, all seven of their piglets were type A. How do you account for this?

21-3 Why do most pairs of cattle twins have identical blood types? Why do they accept transplants of skin from their co-twins?

21-4 Why is it desirable to type the blood of both twins to exclude freemartinism in a heifer born co-twin to a bull calf?

21-5 How can blood typing for red-cell antigens be used to determine the potassium concentration of a sheep's red blood cells?

21-6 In cattle, the gene for white face, S^H, is dominant to its allele for solid color, S, which is dominant to spotted, s. If the mother of Owen's quintuplet calves (two white-faced, two solid-colored, and one spotted) was a solid-colored cow, what does this tell you about the genotype for color pattern of the bull which sired the calves?

21-7 When Crittenden and co-workers compared blood types and susceptibility to Rous sarcoma virus in a group of fowls, they found that all 35 birds possessing the b^s allele for susceptibility (19 $b^s b^s$ and 16 $b^s b^r$) belonged to the R1-positive blood type (R^1-), while all five birds of the resistant genotype $b^r b^r$ were R1 negative (rr). What are possible causes of the association between blood type and susceptibility to the sarcoma virus?

21-8 In pigs, transferrin B is found in high frequency in most breeds, and transferrin A is also relatively common. Transferrin C is quite rare. All are codominant. In Scotland, Imlah crossed related B/C boars and sows, and in eight litters he observed a total of 22 B/B, 40 B/C, and no C/C piglets. The average litter size was 1.4 pigs less than the average for other litters. How might you account for these results?

He then crossed one of the B/C boars with sows of a different breed and produced four litters in which he observed 4 B/B, 32 B/C, and 10 C/C piglets. How can you account for the C/C piglets in these litters when there were none in the first eight litters?

21-9 Højgaard and his associates found that among offspring of bulls homozygous for the Tf^A and Tf^D alleles determining transferrins in bovine serum, the distribution of phenotypes was as follows (using only superscripts for simplicity):

	AA	DD	AD	AE	DE
By sires AA	84	0	70	15	1
By sires DD	0	46	45	1	7

Did any of these calves show phenotypes (which are here also genotypes) not expected? If so, which?

21-10 How would you account for the exceptions in the previous question?

21-11 If we disregard those exceptions, what were the frequencies of the genes Tf^A, Tf^D, and Tf^E among the mothers of the calves?

21-12 With respect to these same transferrins, what phenotypes should be (a) possible or (b) not possible, among calves from cows of the following genotypes (using superscripts only): AA, DD, EE, AD, AE, DE?

21-13 If the number of multiple alleles determining transferrins in the blood serum of sheep were only 5, as first studies indicated, how many different genotypes could there be with respect to those transferrins?

21-14 In this chapter we learned of genetic variations in the hemoglobin, transferrins, and β-lactoglobulins of cattle. Remarkable differences in these have been found in cattle of different breeds in different parts of Europe, Asia, and Africa. How could they help people trying to trace the origins of various breeds?

21-15 Weitkamp and co-workers have described close linkage (2.3 per cent recombination) in humans between the loci for serum albumin types and Gc (group-specific component) protein types. In horses the vitamin-D-binding protein is homologous to Gc protein in human blood, and genetic polymorphism has been demonstrated for Gc in horses. When Sandberg and Juneja classified offspring of a stallion heterozygous for albumin and Gc types and 233 doubly heterozygous mares, they found only 2 recombinant types among 233 foals. How does the linkage between albumin and Gc types in horses compare with that in humans? What do you suppose is responsible for the resemblance?

21-16 When Jorgensen and co-workers classified Danish Landrace pigs for halothane sensitivity and for PHI type, they found that 49 out of 313 BB types were halothane sensitive, but none of 213 AB or 5 AA types was. What are possible reasons for the association of PHI type with halothane sensitivity? Which is most likely? Why?

Selected References

AGAR, N. S., J. V. EVANS, and J. ROBERTS. 1972. Red blood cell potassium and haemoglobin polymorphism in sheep. A review. *Animal Breeding Abstracts* **40**:407–36. (A comprehensive review with many references.)

ASCHAFFENBURG, R., and J. DREWRY. 1957. Genetics of the β-lactoglobulins of cow's milk. *Nature* **180**:649–50. (First report of genetic variation in proteins of cow's milk.)

ASHTON, G. C. 1960. β-globulin polymorphism and economic factors in dairy cattle. *J. Agric. Sci.* **54**:321–28 + P1. (Evidence for effect of transferrin type on yield of milk.)

ASHTON, G. C. 1965. Cattle serum transferrins: A balanced polymorphism? *Genetics* **52**:983–97. (Possible effects of transferrin type on fertility in cattle.)

EVANS, J. V., M. H. BLUNT, and W. H. SOUTHCOTT. 1963. The effects of infection with *Haemonchus contortus* on the sodium and potassium concentrations in the erythrocytes and plasma, in sheep of different haemoglobin types. *Aust J. Agric. Res.* **14**:549–58. (Evidence that hemoglobin type makes a difference in number of intestinal worms harbored.)

EVANS, J. V., and H. N. TURNER. 1965 Haemoglobin type and reproductive performance in Australian Merino sheep. *Nature* (London) **207**:1396–97. (Evidence that hemoglobin type makes a difference in reproductive performance.)

GAHNE, B., R. K. JUNEJA, and J. GROLMUS. 1977. Horizontal polyacrylamide gradient gel electrophoresis for the simultaneous phenotyping of transferrin, post-transferrin, albumin and post-albumin in the blood plasma of cattle. *Anim. Blood Grps biochem. Genet.* **8**:127–37. (Source for Fig. 21-2; shows how to identify variants for four polymorphisms at once.)

GROSCLAUDE, F., J. C. MERCIER, and B. RIBADEAU DUMAS. 1973. Genetic aspects of cattle casein research. *Netherlands Milk and Dairy Journal* **27**:328–40. (Details of biochemistry and genetics of cattle caseins.)

KIDDY, C. A., R. H. MILLER, C. STORMONT, and F. N. DICKINSON. 1975. Transferrin type and transmitting ability for production in dairy bulls. *J. Dairy Sci.* **58**:1501–6. (Evaluation of extensive data on effect of transferrin type on breeding value for milk production.)

McDERMID, E. M., N. S. AGAR, and C. K. CHAI. 1975. Electrophoretic variation of red cell enzyme systems in farm animals. *Anim. Blood Grps biochem. Genet.* **6**:127–74. (A review of many polymorphisms, most of which we have not mentioned.)

OWEN, R. D., H. P. DAVIS, and R. F. MORGAN. 1946. Quintuplet calves and erythrocyte mosaicism. *J. Heredity* **37**:291–97. (Source for Fig. 21-1; detailed descriptions of the quintuplet calves.)

PAULING, L., H. A. ITANO, S. J. SINGER, and I. C. WELLS. 1949. Sickle cell anemia, a molecular disease. *Science* **110**:543–48. (Classic report of the molecular basis for a polymorphism.)

RASMUSEN, B. A. 1975. Blood-group alleles of domesticated animals. In *Handbook of Genetics*, Vol. 4, ed. by R. C. King. New York. Plenum Publishing Corp. Chapter 21. (A review: includes tabulations of blood groups in cattle, sheep, horses, and pigs. Source for data for Tables 21-1 and 21-2.)

RASMUSEN, B. A., and L. L. CHRISTIAN. 1976. H blood types in pigs as predictors of stress susceptibility. *Science* **191:**947–48. (Original report of association between H red-cell antigens and the porcine stress syndrome.)

RASMUSEN, B. A., and J. G. HALL. 1966. Association between potassium concentration and serological type of sheep red blood cells. *Science* **151:**1551–52. (Original report that M antigens were associated with potassium levels.)

RASMUSEN, B. A., and J. M. LEWIS. 1973. The M-L blood-group system and survival of Suffolk and Targhee lambs. *Anim. Blood Grps Biochem. Genet.* 4:55–57. (Source of data for Table 21-4.)

RENDEL, J. 1968. Inheritance of blood characteristics. Basic results and practical applications, In *Genetics and Animal Breeding*, by I. Johansson and J. Rendel. Edinburgh and London. Oliver & Boyd. Chapter 7. (An excellent review; this book, now unfortunately out-of-print, includes much other information of interest to animal breeders.)

SMITHIES, O. 1955. Zone electrophoresis in starch gels: Group variation in the serum proteins of normal human adults. *Biochem J.* **61:**629–41. (Classic report of a new approach to identification of polymorphisms.)

STORMONT, C. 1958. On the applications of blood groups in animal breeding. *Proc Xth Intnl. Congr. Genet.* **1:**364–74. (Source of data for Table 21-3; contains much useful and interesting information about blood groups in cattle.)

Some Inherited Characters of Domestic Animals

It is beyond the scope of this text to catalogue all the genetic variations known in farm animals, but, in addition to those considered in earlier chapters as illustrative examples, there are a few others that merit mention in this one. Some of these last are interesting as distinguishing characteristics of certain breeds. Some have economic value, others are lethal; but, in selecting special cases for consideration here, the principal objective has been to show the wide range of hereditary variations that are known to affect different parts of the bodies of animals.

For that reason, the conditions briefly reviewed in this chapter are not arranged by species, but rather according to the systems or parts of the body that they affect. This prevents the reader interested solely in dogs from skipping blithely over sections on six other animals, and may thus serve the useful purpose of reminding him that, when a mutation occurs in one animal, the same kind of change might be expected in others.

Special problems in the breeding of dairy cattle, sheep, horses, cats, minks, dogs, fowls, and other domestic animals are reviewed in the books listed at the end of this chapter.

THE SKELETON

Breeds distinguished by size or conformation owe their distinctive characteristics mostly to quantitative genes affecting dimensions or proportions of the skeleton. Among these, the developers of breeds selected to produce their ideal animals. As we saw in Fig. 16-5, concepts of what a breed should be can differ greatly. Apart from polygenes, there are many simple mutations that modify the skeleton. Some of these are lethal, others are not. Some are innocuous, but interesting, and a few have even been adopted as distinguishing characteristics of breeds.

Mutations affecting the entire skeleton include the dwarfism found in some breeds of beef cattle, the sex-linked dwarfism of the fowl, and the pituitary dwarfism in dogs, all considered in earlier chapters. In man, there is a rare mutation, apparently always dominant, which causes extreme brittleness of the bones, so that afflicted persons have many fractures. The same condition has been reported as occurring in cats, but further evidence of its genetic basis in that species is desirable.

The Baboon Dogs of de Boom

Shortening of the entire axial skeleton is lethal in cattle (Fig. 9-12) and turkeys, but a similar variation is not so serious in the dog. H.P.A. de Boom of the Veterinary Laboratory at Onderstepoort, in the Transvaal, has studied a mutation which shortens the whole spine but leaves the legs and skull unaffected. The short-spined dogs are known locally as "baboon" dogs because they commonly sit on their hind legs (as do baboons) with their front legs extended forward as props (Fig. 22-1).

When these baboon dogs lie down flat, they can rise only after making special efforts to do so. The mutation is apparently a simple autosomal recessive. Viability of affected puppies is reduced, and, as males have not been raised to maturity (at latest report), it may be lethal to that sex. Affected females reproduce normally, and short-spined puppies are recognizable at birth by their short tails (Fig. 22-1).

Manx Cats

Differences in length of the tail are found among species not distantly related (compare guinea pigs with rats), and it is not surprising that tailless breeds should be established in species that normally have conspicuous tails. Best known of these is probably the Manx cat (Fig. 22-2).

Contrary to the facile explanation sometimes offered that tailless cats are homozygous for a dominant gene and short-tailed ones heterozygous for the same gene, the situation is much more complicated. There is no question about the taillessness being an autosomal dominant mutation. Todd's evidence on that point (including some earlier reports) is given in Table 22-1.

From the significant reduction of litter size in the last of these matings, and the fact that of 29 Manx cats mated only two could have been homozygous, Todd cautiously concluded that his data "lend support to the suspicion" that taillessness may be at least partially lethal. One might suspect, also, that this base suspicion will hardly get Mr. Todd elected as honorary president of the Isle of Man Manx Cat Association, which is doing all it can to foster the production of more and better genuine Manx cats for export to Manx catteries in other parts of the world.

Dunn and his associates found that the back end of a mouse is genetically very complex, and that any one of 16 alleles can cause taillessness in heterozygotes carrying the gene T. Most of them are

FIG. 22-1 The short-spined "baboon" dogs of South Africa. Top: characteristic sitting position; middle: a pregnant bitch showing typical humped back, short tail, and apparent lack of any neck. Below: her litter (by a heterozygous male) showing conspicuous differences between the normal puppy and the three short-spined ones. (Courtesy of H.P.A. de Boom.)

FIG. 22-2 Manx cat, Biddy of MarManx. (Courtesy of Margaret and Bill Kirsten, MarManx Cattery, Spokane, Washington.)

lethal to homozygotes. It would scarcely be surprising if genes with similar effects should be found in animals as closely associated as are the cat and the mouse.

Notail Sheep

The number of caudal vertebrae in the tail of sheep varies in so-called tailless, short-tailed, and long-tailed breeds from 3 to 24. In those parts of the world where tails are usually docked, it would seem

TABLE 22-1
Results in Various Matings of Manx Cats, and Others

Parents	Litters (number)	Litter Size (average)	Progeny Tailless	Tailed
Tailed × tailed	33	3.88	none	all
Manx ♀ × tailed ♂	20	3.15	21	19
Manx ♂ × tailed ♀	6	4.50	15	12
Manx × Manx	39	2.95	63 [a]	27

[a]After excluding litters of two cats considered homozygous.
Source: Data of Todd.

desirable to make most breeds genetically tailless. The feasibility of combining that useful variation with other characteristics of economic value to produce a "Notail" breed was demonstrated at the South Dakota Experiment Station by J.W. Wilson, but the breed associations apparently prefer sheep which, until docking time at least, can wag their tails behind them in the best Bo Peep tradition.

Amputations

Genetic amputation of all four limbs occur in pigs (Fig. 9-2), in cattle, and in man. In cattle the limbs are terminated near the elbow and hock joints, and the skin is neatly patched over the rounded stump (Fig. 22-3). Associated defects include reduction of the jaw, cleft palate, and absence of most teeth. The character has been called *acroteriasis congenita*, or simply "amputated."

The Ancon Sheep

Hereditary shortening of the legs, not lethal, is evident in any Dachshund, and a similar genetic variation was adopted 175 years ago as the distinctive characteristic of the famous Ancon sheep bred in Massachusetts by Seth Wight (or Wright). They were given that name by George Shattuck, the Boston surgeon who dissected one, because their crooked forelegs looked like elbows, the Greek word for which is "ancon." To New England farmers they were better known as "otter

FIG. 22-3 Genetically amputated calf, showing typical abnormality of the head and neat covering of the ends of the amputated limbs. (Courtesy of I. Johansson.)

sheep" because, with their short legs and long backs, a good imagination could see in them some resemblance to otters.

When the first such sheep appeared as a mutation in his flock in 1791, Seth Wight saw in it the progenitor of a new breed, one of particular value in that place and time because short-legged sheep were less able than others to jump over the low walls and fences by which they were restrained. That advantage was offset by their inability to run well, or to walk any great distance.

An identical (or similar) mutation that occurred in Cheviots in Norway in 1919 was studied by Christian Wriedt. After his death, some of the Norwegian Ancons were imported by Landauer and studied at the University of Connecticut (Fig. 22-4). With numbers unusually good for animals as big as sheep, the character was proven to be a simple recessive, autosomal mutation. From matings of Ancon × Ancon, no fewer than 53 lambs were produced—all Ancons.

Polydactyly and Syndactyly

Extra fingers and toes occur in mammals and birds, but in domestic mammals this variation has not met with sufficient favor in the eyes of breeders to be preserved as a breed characteristic. Poultry fanciers have thought otherwise and made "five toes" an indispensable sign of good breeding in Houdans (in France), Dorkings (in England), Silkies (in Asia), and Antokolkas (in Poland). The fifth toe is not really an extra digit, but a variable bifurcation of the first toe, or hallux. The popularity of polydactyly as a breed characteristic is attributable in large measure to the poultry fancier's love for any bizarre variation, but as late as the twentieth century there persisted (in Poland) a

FIG. 22-4   An Ancon ram (left), an Ancon ewe (right), and a normal ewe (center). (From Landauer and Chang in *J. Heredity* as cited.)

legend promoted by Columella in the first one: that birds with five toes lay more eggs than those with four.

Polydactyly is a dominant, autosomal character that is common in cats, and, as in other animals, the expression of it is variable. It may occur in the front feet without showing in the hind ones. Ithaca, New York, abounds with polydactylous cats, which, according to local tradition, are descendants of a particularly glamorous and amorous tom-cat that came to the town in 1917.

Syndactyly, or fusion of digits, has been known in cattle and swine for many years. Herds of swine with hooves not cloven, sometimes called mule-foot pigs, have been established, but they have not persisted to make distinct breeds.

THE INTEGUMENT

We have already considered many variations in the skin and in such outgrowths from it as hair, feathers, and horns, but there are a few additional cases that merit special mention.

Ichthyosis in Norwegian Red Polls

Few genetic defects in the skin are as severe as the extreme ichthyosis studied by Tuff and Gleditsch in Norwegian Red Polls. At birth, skin of the calves appears hairless and to consist of horny plates resembling big scales. These are separated by grooves 2–6 millimeters broad (Fig. 22-5). Hairs are present in these grooves, and also under some of the plates. Knees and metacarpal regions have normal skin and hair. The plates, which are pliable at first, turn darker in color and become rigid within 24 hours of birth.

This condition results from an excessive production of keratin and consequent abnormal cornification of the epithelial layer of the skin. Nine cases were found, of which seven (and possibly all nine) traced back to one cow. This abnormality, like most others, is a simple autosomal recessive character.

Variations in Hair and Feathers

In the previous case the whole skin was abnormal. Much more common are those interesting cases in which the skin is normal, but the follicles in it, which should produce normal hairs or feathers, fail

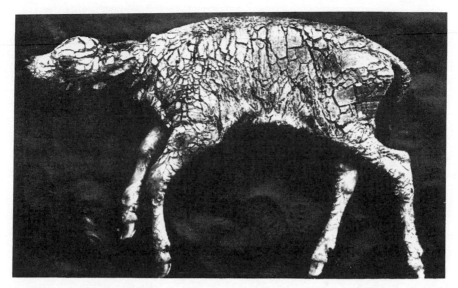

FIG. 22-5 Extreme ichthyosis, a lethal character found in Norwegian Red Polls. (From P. Tuff and L. A. Gleditsch in *Nord, Veterinaermed.*)

to do so. Even within a single species, one mutation causing hairlessness may be lethal and another not, as we have already seen in cattle (Chapters 2 and 9.)

Various kinds and degrees of hairlessness are known in mice, rats, rabbits, cats, and dogs, but the only one to be retained as a breed characteristic is that distinguishing the breeds of dogs known in various parts of the world as Chinese, Turkish, Persian, or Mexican Hairless. From genetic studies with these by Letard, it seems probable that all these hairless dogs are heterozygotes, and that the homozygotes have such gross abnormalities (absence of external ears and malformation of the buccal cavity) that they are not born alive.

In the scaleless fowls studied by Abbott and Asmundson, a recessive autosomal mutation eliminates all but a few straggling feathers, and also other derivatives of scales, such as the spurs and foot pads (Fig. 22-6). Under natural rearing conditions, scaleless chicks cannot survive, but if they are kept at suitably high temperatures, they grow as well as normal fowls and even attain sexual maturity at normal ages.

Genes causing curly hair are apparently considered undesirable by cattle breeders, but not necessarily so by breeders of Percheron horses, and, according to newspaper reports, have even been preserved by one breeder in a western state to make a new breed of horses—the Pendletonian. Dog breeders select genes to make smooth

FIG. 22-6 Scaleless hens at 18 months of age, in full and maximum plumage following completion of their moult. (From U. K. Abbott and V. S. Asmundson in *J. Heredity.*)

coats in some breeds, rough coats in others. Rabbit fanciers have preserved three different mutations that eliminate the long guard hairs to produce the velvety close-cropped fur of Rex rabbits. The cat breeders, having previously utilized a gene that lengthens the hair to develop Persian breeds of various hues, have now made a new Rex breed (or breeds) from genes that eliminate the guard hairs. Two such mutations occurred independently, one in England, the other in Germany, and it remains to be seen whether these are at the same locus or different ones. The rex mutation shortens the whiskers. It reduces some other hairs as well as the guard hairs and thus gives the cat a nicely waved coat (Fig. 22-7).

The Hairy Romney Sheep

Most genetic variation in the fleece of sheep is polygenic, but, in the Romney Marsh flock at Massey Agricultural College in New Zealand, Dry found a single-gene difference which has proven to be of great economic value. The incompletely dominant gene, *N*, causes in the fleece of heterozygotes a certain amount of hairiness, *i.e.*, fibres lacking the normal and desirable crimp. Study of this defect led to the

FIG. 22-7 A rex cat showing waved coat and absence of guard hairs. This one, imported from Germany, was named Christopher Columbus, because he came to America. (Courtesy of Mrs. William O'Shea, Vernon, New York.)

discovery that fleece of the homozygotes, NN, although entirely too hairy for the usual purposes to which wool is put, has exceptional value as a wool for carpets (Fig. 22-8). Moreover, Dry learned to distinguish the NN, Nn, and nn genotypes in newborn lambs by the extent to which they showed "halo-hairs" in various regions of the body. Halo-hairs are large, medullated fibres which project above the coat of NN lambs at birth. Heterozygotes have fewer halo-hairs toward the anterior end of the body, particularly in an area corresponding to the armpit, just behind the shoulder.

COAT COLORS

The cost of making extensive experimental matings in large animals just to study colors and patterns is so great that such investigations are seldom made. As a result, theories about the genes that determine coat colors in large animals are based on observations of limited numbers and are not often subjected to critical tests. This handicap does not apply to laboratory animals like mice and rabbits; hence genes for colors and patterns in those species are well known. To illustrate the complexities of such analyses, let us see what is known about colors in two smaller domestic animals that have been well studied. One is the mink; the other is the dog.

FIG. 22-8   Romney ewes showing (top) the normal woolly fleece and (bottom) the dominant hairy type that proved so valuable for carpet wool. (Courtesy of F. W. Dry, Massey Agricultural College, Palmerston, New Zealand.)

Mutant Color Phases in the Mink

It is probably safe to say that never in the history of the domestication of animals has there been such a concerted search for new mutations and combinations affecting coat color as that staged by the mink breeders of the world since about 1930. Perhaps the scramble was accelerated by the fantastic prices paid for platinum foxes when the platinums were at the zenith of their popularity. At any rate, many new color mutants have been multiplied, either singly or in combination with other mutations, and then hopefully subjected to the critical test of the fur markets. Some have been approved; many have not. From all of this enthusiasm, science has gained greatly. Most of the new shades so assiduously tested are caused by simple mutations, of which as many as three have been combined in some varieties. As a result, more is known about genes affecting color in the mink than about similar genes in most other mammals.

A few of the color varieties currently popular are listed, with their genotypes, in Table 22-2. One of these that has maintained its popularity in the market longer than any other is the Royal Pastel, which differs from the wild mink only in being homozygous for a recessive gene that replaces black with brown. Platinum (pp) is distinguished from Imperial Platinum ($ipip$) only by experts. In some cases different genes that produce practically identical colors in the fur are distinguishable because they cause different colors in the eyes. Sapphires, which are very popular in the fur trade, are double recessives, produced originally by crossing either Platinum ($AlAl\ pp$) or Imperial Platinum ($AlAl\ ipip$) with Gunmetal (then called Aleutian), which is $alal\ PP$, and getting Sapphires in the F_2 generation. (See Problems 5-11 to 5-18). They often show reddish eyes. It should be noted that Pearl and Hope are triple recessives.

Although the mink breeders refer to these color varieties as mutant color phases, those that persist are likely to be known eventually as different breeds. The mink breeders as a group are remarkably well posted on the genetics of their favorite species, and are advised in the publications for their industry just what to expect, and in what proportions, in the progeny from matings of different kinds of parents.

TABLE 22-2
Genotypes of Some Popular Mutant Color Phases in the Mink[a]

Natural Dark, wild type	BB	PP	Al Al	Ba Ba	$C^H C^H$	Bm Bm	Bp Bp
Royal Pastel, light brown	bb	+	+	+	+	+	+
Platinum, grey	+	pp	+	+	+	+	+
Gunmetal, dark grey	+	+	al al	+	+	+	+
Sapphire, bluish-grey	+	pp	al al	+	+	+	+
Heinen Buff, light tan	+	pp	+	ba ba	+	+	+
Regal White	bb	+	+	+	cc	+	+
Lavender, pale grey fawn	+	+	al al	+	+	bm bm	+
Pearl, pale grey	+	pp	al al	+	+	+	bp bp
Hope, pale grey	+	pp	al al	ba ba	+	+	+

[a]The + signs indicate homozygosity for the wild-type genes shown at the top of each column.
Source: Courtesy of R.M. Shackelford.

When the gene (*F*) causing Blufrost (which may be the same as the "Stewart" mutation) proved to be undesirable because it is lethal to most homozygotes and makes the few surviving males sterile, the mink breeders abandoned Blufrost (also called Silver Sable) as a distinct color phase. However, they kept the gene *F* (in the heterozygous condition) as essential in the "Breath-of-Spring" series of minks because it makes the under-fur lighter in color than the guard hairs. The contrast apparently adds considerably to the value of such pelts in the fur markets and results in special prices for BOS Pastels, BOS Platinums, and BOS Sapphires.

Table 22-2 lists only ten mutant color phases in the mink. Many others are known.

Colors in Dogs

To give some idea of the complexity of colors and patterns in domestic animals, a number of the genes found by C.C. Little to affect them in the dog are listed in Table 22-3. From his investigations, Little was able to assign genotypes to many breeds, as for example:

Beagle (with no ticking):	$a^t\ B\ C\ D\ E\ g\ m\ P\ s^p\ t$
Dalmatian (with black spots):	$A^s\ B\ C\ D\ E\ g\ m\ P\ s^w\ T$
Weimaraner:	$A^s\ b\ C\ d\ E\ g\ m\ P\ S\ t$

For simplicity in these examples, only single genes are shown. Pure breeds would be homozygous for each one.

The merle type of dilution (Table 22-3) found in several breeds of dogs leaves heterozygotes with black (undiluted) blotches on a background that is either bluish-grey or dilute tan. Dogs homozygous for this gene are almost entirely white (except Foxhounds, which show some color) and their eyes are very pale blue, sometimes called "china" eyes. Such homozygotes are often more or less sightless, and some lack an eye on one or both sides. They are also often deaf.

Analogies and Homologies

Anyone reviewing what is known about genetic variations of coat color in different mammals is struck by the fact that some variations are much the same in several different species. Even in this text, we have referred to genes in cattle, rabbits, mice, rats, minks, and dogs which

TABLE 22-3

Genes Affecting Coat Colors in Dogs

Locus	Allele	Effect of the Gene	Breed Exemplifying
A	A^s	Distributes dark pigment over whole body	Newfoundland, Chesapeake Bay Retriever
	a^v	Restricts dark pigment; makes sable or tan	Basenji, Irish Terrier
	a^t	Causes bicolor varieties like black and tan	Doberman Pinscher
B	B	Produces black pigment	Black Cocker Spaniel
	b	Produces chocolate brown or liver	Weimaraner
C	C	Induces "full depth of pigmentation"	Irish setter; some black breeds
	c^{ch}	Chinchilla; reduces red more than black	Norwegian Elk Hound
	c^a	True albinism; very rare in dogs	None
D	D	Intensely pigmented	Most breeds
	d	Causes Maltese dilution, or blue	Blue Great Dane, and others
E	E^m	Induces "black mask" pattern	Pug, Norwegian Elk Hound
	E	Extension of dark pigment over coat	Cocker Spaniel, black or brown
	e^{br}	Brindle; interactions not clear	Boston Terrier, Deerhound
	e	Restricts black to eye, leaving coat red or yellow	Irish Setter; Cocker Spaniel, red
G	G	Progressive greying from birth to old age	Bedlington and Kerry Blue Terriers
	g	No such greying	Most breeds
M	M	Causes merle or dapple, often with white areas	Merle Collie, Dunkerhund
	m	Uniform pigmentation	Most breeds
S	S	Solid coat, no white	Any with solid color
	s^i	Irish spotting, little white	Basenji
	s^p	Piebald spotting, up to 80 per cent white	Beagle
	s^w	Extreme piebald, small black patches	Bull Terrier, Sealyham
T	T	Ticking; flecks in white areas	Dalmatian, Setter
	t	No ticking; white areas clear	Most breeds

Source: Condensed from C. C. Little, as cited.

eliminate black pigment and leave the coat some shade of brown or red. A gene with similar effects makes horses chestnut, or sorrel instead of black. As it happens, the seven genes accomplishing this trick in seven species have all been assigned the symbol b, with black designated as B. This was done by the specialists who studied the genetics of color in the species mentioned.

It would be simple to say that one and the same gene, b, changes black to brown in the seven species mentioned, but it is doubtful how far we should go with such simplification. The bb animals are not all of the same color. They are red in cattle, brown in mice and rabbits, chocolate brown in rats and dogs, light brown in the mink, and chestnut or sorrel in horses. It may be safer just to say that in each of these seven species there are genes which prevent the formation of black melanin, that some of these produce different shades of red or brown, and that presumably they exert similar effects on chemical processes in the different species. A gene in the fowl makes Rhode Island Reds much the same color as red Cocker Spaniels, but we can hardly assume that it is the same gene, at the same locus of corresponding chromosomes! Even within the mammals it is perhaps safer, when comparing (for example) minks ($2n = 30$) with dogs ($2n = 78$), to consider that the bb genotypes—liver in the dog, and pastel or light brown in the mink—are analogous, but not necessarily homologous.

At the same time, we should note that some genes affecting color do so similarly in the several species. It seems likely that complete, pink-eyed albinism, which occurs in various species from frogs and fowls through mice to men, may be caused in all of them by genes that block at some one point the formation of melanin from tyrosine. Such genes can be in different chromosomes in different species. Incomplete albinism, which leaves enough melanin to show a "ghost" pattern, is sex-linked in the fowl, but is caused by an autosomal gene in cattle. The effects are analogous, but the genes are not homologous.

A chinchilla phenotype occurs in dogs as in rabbits. Genes causing dilution of pigment are found in most mammals and birds that have been studied. White spotting is recessive to full color in most species, but there is a dominant type in the mouse. In several species a recessive gene, usually designated as e, restricts black to the eye, leaving the coat red or yellow. In fowls, there is a pattern called Columbian in which black is restricted to the extremities (neck, wings, and tail) much the same as in Himalayan rabbits. All of such similarities suggest that genes direct the formation and distribution of pigment along similar lines in different species. Little is lost by considering the resultant phenotypes analogous, but without assuming that the genes causing them are homologous.

Whites and Dilutions

Some kinds of white or white spotting are lethal to homozygotes. Others, not lethal, are associated with faulty development of the reproductive tract. Some genes that dilute melanic pigment in the coat of heterozygotes are also lethal to homozygotes, but there seems to be no general rule applicable to all of these cases. The variability among them is shown by the few examples here considered. Others are listed by Searle (1968).

The Heggedal Mink   We have already considered three dominant dilutions of color in the fox, all of which are lethal to homozygotes (Chapter 11). Another interesting case is that of the Heggedal white mink, studied by Nes in Norway. Genotypically, it is a pale blue from which the melanin is eliminated by the combined action of two dominant, allelic genes, S and S^H. The Heggedal gene S^H (known to breeders as the Shadow factor) is lethal to most homozygotes before implantation and to the remainder during the last third of the gestation period. Its action is thus very similar to that of the gene causing Belyaev's Georgian whites. All living Heggedals must be heterozygous. Among 136 sterile females carrying S^H, Nes found that about 40 per cent had abnormalities of the reproductive tract.

Others   Dominant white is also lethal to homozygotes in canaries (Duncker) and in horses (Pulos and Hutt). In both species the lethal action apparently occurs at early embryonic stages. With all the evidence that dominant white can be deadly, we must remember that the White Leghorn, one of our most productive domestic animals, is also dominant white, and none the worse for it. Presumably, there are different kinds of dominant white. Other white varieties of the fowl are recessive whites, but none the better for that.

Silver Sable (Blufrost) minks exemplify a dilution of coat color that is lethal to homozygotes (see Problem 9-3). In silver-white guinea pigs homozygous for two recessive genes, each of which causes (separately) some dilution of coat color, about half the females are sterile but show no consistent abnormality of their reproductive tracts. The males are all sterile, having testes only a quarter of normal size, in which there is no spermatogenesis (Wright).

In Swedish Highland cattle, the incidence of a hereditary hypoplasia of the gonads that affected both sexes was reduced by selection from about 25 per cent in 1935 to 8 per cent seven years later. Settergren found that, among 882 females, those 90 per cent or more white had 13.5 per cent showing that hypoplasia and 10.3 per cent more were suspect. In those less than 90 per cent white, there were only two possible cases, both uncertain.

White-heifer Disease

This condition, long known to stockmen, is of special interest to animal geneticists because, as yet, no satisfactory genetic explanation is known for it. The name comes from the fact that it is most commonly found in white heifers of the Shorthorn breed, or in breeds related to them, but the same kind of anatomical abnormality has been found in several other breeds.

The affected heifers are sterile, but their abnormality is not the same as in freemartins, which are also sterile. In white heifer disease the ovaries and external genitalia are normal, but the uterus and the vagina are not completely developed. Spriggs estimated that about 10 per cent of white Shorthorn heifers in England are affected. In a large herd studied by Rendel it was found that during 15 years the frequency of this abnormality in red, roan, and white heifers was 1.1, 3.5, and 39.1 per cent, respectively. From calculations of the expected frequencies of these three colors among registered Shorthorns and the deficiency of white females below expectation, Rendel estimated that about 20 per cent of the white heifers had been eliminated for sterility. The proportions of affected heifers found by Hanset among large numbers of Belgian White and Blue Pied cattle (derived from Shorthorns) show a significantly higher frequency among whites than in the others:

	Heifers (number)	Affected	
		Number	Per cent
White	1,518	237	14.9
Blue Pied	1,250	53	4.2

It is not possible to review here all the theories that have been advanced to account for the association of this abnormality with the white color. None has proven satisfactory. After reviewing them all, and rejecting one of his own earlier theories, Hanset suggests that the condition may be caused by the action of the gene for white in conjunction with an unknown number of "auxiliary" genes at other loci. It is of interest to note that this kind of abnormal development of the reproductive tract affects only females—not males.

JOINTS

This is a convenient heading under which to consider some abnormalities for which the exact cause may lie in a tendon, perhaps a muscle, or even in the skeleton, but which are manifested chiefly by malfunction of one or more joints. For example, some people have "little fingers" (one or both) which cannot be placed flat on a table because the first joint is bent upward and rigid, so that the finger cannot be straightened out. It has been ascribed by some investigators to shortening of a tendon, and by others to abnormal attachment of muscles controlling that joint. Whichever is right, it is a dominant, autosomal mutation, not always manifested by those carrying the gene (*i.e.*, incomplete penetrance) and often evident only in one hand, but more evident as a peculiarity of a joint than anything else.

In lambs, calves, and piglets, hereditary rigid joints at birth have been attributed to muscle contractures, and found to be caused by recessive lethal genes. In calves, a less extreme stiffness of a single joint, called flexed pasterns, is also hereditary, but the congenital weakness is commonly overcome as the calf gets older, a process sometimes accelerated by the temporary use of splints. Obviously affected animals should be salvaged for market, but not for breeding.

Hip Dysplasia in Dogs

This abnormality has long been a source of trouble, particularly in German Shepherds, but it has been found in many other breeds. It is sometimes called subluxation of the hip or dislocation of the femur, among other names. The condition provides an almost perfect example of a defect that is hereditary but for which it is difficult to get good genetic evidence. Data from questionnaires to breeders have little value, not for the usual reason that few breeders are willing to admit that their animals have any bad genes whatever, but more because the expression of the dysplasia is so variable that accurate recognition of it cannot be made without diagnosis by X-rays. The basic abnormality may be in the shape of the acetabulum, which is shallow and becomes somewhat oval in shape, instead of being deep and round. As a result, in older dogs, the head of the femur becomes somewhat flattened.

Hip dysplasia is clearly inherited. In an extensive study of this abnormality in German Shepherds of the Swedish Army, Henricson *et al.* found (by radiographic examination) that, among 462 dogs from parents that were both normal, the proportion with dysplastic hips was

TABLE 22-4
Dysplasia in German Shepherds

Parents	Progeny	
	Number	With Dysplasia (per cent)
Both normal	462	37.5
Normal ♂ × dysplastic ♀	213	41.3
Dysplastic ♂ × normal ♀	89	52.8
Both dysplastic	88	84.1

Source: Data of Henricson *et al.*

37.5 per cent. Males and females did not differ significantly in the proportion affected. Their data on the incidence of the defect in litters from different kinds of matings are given in Table 22-4.

It is now clear that hip dysplasia in dogs is a polygenic character, and that, as with other polygenic traits, there may be environmental influences (as yet not known) which influence it. The range of expression is so great that most radiologists now recognize four or five different grades of severity. Henricson *et al.* showed that dogs having dysplasia graded as borderline (or Grade 1) had no more of it in their progeny than did those with normal hips.

MUSCLES

Several kinds of inherited muscle dystrophies are known in man, and at least one of these has a counterpart in mice, but the extent to which such afflictions occur in domestic animals has yet to be determined. Although congenital paralysis (particularly of the hind limbs) has been found in calves, lambs, and piglets, the basic abnormality in these cases may not be in the muscles, but in the nerves that should activate them.

Hypertrophy of muscles is inherited in beef cattle, but not as any simple Mendelian trait. Affected animals have unusually thick thighs and loins, and are sometimes called (erroneously) double-muscled. According to some reports, the condition is incompletely recessive but variable in expression. It has been suggested that stockmen might select the heterozygotes as breeders because of their better fleshing, but dystocia is a problem in these animals. Flesh of the hypertrophied muscles has less fat than butchers consider "normal" and the affected animals are apparently disliked by some meat packers, but one wonders to what extent such dislike is directed more against anything not "normal" than to any objectionable qualities in the meat.

The remarkable development of the hams and loin characteristic of the spotted Piétrain pigs of Belgium may possibly depend on some genetically induced hypertrophy of muscles. At any rate, those pigs command substantial premiums in the markets because of the superior fleshing known to be characteristic of the Piétrain breed.

DIGESTIVE SYSTEM

The pendulous crop in turkeys, the digestive troubles of grey Karakuls, and the *atresia coli* of horses, all considered earlier, are good examples of genetic abnormalities affecting the digestive tract at different levels and in different species. To them we add here only one more example, at another level, in a different animal.

Hereditary Hyperplastic Gingivitis in the Fox

This remarkable abnormality was studied in silver foxes by Dyrendahl and Henricson of the Royal Veterinary College in Stockholm. It becomes evident without any special search for it at two to three years of age, but, as early as six months after birth, swelling of the gums around the posterior cheek teeth can be recognized in both jaws. Growth is slow, but a tumorous mass gradually spreads forward and proliferates to involve both upper and lower jaws completely. At two to three years the lips cannot cover the mass and only the highest crowns of the teeth are visible above the "cauliflower-like" mass covering both jaws. Older foxes are sometimes unable to close their mouths.

From studies of the frequency of this hyperplastic gingivitis at a large fox farm during six years, it was concluded that the defect is probably caused by an autosomal recessive gene with rather low penetrance in homozygotes. Altogether 901 foxes were found to show the abnormality. As matings of parents both affected yielded 143 with the gingivitis among 465 progeny, penetrance (if a single gene is responsible) was only about 30 per cent.

The gingivitis was more common among animals with first-class fur (*i.e.*, a dense undercoat) than in those with fur of poor quality. Coats were examined when animals were either pelted or selected for breeding, *i.e.*, when the foxes were about nine months old, and before the gingivitis could have affected their health. Reasons for the association between good fur and bad gums were not known.

NERVOUS SYSTEM

A number of hereditary abnormalities in domestic animals have been shown to be caused by defects in the nervous system. As many of such defects cause paralysis in some degree, or faulty locomotion, they are easily recognized. Disorders of the nervous system considered in previous chapters include ataxia in dogs, the partial paralysis and muscular atrophy studied by Stockard in the same species, internal hydrocephaly in cattle, posterior paralysis in cattle, paroxysm in chicks, and probably also the lethal in poults and chicks that has been called (for want of a more accurate name) congenital loco. A glance through the list of genes mapped in the mouse (Fig. 10-5) will show that a remarkable proportion of the mutations identified in that species affect balance or movement, and thus give rise to such graphic appellations as reeler, fidget, tipsy, and others.

Epilepsy

This condition occurs in several species of mammals, but for various reasons, genetic studies were long limited mostly to analyses of pedigrees. In dogs, experimental investigations are complicated because the condition is uncommon during the first year of life, and some dogs may not show it until over six years of age. In contrast, the epileptic chicks of Crawford (1970) proved to be ideal models for the study of epilepsy. Not only were their seizures similar to those of epilepsy in man (*grand mal*), but so also (as later studies showed) were their encephalograms and reactions to anticonvulsant drugs. They were excellent for genetic studies because the condition was evident at hatching or (when provoked by excitement) within 24 hours. It had no adverse effects on viability. Because any attempt to copulate usually triggered a seizure, the epileptic birds had to be reproduced by artificial insemination.

For the usual Mendelian analyses, Crawford hatched 2,282 chicks and learned that the epilepsy was a simple, autosomal, recessive character, with penetrance incomplete in chicks from heterozygous dams. From parents both epileptic, all chicks were epileptic. The mode of inheritance may be more complex in other animals, but there is no longer any doubt that epilepsy can be a genetic defect. The same conclusion must be drawn from the high concordance of epilepsy in Conrad's monozygotic twins (Table 15-3).

Disorders Attributed to Cerebellar Hypoplasia

These are known best in cattle and sheep, in which species the disorder is a simple autosomal recessive trait, and in cats and chickens. Affected calves were first reported in Herefords, but the same condition has since been found in other breeds. The calves are born alive but cannot stand, and lie with their limbs rigidly extended. The cerebellum is greatly reduced in size and shows pathological changes in the cortex. Less severe disorders found in Jerseys have been described as congenital ataxia and congenital spasms. Some of the calves affected in one herd showed no symptoms until a week or two old, when movements of their legs became incoordinated so that they could not stand without help. Spasmodic movement of the head was characteristic of those affected, and none survived.

A similar abnormality has been found in several breeds of sheep. Some of Rasmusen's affected Corriedale lambs could walk, but only with difficulty, and others could not. Those last turned the head when lifted and had tremors. In this case the cerebellum may be normal in size but on histological examination it shows a deficiency of certain cells in the cortex known as Purkinje cells.

Fowls that develop the sex-linked, recessive abnormality dubbed "shaker" are normal until 18 to 35 days of age, when their heads vibrate rapidly, but with amplifications so small that the vibration might be overlooked. Locomotion is impaired, and by three months of age any that have survived are usually unable to stand. As in the sheep (and probably in some of the other cases as well) the cells of Purkinje are degenerate.

Hydrocephaly

The condition shown in Fig. 9-5 is internal hydrocephaly, in which the excess fluid accumulates in the ventricles of the brain. Sometimes fluid accumulates outside the brain, in spaces between the arachnoid membrane and the *pia mater*. Such an external hydrocephaly has been found to be caused in pigs by a recessive autosomal mutation. The case is rather interesting because it has apparently been found only in Durocs, but in two unrelated herds of that breed. Pressure of the fluid causes enlargement of the cranium, and the overlying parietal and frontal bones are thin, often with holes in them. The gene responsible has other effects, for most of the affected pigs are paler in color than normal. Oddly enough, the fact that they have a disorder in the head is best indicated at birth by a defect at the other end of the body—they have extremely short tails. Among 178 piglets from parents both heterozygous, Blunn and Hughes found 42 with hydrocephaly. All died within two days.

Some kinds of hydrocephaly are not necessarily lethal, and in the fowl that condition provides the distinguishing characteristic of Houdans, Polish, and other breeds (Fig. 4-5). The visitor to the poultry show sees in these only a beautiful crest of feathers, or top-knot, sometimes so good that it obscures the eyes and face, but the elongation of the feathers in that crest is caused by internal hydrocephaly. Under the best crests the cranium has a great dome, the result of pressure from the enlarged cerebral hemispheres beneath. This condition is caused by an incompletely dominant gene, Cr, the position of which has been spotted in the chromosome map of the fowl (Fig. 10-6).

Somewhat similarly, an *encephalocele*, or hernia of the brain, in the parieto-occipital region is responsible for the top-knot which is the sole distinction of a breed of ducks—Crested White—from other white ones. In this case the causative gene is lethal to the homozygotes, and only about half the heterozygotes show a crest.

Other Disorders of the Nervous System

The posterior paralysis of Red Danish calves is attributed to a lesion in the forebrain. An autosomal recessive paralysis of the hind legs of pigs has been reported in different breeds in three countries with similar symptoms in all three. The newborn piglets lie on their right sides, and if placed on their left sides, they try to turn over to their right and frequently manage to do so.

Some mutations make animals so sensitive to sudden alarms (such as a loud noise or the sudden appearance of some person) that they undergo a temporary seizure similar to that in tetany. In this class are the famous "fainting goats," which might more appropriately be called the goats scared stiff. When suddenly frightened, they become completely rigid, and some even fall to the ground. Others can easily be pushed over. After the seizure, which lasts 10 to 30 seconds, the goats start walking, but their hind legs drag. Although the condition, sometimes called *myotonia congenita*, has been known for many years, and has generally been assumed to be hereditary, any good evidence of its genetic basis has yet to be adduced. Similar audiogenic seizures are hereditary in mice, and Cole's "paroxysm" in the fowl (Fig. 10-3), which resembles closely the condition in goats, is a sex-linked lethal character.

Other disorders of the nervous system in domestic animals that have been ascribed to heredity are given in the useful check-list of Saunders (1952).

REPRODUCTION

We have read in earlier chapters quite a bit about lethal genes, freemartins, impotency, genital hypoplasia, atresia of the oviduct, intersexes, and sterility. Additional cases of genetic abnormalities affecting reproduction are discussed on pp. 525–26. No more need be added here.

RETROSPECT AND PROSPECT

The purpose of this chapter (and, in fact, of others preceding it) is to show that genetic variations can affect any or every part of the animal body, and that even a single gene in the homozygous state can determine whether the animal will live or die, be ill or well, fertile or sterile, valuable or useless, a respectable representative of its breed or an outcast. The few examples cited in each section are only small samples of the variation already known. That, in turn, is only a small part of the total variation which remains to be explored.

It is hoped that recognition of the diversity of genetic variation in domestic animals may serve as some incentive for more study of it in the future than we have had in the past. Fortunately, the great range in that variation is matched by an almost similar range in the interests of people who breed domestic animals, or multiply their numbers, or minister to their welfare. Few of us alone are likely to breed sapphire minks, Manx cats, and lambs with halo-hairs, but, whatever our pet species, and whether we prefer to look at whole animals, or only those parts of them that can be put in a test tube or under a microscope, each has a chance to transfer some mite of knowledge about the genes of domestic animals from the realm of the unknown to that of science.

Problems

22-1 Wriedt and Mohr found that four sires of amputated calves mated to daughters of other carrier bulls yielded 13 amputated calves among 115. Was that in accord with expectations, or not?

22-2 From matings *inter se* of cats heterozygous for taillessness, Todd obtained 63 tailless and 27 tailed kittens. Can you tell from these figures whether or not homozygosity for taillessness is lethal?

22-3 If further data should confirm indications that most genetically tailless cats are heterozygotes, how would you advise breeders of Manx cats to select, in order to reduce the proportion of unwanted "stumpies," or bob-tailed cats?

22-4 Referring to Table 22-1, what size of litter would be expected from the Manx × Manx matings if the gene for taillessness were lethal, and if the normal, average litter size for cats were 3.88?

22-5 Does your answer affect your answer to Problem 22-2?

22-6 Sooner or later, some sheep breeder will find another mutant of the Ancon type, and will want you to breed a flock of them. If he should appeal to you for advice, how would you advise him to proceed? Remember that close inbreeding must be avoided at early stages if reproduction is to be maintained.

22-7 For the situation of Problem 22-6 would you rather have the original mutant a ram or a ewe?

22-8 After the breeder of the new Ancons has a flock large enough to permit some selection among them, he will want to reduce the proportion so crippled by their short bent legs that they have difficulty in walking. How should he select to that end?

22-9 Do you know any breed of domestic animals in which this trick has been done? There is one; it begins and ends with "d."

22-10 If a breeder of rex cats, worried a little about inbreeding in her stock of them, wants to introduce blood of rexes that originated in another country, by what matings could she determine whether the two kinds are genetically identical or different, and what results could she expect in first crosses, backcrosses, and F_2 generations if the two mutations are (a) identical? (b) not at the same locus?

22-11 Dry was able to identify his *NN* hairy Romneys at birth by their halo-hairs. Do you know any other cases in domestic animals in which those likely to be most (or least) desirable for economic purposes can be identified at birth, or soon after?

22-12 Among all the variations considered in this book, where would you consider it best to seek for indicators at early ages that might identify animals likely to be most productive of milk, eggs, or meat?

22-13 Adding the new examples in this chapter, how many breeds or color varieties of domestic animals do you know in which the type preferred by the breeder is heterozygous for a dominant gene?

22-14 Sort your list from Problem 22-13 into two classes: (a) those in which the homozygote is not viable or fertile and (b) those in which it is both viable and fertile, but of a type not wanted.

22-15 In dogs, progressive atrophy of the retina, which begins in the rods of that structure, causes first, night blindness, and later, defective vision in daylight. It is known to occur more frequently in Irish Setters than in other breeds. Veterinarians say it is inherited as a simple recessive autosomal character, but actual evidence is scanty. Assuming that they are right, if the owner of a kennel should appeal to you for help in order to eliminate it with the minimum loss of valuable dogs, what would you advise?

22-16 In what two ways could the hereditary gingivitis of foxes be genetically associated with the denser under-fur noted in the affected animals?

22-17 If a fox breeder thought it worth while to try to separate the two traits, and to retain the dense under-fur without the gingivitis, could that be possible if either of your answers to Problem 22-16 were correct? Which one?

22-18 Under the circumstances would it be not worth while to try to separate the gingivitis from the good fur?

Selected References

CRAWFORD, R. D. 1970. Epileptiform seizures in domestic fowl. *J. Heredity* **61:** 185–188. (Records a fine study.)

DRY, F. W. 1955. The dominant *N* gene in New Zealand Romney sheep. *Austral. J. Agric. Res.* **6:** 725–69. (See also pp. 608–23 and 832–62 in the same volume for further information.)

EVANS, J. W., A. BORTON, H. F. HINZ, and L. D. VANVLECK. 1977. *The Horse.* San Francisco. W. H. Freeman & Co. (With chapters by other specialists; all you want to know about horses.)

HANSET, R. 1960. La "maladie des génisses blanches": son aspect génétique. III. Hypothèses en présence—Discussion générale. *Ann. Méd. Vét.* **104:** 49–93. (See also extensive original data in his earlier papers, *Ibid.* **103:** 281–98, and **104:** 3–33, on the same subject.)

HENRICSON, B., I. NORBERG, and S.-E. OLSSON. 1966. On the etiology and pathogenesis of hip dysplasia: A comparative review. *J. Small Anim. Pract.* **12:** 171–77. (Reports extensive studies with German Shepherds of the Swedish Army.)

HUTT, F. B. 1979. *Genetics for Dog Breeders.* San Francisco. W. H. Freeman & Co. (Simple principles and a survey of genetic disorders.)

HUTT, F. B. 1949. *Genetics of the Fowl.* New York. McGraw-Hill Book Co., Inc. (Covers what was known about heredity in this species up to 1948.)

JOHANSSON, I. 1961. *Genetic Aspects of Dairy Cattle Breeding.* Urbana, Ill. Univ. of Illinois Press. (See particularly Chapter 8 on genetic and environmental variations and Chapter 12 on progeny-testing.)

JOHANSSON, I., and J. RENDEL. 1968. *Genetics and Animal Breeding.* Edinburgh and London. Oliver and Boyd. (Deals particularly with the larger domestic animals.)

LITTLE, C. C. 1957. *The Inheritance of Coat Color in Dogs.* Ithaca, N.Y. Cornell Univ. Press. Later, New York. Howell Book House. (Summarizes the author's studies of color in dogs; genotypes for many breeds. Source of data in Table 22-3).

NES, N. 1964. The homozygous lethal effect of the Heggedal factor (Shadow factor). *Acta Agric. Scand.* **14:** 208–28. (With evidence that most die before implantation.)

NES, N. 1965. Abnormalities of the female genital organs in mink heterozygous for the Heggedal factor (Shadow factor). *Acta Vet. Scand.* **6:** 65–99. (With illustrations of the different kinds found.)

RAE, A. L. 1956. The genetics of the sheep. *Advances in Genet.* **8:** 189–265. (Review of inheritance and of special problems of sheep breeders.)

ROBINSON, R. 1978. *Genetics for Cat Breeders.* 2nd ed. Oxford. Pergamon Press. (An up-to-date survey.)

SAUNDERS, L. Z. 1952. A check list of hereditary and familial diseases of the central nervous system in domestic animals. *Cornell Vet.* **42:** 592–600. (A useful check-list, with references.)

SEARLE, A. G. 1968. *Comparative Genetics of Coat Colour in Mammals.* London: Logos Press. New York: Academic Press. (See Chapter 12 for pathological conditions associated with certain colors in 16 species.)

SHACKELFORD, R. M. 1957. *Genetics of the Ranch Mink.* 2nd ed., New York. Black Fox Magazine and Modern Mink Breeder. (With genotypes and colors.)

TODD, N. B. 1961. The inheritance of taillessness in Manx cats. *J. Heredity* **52:** 228–32. (Source of useful new data on this problem, including those in Table 22-1.)

ZIMMERMAN, H. 1963. *Genetische Grundlagen der Mutationsnerzzucht.* Jena. Gustav Fischer Verlag. (Descriptions and genotypes for 101 color varieties in the mink; well illustrated.)

CHAPTER

23

Genetic Resistance
to Disease

For two reasons it seems appropriate to relegate this subject to the last chapter. In the first place, it is one of the senior author's special fields of interest, and, therefore, some slight sense of modesty has prevented undue clamor about it in the earlier chapters. Secondly, genetic resistance to disease seems to be about the last and least among the varied interests of animal pathologists and animal breeders; hence its appropriate place seems to be in the final chapter.

Disease-resistant Plants

To begin with, it seems desirable to point out that here is a field in which geneticists and pathologists interested in domestic animals lag far behind their opposite numbers who deal with cultivated plants. Both animal breeders and plant breeders select for greater yields, or for specimens closer to their ideas of perfection. To attain their objectives, the plant breeders, co-operating with plant pathologists, have utilized genetic differences in resistance to various pathogens to develop highly productive, disease-resistant varieties of most crops and plants of economic importance. In so doing, they have not abandoned other means of controlling disease, but, rather, have sought to raise the efficacy of their fungicides and insecticides by supplementing their action with aid from helpful genes. Sometimes the genes bear the brunt of the battle alone, as in wheats resistant to stem rust. At other times, sprays and dusts are indispensable, as with many orchard fruits. Often both kinds of control together make the difference between a profitable crop and disaster.

Without any lengthy listing of varieties of cultivated plants economically valuable because of their genetic resistance to disease, a few examples can be cited. In North America the outstanding example, from the standpoint of the value of the crop in dollars and calories, is probably that of the spring wheats resistant to stem rust. In the first 20 years of this century, yields of wheat on the prairies depended in large measure on weather conditions, and on the rapidity with which spores of the causative fungus spread northward from the wintering ground of the organism in Mexico and Central America. During the next 40 years, many resistant varieties were developed in the United States and Canada. The problem was complicated by the fact that, to be useful, the new varieties had to be not only resistant to stem rust, but also good in yield, stiff in the straw, able to mature before frost, and, withal, to produce a flour that would make good bread. All these were combined in such outstanding varieties as Thatcher and Rival in the United States and Renown and Regent in Canada.

The problem of the wheat breeders was further complicated

because there are many biological races of the rust fungus, and wheats resistant to one may not be resistant to others. A new race called 15B became a serious problem in the 1950's because none of the varieties previously satisfactory could withstand it. New varieties that could do so were eventually developed.

Here is an essential crop, and a pathogen that cannot be eradicated, or economically controlled by sprays, dusts, or treatment of the seed. Thanks to alert and dedicated co-operation between plant breeders and plant pathologists, stem rust of wheat has been effectively controlled in North America (also elsewhere by similar procedures). As a result, we have not only cheap bread, but also a great surplus of wheat to store in old ships or to feed the hungry in over-populated countries elsewhere.

One genetically resistant plant that has maintained its resistance to rust (of a different kind from that of wheat) for over half a century is the Martha Washington variety of asparagus. Rutgers, Marglobe, and other tomatoes are resistant to fusarium wilt, and one can hardly pick up a farm magazine without reading of some new raspberry resistant to mosaic, a new bean to withstand anthracnose, or some other resistant variety in other plants.

Some of the new varieties are valuable because of genetic resistance, not just to plant diseases, but also to insects. A plant louse that was a serious pest for years in vineyards was finally overcome by the discovery of resistant varieties of grapes, some of which are said to have retained their resistance for over 70 years. The Hessian fly, the worst insect pest of wheat, has long been partially controlled by strict observance of fly-free dates for sowing. The efficacy of that method has been greatly enhanced by the development of wheats that are genetically resistant to some of the several biotypes of the fly. Similarly, aphids besetting alfalfa have been controlled in recent years by the development and use of resistant varieties of alfalfa. Many more such cases are known in which the plant breeders have foiled invaders from the animal kingdom by breeding genetically resistant varieties.

Resistance to Disease in Animals

By contrast, cases in which genetic resistance to some disease has been deliberately sought and utilized for the development of superior strains of domestic animals are few. That does not mean any lack of such genetic resistance in animals, but rather that it has not been utilized. Anyone observing a flock of hens stricken with Newcastle disease, or leukosis, or anything else, must see that while the disease is fatal to some birds, others are entirely unaffected, and continue unconcernedly to do their duty. There can be little doubt that they have been exposed. One cow may develop mastitis when her neighbors on either side do not; some cows, in fact, resist deliberate attempts to infect them, when others of the same age cannot. Similarly, over 99 per cent of us (adults) have successfully withstood poliomyelitis without the protection of the Salk vaccine, or any other, because we are genetically resistant to the viruses which cause that disease.

The reason why genetic resistance to disease has been so little utilized in domestic animals is not any lack of such resistance, but, rather, inability or reluctance on the part of most official controllers of disease to recognize its potentialities. Disinfectants, vaccination, sulfa drugs, and antibiotics are orthodox; genes are not. Admittedly, animals are not plants, but while animal breeders must envy the comparative ease with which plant breeders develop varieties resistant to disease, the possibilities of doing the same with animals are not so hopeless as one might think from the unfortunate lack of research in this field. It is not suggested that genetic resistance should replace other measures of control where these are adequate, but some ubiquitous pathogens and parasites cannot be eradicated by any known means, and others can be banished only temporarily. Under such circumstances it would seem only logical to multiply the kind of animals that can either vanquish those organisms or tolerate them without any ill effects.

We have already considered some examples of genetic resistance to disease, but without referring to them as such. The female dog that is heterozygous for hemophilia is genetically resistant to that disease because the dominant allele prevents the recessive one from interfering with the normal clotting of the blood. The hemizygous male carries the same recessive gene, but lacks the protection against it which is provided in the female by the dominant allele. Here a difference in one gene makes the female genetically resistant and the male genetically susceptible. Similarly, animals heterozygous for lethal genes have resistance to the pathological conditions that are fatal to homozygotes.

Any reader unsatisfied with these examples because they do not deal with infectious diseases should turn again to Chapter 16, and recall how quickly the hardy oysters of Malpeque Bay made their own race resistant to the mysterious infectious disease that once nearly wiped them out. They did it by simple mass selection, a comparatively slow process at best, but effective in their case because of the enormous numbers of progeny produced by any one oyster. Similarly, but more slowly, natural selection bred Romney Marsh sheep that are comparatively resistant to trichostrongyle worms, and is now breeding rabbits resistant to myxomatosis. Breeders of domestic animals need not be restricted by the slow course of mass selection (whether natural or artificial) but can accelerate development of disease-resistant stock by using the progeny test. Let us consider a few examples of useful genetic resistance to infectious disease or parasites in domestic animals.

Honey-Bees Resistant to American Foulbrood

Among all domestic animals, none has demonstrated to its masters as well as has the honey-bee the commendable virtues of self-reliance, industry, thrift, co-operation, and monogamy in the male. Two more remarkable examples of the better way of life should be added, one in the field of public health, the other in eugenics. Some years ago, O. W. Park and others deliberately infected colonies of bees with *Bacillus larvae,* the organism causing American foulbrood, which is one of the worst diseases against which bee keepers have to contend. Some colonies withstood the infection better than others, and, by selection from that kind, the proportion of resistant colonies was raised in three years from the original 28 per cent to 75 per cent.

It was assumed that the selection practised had raised genetic resistance of those bees to the disease, but within a few years this supposition was found to be entirely erroneous. The superiority of the apparently "resistant" colonies resulted entirely from assiduous devotion to duty on the part of their public-health officials charged with responsibility for maintaining the sanitation of the hive and the health of the community. In the "resistant" colonies, these officials were particularly adept in detecting the disease at early stages, and diligent in cleaning out the infected cells, so that the diseased brood was removed before the causative bacteria reached the dangerous stage of infectious spores.

These activities, it will be recognized, are much the same as those of other public-health officials in other communities. As the bees were thus effectively guarded from any exposure to the disease, no selection for genetic resistance to it was possible. What selection had accomplished was the breeding of public-health officials more intelligent, diligent, and devoted to duty than those in colonies where disease flourished. It seems unlikely that we may be able to do as well in our own species. In any case, we are indebted to the honey-bee for showing us that differences in behavior can be genetically determined.

In later years, when new techniques made it possible to infect individual larvae, true genetic resistance to American foulbrood was revealed by Rothenbuhler and Thompson, who tested large samples of three different strains and found results as follows:

Strain	Previous Selection for Resistance	Survived (per cent)
V	None	25
C	One generation only	47
B	Many years of natural selection	67

Resistance to Mastitis in Dairy Cattle

No better example can be given of genetic variation already demonstrated, but still waiting to be utilized, than that responsible for mastitis in cattle. It is a sad commentary on our times that, after years of sanitation and in the midst of the glorious era of antibiotics, a committee of the American Dairy Science Association concluded in 1956 that "mastitis is the most costly disease of dairy cattle *not under satisfactory control*" (italics supplied). Few cows are now retained in dairy herds for more than four lactation periods, and, since many of those culled out are eliminated because of mastitis, it would seem that the disease is still *not under satisfactory control*. Blosser (1979), after an extensive survey of 9.5 million cows in 33 states of the United States, concluded that in 1976 mastitis caused (in that country) a loss of $1,294 billion. His review of reports from other countries tells of similar losses there (in proportion to the number of dairy cattle). One report from the United Kingdom stated that the incidence of the disease had not changed in 25 years. It seems clear that, while the magical antibiotics have temporarily alleviated problems in many cases, they have contributed little or nothing to the eradication of mastitis.

Mastitis is a general term for infection of the udder, which is caused in large measure by *Staphylococcus aureus*, or *Streptococcus agalactiae*, but also by other bacteria. These can be banished temporarily by antibiotics, or suppressed temporarily by "reverse flushes" and other measures to promote extreme sanitation, but are not likely to be eradicated.

Apart from several cases in which certain families have proven unusually resistant or susceptible to mastitis, significant data have come from investigations in New Zealand and in North Carolina, where cows in many herds were classified as resistant or susceptible, and the frequency of these two kinds then determined in the daughters from each maternal class. In North Carolina, Legates and Grinnells made their classifications by a method differing from that used by Ward in New Zealand, but the two studies yielded almost identical results (Table 23-1). For a statistical test of the significance of Ward's findings in the 20 herds at Canterbury, see the last part of Chapter 4.

In all three investigations, resistant cows had 33 to 38 per cent fewer daughters susceptible than the susceptible cows had. It should be noted that the difference results from the equivalent of a single generation of mass selection among females only, and without consideration of genetic differences among sires in susceptibility of their daughters. Although that is little more than a beginning in any selection program, the data in Table 23-1 suggest that it is more effective in reducing mastitis than the sanitary procedures commonly used.

One wonders how much more the incidence of mastitis could be lowered with further selection, and particularly if differences among sires were to be considered as well. Officials supervising the numerous centres for artificial insemination could find such differences with little further effort. There is no doubt that they exist. Among Jerseys at Pennsylvania State University, Reid found that, while the incidence of

TABLE 23-1

The Relation of Mastitis in Dams to That in Daughters

| Investigator and Place | Herds (number) | Susceptible Dams | | Resistant Dams | | |
		Number	Susceptible Daughters (per cent)	Number	Susceptible Daughters (per cent)	Effective Reduction (per cent)
Ward; Canterbury, N. Z.	20	86	89.5	109	56.0	38
Ward; Manawatu, N. Z.	15	128	81.3	171	54.4	33
Legates and Grinnells; North Carolina	11	114	53.0	82	35.0	34

mastitis among heifers from certain sires was below 14 per cent, the corresponding figure among daughters of another was 55 per cent.

Readers of this book who are trained in orthodoxy might relish some of the delightful heresy to be found in the following headings of sections in Reid's report of his studies:

Sanitation fails to reduce incidence.

Disinfection of udders, teats, and teat cups ineffective.

Bactericides useless in udder washes.

Cleaning and disinfecting stable floors of little avail.

Segregation of mastitic cows helps very little.

At the same time, Reid did not decry sanitation, and pointed out its importance in the production of milk of good quality. He also emphasized the fact that, as one might expect, susceptibility was aggravated by poor housing, lack of adequate bedding, exposure to inclement weather, and other adverse environmental conditions.

Organisms causing mastitis are not likely to be eradicated; hence the disease must be considered as enzoötic.[1] Obviously, some cows are genetically able to withstand it. It would seem sensible to multiply that kind. This has not yet been done to any extent in practice. Here is an opportunity for people who supervise the breeding of cattle, perhaps for some reader of this book. Further evidence of genetic resistance to mastitis has been reviewed elsewhere (Hutt, 1958).

More recently, other investigators have found significant differences between breeds of dairy cattle in susceptibility to mastitis, and, within breeds, differences among sires in susceptibility of their daughters. Among 64 large sire-families studied over three years by Skolasinski *et al.* (1977), the incidence of mastitis in 50 families ranged only from 20 to 30 per cent, but there were eight families in which the incidence varied from 30 to 70 per cent, and six in which it was no higher than 20 per cent. Differences among sires in susceptibility of their daughters to mastitis were significant in the 10 large sire-families studied by Grootenhuis (1976) during their *third* lactation, but were only "on the border of significance" in the 21 sire-families studied by Ryniewicz (1972) during their *first* lactation. It is reasonably certain that her differences among sires would have crossed that border and become highly significant if the daughters had been investigated at a later lactation.

The fact that susceptibility to mastitis increases with age has led some animal breeders to suggest that genetic resistance to that

[1]Let us leave the terms *epidemic* and *endemic* for people, not cows.

disease, while clearly evident, may be difficult to utilize because of the difficulty of progeny-testing sires for a condition not fully manifested in the first lactation. That would be true if we were dependent solely on progeny tests. There is some possibility, however, of utilizing an indicator that could reveal resistance and susceptibility during the first lactation, or even before it. (See Indicators of Genetic Resistance, pp. 555–62).

Resistance to Pullorum Disease in the Fowl

The cycle by which pullorum disease is passed through the egg from one generation to the next was explained in Chapter 14. An obvious method of control is to break that cycle by eliminating from the breeding pens all hens that carry the causative organism, *Salmonella pullorum*. Fortunately, most of such birds are readily detected by an agglutination test, and, by routine tests of all breeding flocks and elimination of reactors therefrom, pullorum disease has been practically eliminated in some parts of the world.

For animal geneticists, pullorum disease has some merits that entitle it to more than passing interest. For one thing, there are natural, inherent differences among breeds in susceptibility to it. Heavy breeds like Rhode Island Reds and Plymouth Rocks are three to five times as susceptible to it as are White Leghorns. This was demonstrated by the proportions of reactors in the two classes among flocks tested for the first time in programs aimed at eradication of the disease from entire states or provinces. It was confirmed by experimental inoculation of chicks. It has been shown, moreover, that the superior resistance of White Leghorns is independent of their dominant-white color, and, hence, more likely an attribute of the whole Leghorn breed rather than of just one variety.

In experiments which began in the 1920's at the University of Illinois, Roberts and Card demonstrated the feasibility of raising the innate resistance of White Leghorns still higher by selective breeding for resistance. In this case all chicks under test were given large oral doses of *S. pullorum*, a method of infection which (apart from dosage) is the same as the natural channel of infection for most chicks. In two strains of Leghorns selected for only four years, the proportions surviving the standard dose were 61 and 70 per cent, against only 28 per cent for controls. A strain selected for 9 years had 74 per cent surviving the standard dose. In reciprocal crosses between resistant and susceptible stocks, it was found that chicks of the former strain did not owe their resistance to any passive immunity transmitted from dam to chick through the egg. Resistant sires produced resistant offspring just as well as resistant dams. Clearly, the ability to resist was transmitted by the genes.

This was the first demonstration in domestic animals bigger than the rat of the feasibility of raising resistance to infectious disease by breeding. The method has not been used in practice, partly because control by elimination of reactors is easier, and more orthodox, but also because the rising proportion of Leghorns in the poultry world has automatically multiplied the breed endowed with natural resistance.

Chicks and pullorum disease together provide a good example of the influence of environment on heredity. Chicks are normally brooded during their first week of life at about 35° C. Some of those genetically resistant to *S. pullorum* at that temperature will succumb if brooded at only 30° C. Conversely, many of those susceptible at 35° C. are resistant when brooded at 40° C.

Resistance to Marek's Disease and to Lymphoid Leukosis in the Fowl

For about fifty years, one of the worst diseases afflicting the domestic fowl was known as the avian-leukosis complex. Poultrymen simply

FIG. 23-1 A White Leghorn pullet showing at six months of age a characteristic position of paralyzed birds stricken with neurolymphomatosis (Marek's disease). (Courtesy of Cornell University, and of Cornell University Press; from *Genetic Resistance to Disease in Domestic Animals*, by F. B. Hutt.)

called it fowl paralysis and big-liver disease. Affected pullets could show (at any time after six weeks of age) complete paralysis of one or both legs and wings (Fig. 23-1). That paralysis was caused by the formation of small lymphocytic tumors in the sciatic or brachial nerves. Birds that escaped such *neurolymphomatosis* might subsequently die with tumors in the liver or in other visceral organs.

Some pathologists believed that all these lesions (and others) were expressions of one and the same disease, and that they were all caused by a virus. Others thought that two different viruses, or more, were responsible for the differing manifestations of the complex. Eventually, it was recognized that the complex did consist of two distinct diseases. One of these (neurolymphomatosis) has been named *Marek's disease*, after the distinguished Hungarian pathologist who first described it in 1907. The other, which is now called *lymphoid leukosis*, has tumors, not only in the visceral organs, but also in the bursa of Fabricius.

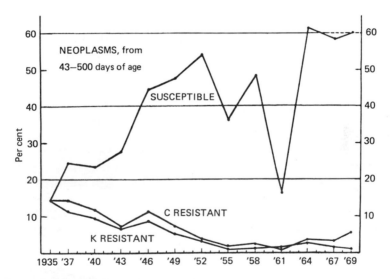

FIG. 23-2 Mortality from all causes in 32 annual flocks of three strains of White Leghorns at Cornell University. Two strains (C and K) were bred for resistance to neoplasms and other diseases. A third strain was selected for susceptibility to Marek's disease and leukosis. Birds of all three strains were intermingled throughout the test each year.

Fluctuations from year to year are smoothed out by three-year moving averages except for the terminal points. The sharp decrease in mortality of the susceptible line (1961) resulted from inadequate exposure to the disease that year.

Source: Data of F. B. Hutt and R. K. Cole, Cornell University.

Until about 1970, only two effective measures for control of Marek's disease were known. One of these is to isolate the chicks at hatching and to keep them away from older birds (known to be the reservoir of infection) until the young pullets are four or five months old. By that age, they are nearly all resistant and can safely be returned to infected premises. The effectiveness of this practice is shown in Fig. 15-5.

The other method of control is to breed strains resistant to Marek's disease. The feasibility of doing so has been demonstrated at Cornell University and elsewhere (Fig. 23-2). Details of the 30-year experiment at Cornell were reported by Cole and Hutt (1973).

After the virus responsible for Marek's disease was identified, it was not long before effective vaccines became available, and the disease is now controlled fairly well by routine vaccination, usually at one day of age. We shall read more about genetic resistance to Marek's disease in the section on "Indicators" later in this chapter.

Resistance to Internal Parasites

Breeds of sheep, and individuals within breeds, differ in susceptibility to the trichostrongyle worms that beset them. In a remarkable experiment, Euzéby et al. (1961) tested six lambs of each of four breeds by exposing them (all together) in a field known to be infested with trichostrongyle worms. To ensure continued exposure, five ewes of each breed were kept in the same field during the course of the experiment.

The degree of resultant infestation of the lambs was measured by (a) the number of worm eggs per gram of feces in samples taken during the experiment and (b) the number of worms per breed found at necropsy after exposure of the lambs for four or five months. The total number of worms was estimated from 10-ml. samples of the contents of the digestive tract, and in those samples the numbers of each of five different kinds of worms were counted.

The results showed remarkable differences among breeds (Table 23-2). These were not related to differences in size. Among the four breeds, the Southdowns (small) were most susceptible, but the Charmoise (also small) were most resistant. In the two large breeds, the Bleu de Maine were highly susceptible, and the Texel comparatively resistant. Resistance of some breeds to certain worms was evident. For example, the most abundant parasite, *Haemonchus contortus*, numbered 2,650 in the Southdowns but only 15 in the Charmoise.

The report does not state whether the six lambs of each breed

TABLE 23-2

Differences Among Breeds in Resistance to Trichostrongyle Worms

	Southdown	Bleu de Maine	Texel	Charmoise
Eggs per gram of feces	500	530	350	330
Worms per breed	6,830	6,738	2,756	1,895

Source: Data of Euzéby et al. as cited.

came from several different rams of that breed, or from one only. In the latter case, the differences observed could be differences among four sire-families, and not necessarily among four breeds. In either case, those differences would be genetic in origin.

In Chapter 16 we read briefly of the special resistance of Romney Marsh sheep to trichostrongyle worms. At the University of California, Stewart and his associates tested four Romneys against samples of other breeds (all maintained in one flock) for resistance to one of these worms, *Ostertagia circumcincta*. Severity of infestation was measured at intervals of two weeks for a full year by counting the number of worm eggs per gram of feces. (Fig. 23-3).

Apart from the consistently low count for the Romneys, and hence the obviously greater ability of that breed to withstand the parasite, a significant finding in this study was that within other breeds individuals differed greatly in resistance. Following up that lead, Gregory and other investigators have shown that some rams can sire lambs more resistant to trichostrongyle worms than those from other rams. Similarly, Warwick and his associates selected in goats for resistance to the worm *Haemonchus contortus*, and, on comparing progeny of resistant sires with those of unselected controls, found the proportions resistant to be:

In kids from resistant ♂♂ × daughters of resistant ♀♀	83 per cent
In kids from resistant ♂♂ × unselected controls	71 per cent
In kids from unselected controls	31–33 percent

It would seem that genetic ability to resist intestinal parasites in sheep and goats could be utilized to great advantage. The phenothiazine and other drugs on which chief reliance has been

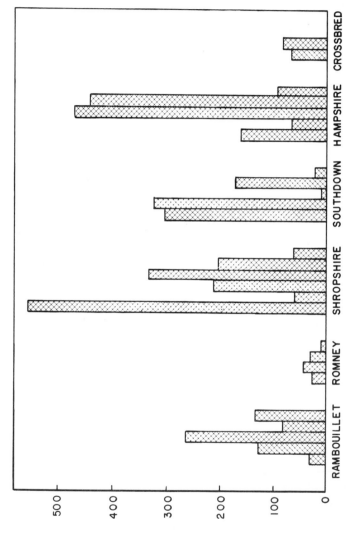

FIG. 23-3 Differences among sheep in infestation with the worm *Ostertagia circumcincta* as measured by average number of eggs per gram of feces during one year. Each column represents one animal. Uniformly low counts in the Romneys are conspicuous. (After P. W. Gregory in *Proc. Amer. Soc. Animal Prod.*)

placed for controlling worms in these species do not kill all the worms, but merely reduce them to a number that the animal can tolerate without becoming unthrifty. Who can assure us that the survivors will not multiply and eventually produce races resistant to such drugs, just as the houseflies did with DDT?

Trypanosomes carried by tsetse flies are a serious threat to the production of cattle in many parts of Africa, but some breeds are so resistant to trypanosomiasis that they have a special value in areas where the flies abound. West African Shorthorns and the N'Damas, which originated in French Guinea, have proved to be more resistant to these protozoan blood parasites than breeds from areas where the flies are not prevalent. In Nigeria, Chandler found the N'Damas to be resistant to two species of trypanosomes and, within those species, to strains of different geographic origins. Such a general resistance could make the N'Damas useful over wide areas where the tsetse fly is common.

Other protozoan parasites to which genetic resistance is known include coccidia of the genus Eimeria, which are ubiquitous parasites of fowls. The feasibility of breeding strains highly resistant, and of doing so in a few generations of selection, was demonstrated by Rosenberg and others. Crosses of the resistant lines with unselected stock showed that, as with most genetic resistance to other diseases, resistance to coccidiosis is a quantitative character.

Resistance to External Parasites

Animals differ genetically not only in resistance to internal parasites, but to those on the outside as well. Resistance to blood-sucking arthropod parasites of one kind or another is vitally important in the many parts of the world where insects and ticks act as vectors, *i.e.*, carriers of pathogenic protozoa, Rickettsia, and other organisms. Diseases thus transmitted include the yellow fever and malaria of man, and trypanosomiasis, anaplasmosis, and piroplasmosis (one kind of which is better known in the United States as Texas fever) of cattle. It has been found in various parts of the world that zebu cattle are much more resistant to ticks, and, hence, appear more resistant to tick-borne diseases than are cattle of European origin. This was well illustrated by Bonsma's studies in the Transvaal of mortality from "heartwater," a disease so-named because one characteristic of it is an accumulation of fluid in the pericardium. It is caused by a Rickettsia which is carried by the "bont" tick, *Amblyomma hebraeum*. Mortality from this disease among calves with differing degrees of

Africander (zebu) blood, but all born and raised at one experiment station, is shown in Table 23-3.

In this table one should note (a) the great difference between the zebu breed and the European ones, (b) that resistance to heartwater in these two groups was shown, not only by lower mortality, but also by longer survival among those that died, and (c) that the crossbred cattle, even those only half Africander, resembled the resistant parent more than the susceptible. A similar dominance of resistance to tick-borne disease has been found in other F_1 crosses from other zebu breeds, in other parts of the world. That this advantage of the zebus can be credited to some innate ability to repel ticks is generally recognized. Actual counts of bont ticks on Bonsma's cattle showed that in European breeds the ticks on three regions of the body (with measured, uniform areas in each region) outnumbered those on Africanders by 2.2, 2.9, and 7.5 to 1.

In various parts of the world investigators are now seeking to find how the zebus repel ticks, and are studying the possible roles of hypersensitivity of the skin, its thickness, and its secretions. Whatever it may be, this valuable attribute is not restricted to zebus for it has been found also in "criollo" cattle of Nicaragua and Honduras. In nine monthly counts of two kinds of ticks (neither the same as Bonsma's bont tick), and with each cow sprayed after each count so that she could start clean for the next month, Ulloa and de Alba found average counts for the criollos to be far below those of Jerseys. The 23 cows from Honduras were particularly resistant, with average monthly counts in the standard areas being only 4.2 for one species and 1.5 for the other, against corresponding figures of 21.8 and 10.0 for the ten Jerseys. One of the Honduran cows even kept her score at zero for the whole nine months. It would be nice to know how she did it.

TABLE 23-3

Mortality from Heartwater to 30 Months of Age in Cattle Born at Mara in the Transvaal

Breed	Calves Born (number)	Died (per cent)	Average Age at Death (months)
Pure Africander (zebu)	246	5.3	11
¾ Africander, ¼ European	86	7.0	8
½ Africander, ½ European	397	10.2	6
Pure European	28	60.7	5

Source: From data of Bonsma.

Klendusity

We should recognize that animals which escape death from tick-borne diseases because of their ability to repel ticks are not necessarily genetically resistant to the protozoa, Rickettsia, or other organisms which cause those diseases. By keeping the ticks off, or by reducing their numbers, some animals may escape infection entirely, and others get such a light dose that their natural mechanisms of defence can cope with it. Similarly, some plants escape infections by repelling insects that carry them. The Herbert and Lloyd George varieties of raspberries seldom get mosaic disease because fine hairs on the leaves prevent the aphids which carry it from getting close enough to pierce the leaf. Plant pathologists call this property of escaping disease *klendusity*. They have found that the apparently resistant raspberries are actually quite susceptible when they do become infected. They are klendusic, but not resistant.

Accordingly, until more is known about them, it would seem appropriate to attribute the low incidence of tick-borne diseases in the zebus more to klendusity than to genetic resistance to the organisms carried by the ticks. They may have both. The fact that the Africander calves which died of heartwater withstood that infection longer than susceptible calves of European breeds (Table 23-3) suggests greater resistance *after* infection, but that, in turn, might have resulted from a lighter dosage, or a later one, as could happen if their ticks were fewer in number than those on calves of the European breeds.

Specific Resistance to Specific Diseases

The utilization of genetic resistance to disease in domestic animals would be of more interest to animal breeders and veterinarians, perhaps, if resistance to one disease of one species automatically carried with it resistance to all others. Unfortunately, that is not the case. It is true that some genetic weakness, such as impaired ability to form antibodies, can make animals susceptible to any of several different kinds of infection; hence the normal individuals not homozygous for that defect are more resistant to those same infections. It has also been shown that animals genetically resistant to one pathogen may in some cases be similarly resistant to another that is closely related to the first. Thus, White Leghorns are comparatively resistant both to *Salmonella pullorum* and to *S. gallinarum* (which causes fowl typhoid).

Apart from such exceptions, the more general rule seems to be that genetic resistance to one disease is specific for that disease, and does not often extend to others that are entirely different. For example, the remarkable ability of zebu cattle to withstand extreme heat, and to repel ticks, is matched by their equally remarkable resistance to bovine tuberculosis (Carmichael), but most of the evidence available indicates that in Africa they are more susceptible than some other breeds to trypanosomiasis. Similarly, when a disease commonly called "blue comb"[2] struck Leghorns at Cornell one year, mortality from that disease (at about three or four months of age) and from leukosis (to 500 days of age) ran as follows in three different strains:

Strain	Blue Comb (per cent)	Leukosis (per cent)
K Resistant (to leukosis)	0.99	6.5
C Resistant (to leukosis)	7.40	8.7
Susceptible (to leukosis)	0.00	65.0

Clearly, the strain genetically most susceptible to leukosis was most resistant to blue comb. Although the two resistant strains were about equally resistant to leukosis, they differed significantly in susceptibility to blue comb and to some other diseases as well. Similarly, Gowen's mice most resistant to typhoid disease proved highly susceptible to a virus causing pseudorabies.

From these examples it is evident that the physiological basis for genetic resistance to one disease can be quite different from that for resistance to some other. To breed domestic animals that can better cope with adverse environments of different kinds, which include pathogenic organisms of assorted sizes and species, it would seem desirable to learn more about the divers physiological mechanisms to which genetically resistant animals owe their biological superiority over those genetically susceptible. Examples of such mechanisms are considered in the next few pages.

Genetic Variation in the Effectiveness of Antibodies

Studies in this field with laboratory rodents give useful leads to possible ways in which genetically resistant domestic animals differ from susceptible ones. At the Vermont Experiment Station, F. A. Rich found that some of his guinea pigs were deficient in blood complement, which is necessary for the functioning of some antibodies. When

[2]Later called monocytosis.

tested by inoculation with the bacterium now called *Salmonella choleraesuis*, mortality in the deficient animals was 77 per cent, against only 20 per cent in their normal litter-mates that did have blood complement. The mutant guinea pigs also proved highly susceptible to other infections. Their peculiar defect was shown to be caused by homozygosity for a recessive autosomal gene.

At the State Serum Institute in Copenhagen, Scheibel noticed that her guinea pigs differed in ability to produce diphtheria antitoxin. By selection in two directions she quickly differentiated two distinct strains, in one of which the proportion of all animals classified as non-producers was reduced to 2.5 per cent in the first selected generation and changed little in the next five. In the other strain, the first generation from parents that were poor producers contained 47 per cent classified as non-producers, and in four more generations of selection the proportion of such animals was raised to 88 per cent.

Other investigators working with diseases in mice have found cases in which genetically determined resistance seems to depend upon the rapidity with which antibodies can be produced in response to infection.

Breeding Resistant Animals Without Exposure to Disease

In most cases, to distinguish animals that are resistant to disease from those that are susceptible, both kinds have to be exposed. Sometimes, as with leukosis in fowls and mastitis in cattle, exposure is almost inevitable and is accepted with as much good grace as we can muster. Deliberate exposure, the introduction of pathogenic organisms to facilitate the breeding of resistant stock, is not likely to be practised by many breeders, and those who do it should hardly expect the endorsement of their veterinarians. If there were ways by which genetically resistant animals could be identified without exposure to the disease, the multiplication of such stock would be more popular. The only thing necessary is to discover enough of the mechanisms by which resistant animals cope with disease to permit identification of such animals.

INDICATORS OF GENETIC RESISTANCE

An Indicator of Resistance to Pullorum Disease

In a series of studies too long to be reviewed here, it was found that the chicks most resistant to *Salmonella pullorum* have superior control of the processes that maintain their body temperatures. The degree of such control can be measured by determining (from the average of several readings) how quickly the chicks can raise their temperatures from about 102° F. (when they hatch) up to the 105° to 107° F. that is normal at ten days of age. The differences among them in that ability are evident by six days after hatching. Chicks most efficient in making the transition are most resistant to pullorum. For one thing they respond to infection, as do other animals, by running their temperatures up to fever heat, and maintain that high temperature better than susceptible chicks. That in turn is believed to accelerate the production of antibodies and other mechanisms for defence, such as *phagocytosis*, or the ingestion of invading organisms by cells specially assigned that task. This tale has all been told elsewhere (Hutt, 1958).

To test the feasibility of using such an indicator of resistance and of breeding resistant stock in birds not infected, selection to differentiate two lines was carried out for two generations. One strain was bred for high temperatures, and the other for low ones. As a result, in the first generation, average temperature during the first six days in the low-temperature line was 105.54° F., or 0.42° F. lower than in the high-temperature line. In the next generation the difference was 0.59° F. When chicks of the two lines were tested by oral inoculation with a standard dose of S. *pullorum*, those of the high-temperature strain were consistently more resistant than chicks of the low-temperature line. Results in two of these trials are shown in Fig. 23-4.

While it seems clear that, by using the early indicator of resistance, it is feasible to raise the genetic resistance of fowls to pullorum disease, and to do so without exposing them, that procedure is not likely to be followed because the extensive programs to eradicate S. *pullorum* have succeeded so well (in some parts of the world) that the disease is no longer considered serious.

Facile control of body temperature can be a useful indicator of resistance to other diseases. Locke found long ago that resistance of rabbits to virulent pneumococci could be measured by their "warming time," *i.e.*, the number of minutes required for a rise in body temperature of three degrees after they had been chilled to temperatures of about 95°F. At each of the three different dosages, rabbits previously found to have warming times under 35 minutes were more resistant than those requiring more than 45 minutes.

Indicators of Resistance to Mastitis

The feasibility has been demonstrated of distinguishing between cows resistant to mastitis and those susceptible to it by experimental

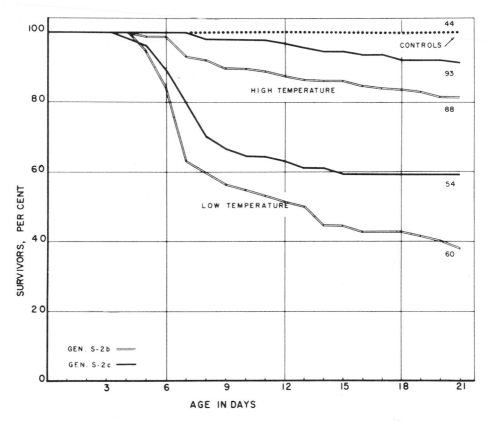

FIG. 23-4 Survival rates for samples of chicks from two lines following inoculation with *S. pullorum* in moderate dose (open line) or lighter dose (black line) showing consistently greater resistance in the high-temperature line than in the other. The numbers tell how many chicks were in each group. Controls were not inoculated. (From F. B. Hutt and R. D. Crawford in *Canad. J. Genet.* and *Cytol.*)

inoculation of the teat canals with *Str. agalactiae*, and then clearing up resultant infections with massive doses of penicillin. Some animals are easily infected, even in their first or second lactation, and others not at all. While such a procedure is an early measure of resistance, it is not likely to be used because of the objections to deliberate infection with pathogenic organisms.

There are some signs that a better indicator of genetic resistance—one not requiring deliberate exposure to pathogens—may be available in the future. In 1963, Adams and Rickard tested *in vitro* the bactericidal potencies of the lipid, sebum-like material from the teat canals of quarters of the udder that differed in susceptibility to *Str. agalactiae,* and they determined by chromatography the corresponding composition with respect to fatty acids. The bacteria proved most susceptible to myristic, palmitoleic, and linoleic acids. Material from quarters resistant to experimental infection contained a higher proportion of these fatty acids than did the material from susceptible quarters. The latter kind contained significantly more palmitic acid than that from resistant quarters.

More recently, Štavíková and Lojda (1975), in Czechoslovakia, studied the composition of the lactosebum in a large number of cattle and found that, of six fatty acids making up the neutral lipid fraction of the lactosebum, levels of lauric and myristic acids were highest in cows resistant to mastitis. They determined the heritability of the lipid composition in the lactosebum to be 0.38 to 0.58, but Edwards *et al.* (1973), who studied variations in the fatty acids of the milk in 25 pairs of Ayrshire twins, found heritabilities of 15 different fatty acids to range mostly from 0.82 to 0.98.

With these leads, to the earlier evidence that heredity determines resistance or susceptibility to mastitis, we can now add evidence that resistance is associated with a high level in the lactosebum of fatty acids that excel in bactericidal potency. Eventually, quantitative assays of fatty acids in secretions of the skin (including the lactosebum) may provide indicators that will greatly facilitate the breeding of cattle resistance to mastitis.

An Indicator of Resistance to Eye Cancer in Hereford Cattle

One disease for which an indicator of resistance can be easily seen at an early age is that in cattle known to stockmen as eye cancer. It is found most commonly in areas of the world where white faces have to cope with brilliant sunlight. The disease begins with a slight conjunctivitis, or with inflammation of the eyelid, and progresses through successive stages to carcinoma, which may involve the eyelids, the eyeball, or both. Carcinoma is seldom seen in animals under four years of age, but the incidence rises steadily in older animals. Some investigators have been reluctant to admit that the disease is more prevalent in Herefords than in other breeds, but the breeders of Aberdeen Angus boast proudly that their breed is exempt. In Australia, French found the incidence of characteristic lesions (including earlier stages) at one experiment station to be as follows:

	Per cent
In 189 Herefords aged 7 to 8 years:	39
In 71 Herefords aged 4 years:	13
In 150 Shorthorns aged 7 to 8 years:	0

Within the Hereford breed, genetic differences in susceptibility have been demonstrated by Anderson in Texas. The frequency of eye cancer in progeny at least four years old was found to be only 11 percent when both parents had been free of it, but 36 per cent in those from parents both affected. Among animals having only one parent with eye cancer, 22 to 25 per cent developed the affliction.

Evidence from South Africa, the United States, and Australia has shown clearly that animals with pigmented eyelids have far less eye cancer than those with white faces and lids not pigmented. It has been shown, moreover, that as the extent of that pigmentation increases, the incidence of lesions decreases steadily. A solid ring of pigment at least half an inch wide around the eye will ensure pigmentation of the lids (see Fig. 4-3).

Here is a disease to which genetic susceptibility and resistance have been demonstrated, and for which an indicator of innate resistance is clearly evident by three months of age, and possibly earlier. It has been shown that the amount of colored skin around the eye is influenced by heredity, and is therefore amenable to selection. Some studies indicate that there is also genetic variation in resistance to eye cancer apart entirely from the pigmentation of the eyelids, but to utilize it would make breeding for resistance a rather slow process because of the late manifestation of the disease. It would seem somewhat simpler to breed an animal with a ring of protective color around its eye.

An Indicator of Resistance to Marek's Disease

Some years ago, R. K. Cole differentiated (from the Cornell Random-bred White Leghorns) two strains, one of which was extremely resistant to Marek's disease and the other extremely susceptible. His procedure was to inoculate fully pedigreed chicks at two days of age with a standard dose of an inoculum carrying the JM strain of virus, which was known to cause Marek's disease. All that died, or showed paralysis, and all that survived to eight weeks were examined for gross lesions of Marek's disease. It was thus possible to identify the most resistant families in the resistant strain (N) and the most susceptible families in the other one (P). The sires that had produced these desirable families were then remated to the most desirable dams. The resultant chicks, which were *not* inoculated, were kept to beget the next generation.

In the fourth generation, susceptibility in the N strain was only 3.6 per cent, but in the P strain it was 96 per cent (Fig. 23-5). The rapid differentiation of these two strains suggested that the genes determining resistance or susceptibility to Marek's disease must be few in

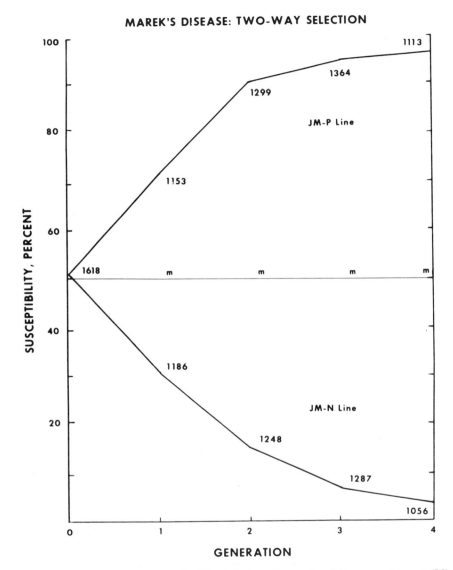

MAREK'S DISEASE: TWO-WAY SELECTION

FIG. 23-5 Differentiation by R. K. Cole (at Cornell) of lines resistant (N) or susceptible (P) to the JM virus (one of those causing Marek's disease). In the fourth selected generations, susceptibility was only 3.6 per cent in the N line but 96 per cent in the other. The rapidity with which this was achieved led eventually to the discovery that the N line owed its resistance to almost complete homozygosity for the B^{21} blood antigen. The numbers by the lines give the numbers of chicks tested in each generation. (Courtesy of R. K. Cole, Cornell University.)

number. Eventually, Briles et al. (1977) found that, by the sixth generation. birds of the N line (resistant) were nearly all homozygous for the B^{21} red-blood cell antigen, whereas those of the P line (susceptible) lacked that antigen, but most of them carried another one, B^{19}.

In further critical tests, backcrosses of F_1 heterozygotes ($B^{19}B^{21}$) to the P line showed that chicks getting the B^{21} allele were remarkably resistant and those of the genotype $B^{19}B^{19}$ were very susceptible. This evidence that even a single dose of the B^{21} allele conveys resistance was confirmed by additional tests. These were not made with the N and P lines, but in another flock of the Cornell Random-bred stock (from which those N and P lines had been derived). Heterozygotes carrying B^{21} (and one of three other B alleles) were mated to birds that lacked it. The progeny were inoculated and blood-typed as before. Again, the birds getting the B^{21} allele proved to be far more resistant than those that lacked it (Table 23-4).

It was still not certain that the B^{21} allele could exert its magical resistance to Marek's disease in flocks other than the Cornell Random-bred, but evidence that it could do so was adduced by Longenecker et al. (1976) who tested its efficacy in other lines. Some of their chicks were inoculated; others were tested with natural

TABLE 23-4

Reduction of the Incidence (per cent) of Marek's Disease by the
Red-Blood-Cell Antigen B^{21}

	Chicks Tested					
Matings or Tests	Geno-type	Num-ber	Marek's	Geno-type	Num-ber	Marek's[b]
Briles et al.						
(a) Within strains N and P						
$B^{19}B^{21} \times B^{19}B^{19}$	$B^{19}B^{21}$	81	8.6	$B^{19}B^{19}$	99	69.7
(b) In another flock						
$B^{21}B^a \times B^aB^a$	$B^{21}B^a$	189	5.8	B^aB^a	191	77.0
Longenecker et al.						
(c) Natural exposure after one day old	$B^{21}B^{21}$ $B^{21}B^a$	141	13.5	B^aB^a	56	31.2
(d) Inoculated at 21 days	$B^{21}B^{21}$ $B^{21}B^a$	273	0.4	B^aB^a	106	19.8

[a]Other B alleles—not 21.
[b]To 20 weeks in (a) and (b), to 109 days in (c), and to 80 days in (d).
Source: Data of Briles et al. and of Longenecker et al., as cited.

exposure, assured by proximity to inoculated chicks. In both cases, those carrying B^{21} had much lower incidence of Marek's disease than those that did not (Table 23-4).

From all of this work, it seems clear that poultry breeders have in the B^{21} allele a useful indicator of resistance to Marek's disease. While the experiments that revealed the value of the B^{21} allele necessitated exposure of the chicks under test to the virus causing the disease, any such deliberate exposure should no longer be necessary. The chief problem for some breeders may be to find laboratories in which the presence or absence of the B^{21} allele can be detected.

Crittenden *et al.* (1970) have shown (see Problem 21-7) that embryos lacking the antigen R_1 are resistant to infection by certain viruses causing avian leukosis. In contrast, those that do carry the dominant allele for R_1 are all susceptible.

It is to be hoped that the continuing studies of the blood groups and protein polymorphism in domestic animals will reveal further cases in which some of them can be utilized as indicators of either resistance or susceptibility to disease.

Applications Not Applied

It should be unnecessary to belabor the point that productivity of domestic animals can be enhanced by prolonging their useful lives as well as by striving to raise the amount of milk, eggs, or meat produced within a given time. Enough evidence of genetic resistance to bacteria, viruses, and parasites has been cited in this chapter to suggest that some genetic variation in resistance to disease is to be expected in every species and to every disease. Lest one assume too readily that controllers of disease would not fail to utilize that genetic resistance whenever other methods are unsatisfactory, it seems desirable to cite an example in which such obvious applications have not been applied.

From 1948 to 1962, official veterinarians in England tried to eradicate the mild form of Newcastle disease, or "fowl pest." This respiratory disease is caused by a virus which is widespread over the face of the earth. It can cause mortality in some laying flocks, but in others it causes only a temporary drop in egg production. Some flocks in full lay and in perfect health are found by blood test to have been exposed, even though they have never shown it. During 14 years of futile slaughter, the number of birds killed annually in England increased steadily. The total numbers of fowls, turkeys, ducks, and geese slaughtered in two years alone (1959-1961) exceeded 11.6 million. Finally, the compensation that was paid for slaughtered flocks reached such figures that the government refused to pay it any longer. Control by vaccines was undertaken in 1962, but slaughter of infected flocks was continued.

During all those years of attempts to eradicate the disease, in spite of the evidence that the great majority of the birds were able to withstand it easily, no attempt was made to see how much the small proportion of susceptible birds could be reduced still further by multiplying the desirable genes of the resistant ones. Instead, those biologically efficient birds were wiped out with the others. Newcastle disease is no great problem in the United States. Most poultrymen control it cheaply and easily by putting a live vaccine in the drinking water.

Resistance Utilized

After reporting in this chapter so many cases in which genetic resistance to disease has been demonstrated but has not been utilized to breed resistant stock, it seems desirable to cite one case in which such resistance was recognized and eventually utilized to the full.

In Chapter 16, we read about the hardy oysters of Malpeque Bay that set out to overcome the peculiar disease that once almost wiped them out. Against it they employed their best weapons—some useful genes and the ability to produce from a single oyster up to 60 million offspring[3] in a year. Resistance was raised rapidly; the disease was vanquished, and within 20 years, the yield of oysters was higher than ever before (Fig. 16-3).

As the disease spread around Prince Edward Island, resistant stock from Malpeque Bay was transferred to areas newly stricken to accelerate the development of resistant oysters. That it did, and so effectively that the time required to develop resistant oysters was cut in half to 10 years. Eventually, in spite of all efforts to prevent spread of the disease, it was found (about 1955) in oyster beds in nearby New Brunswick and Nova Scotia. Again, by transferring many barrels of the genetically resistant stock from Prince Edward Island to the newly stricken areas, the development there of stocks resistant to the Malpeque disease was accelerated. For example, at Malagash, Nova Scotia, mortality in 1964 was only 14 per cent in the introduced resistant stock and had been reduced to 29 per cent in the Malagash oysters, but among hitherto unexposed, unselected oysters brought from Cape Breton, it was 66 per cent (Drinnan and England, 1965).

It is encouraging to reflect that even though man has been slow

[3]In oysters, they are called (collectively) "spat."

to breed for genetic resistance to disease, he has not been slow to utilize such resistance when Nature and natural selection make it available.

Genes Determining the Duration of Life

The sum total of ability to resist diseases of divers kinds can best be measured by the duration of life. Unfortunately, while press reports of venerable horses and hens are common enough, they all tell of exceptional cases, not of averages. Most domestic animals are allowed to live only so long as they are profitable, and, for that reason, genetic studies of longevity in domestic animals have not been made.

It is different with our own species. Barring accidents and wars, in most countries we can count on some effort being made to keep us alive as long as possible, and whether we like it or not. That, in turn, has made possible studies to determine to what extent duration of life in man is determined by heredity. Without reviewing the whole field, four analyses, each with a different procedure, are here briefly summarized:

1. Alexander Graham Bell traced down the ages at death of some 2,300 people of the Hyde family, and sorted them into three classes according to the number of their parents that lived to 80 years or more. Among those having 0, 1, or 2 parents attaining that age, the proportions reaching 80 were 5.3, 9.8, and 20.6 per cent, respectively.

2. The Pearls (R. and R. De W.) determined for 365 people still living at 90, and for a suitable control group, what they called the "total immediate ancestral longevity" (TIAL)—the total ages of two parents and four grandparents. For the nonagenarians, TIAL was more than 60 years higher than for the controls, or about ten years for each of the six persons.

3. Louis Dublin of the Metropolitan Life Insurance Company compared the actual mortality rate with the expected rate (or average) in three groups of policyholders. Among those that were of comparable ages when the policies were issued, for the ones who had both, one, or neither parent living at that time, subsequent mortality was 88, 98, and 111 per cent, respectively, of the average rate.

4. Jarvik *et al.* (1960) used the twin-study method and determined how long the second twin outlasted the first to die in monozygotic and dizygotic twins. Comparisons were made in

six groups differentiated according to age at death of the first twin to die. Data for 84 pairs of MZ twins and 92 pairs of like-sexed DZ twins were available for comparison. In every one of the six groups, the ages at death were closer together for the MZ twins than for the others. As MZ twins show more resemblance than DZ twins in anything influenced by heredity, the role of the genes in determining duration of life is evident.

These four procedures were all different, but their findings were essentially the same. The length of our stay on earth is in large measure decided by the genes that we received from our parents, but here, as with so many other inherited traits, the environment will also play a significant role in determining the outcome. For ourselves, most of us will just have to be content with the genes that fell to our lot, but we can do our best to help them cope with the environment.

For our cows and chickens, cats and dogs, and all the rest of our domestic animals, we can not only provide, as stockmen, the best environment possible, but can also endeavor to ensure, as breeders, that future generations of them get the genes most conducive to maximum health and productivity. It is to be hoped that the principles outlined in this book will be found helpful in that process, and that some readers, at least, may be inspired to put them into practice.

Problems

23-1 A. B. Sabin crossed a strain of mice called PRI, in which all adults were completely resistant to the 17 D strain of yellow-fever virus, with Swiss mice, all of which were susceptible. All of the F_1 were resistant, as were backcrosses of $F_1 \times$ PRI, but when the F_1 were backcrossed to Swiss, exactly half of the 90 progeny proved to be susceptible. In the F_2 generation 28 per cent were susceptible, and some of these crossed to Swiss produced only susceptible progeny. What genetic basis for resistance is indicated?

23-2 It was found that the PRI mice did not vanquish the virus, but that it multiplied in them at a rate only 1/10,000 to 1/100,000 of the rate in susceptible mice. Meanwhile, specific antibodies were developed against the virus. What does this suggest about the physiological basis for resistance?

23-3 Later, Sabin found that PRI mice under three days old were susceptible, that some were the same at three to five days, but after five days all were resistant. In the young susceptible PRI mice, multiplication of the virus was no faster than in resistant adults. What modification of your answer to Problem 23-2 does that require?

23-4 The PRI mice resistant to the virus of yellow fever were also completely or partially resistant to six other pathogenic viruses, but were fully susceptible to ten others. What do such findings suggest about genetic resistance to viral diseases in other animals?

23-5 If there were no other evidence, how would you employ the twin-study method to determine whether or not resistance of dairy cattle to mastitis is genetically determined?

23-6 In making such tests, how would you reduce variation caused by (a) the environment and (b) the fact that susceptibility rises with increasing age?

23-7 It has been demonstrated that in their first or second lactations some cows can be artificially infected with *S. agalactiae,* but others can not. Infections thus induced can be eliminated with large doses of penicillin. In what ways could such tests be useful in the experiment of Problem 23-5?

23-8 A strain of mice called C57 Bl is highly resistant to a viral infection called mousepox, which is believed to be contracted through small lesions in the skin. Schell found that strain highly resistant when inoculated in the footpad, but no more so than others when inoculated intraperitoneally. What bearing have such findings on breeding for resistance to disease in domestic animals?

23-9 Investigators breeding for resistance to disease in laboratory animals try to use a dosage that is lethal, or visibly pathogenic, to about half of the animals in the initial tests. Why are doses more virulent, or less so, not equally useful?

23-10 Pimentel and his associates bred a strain of houseflies so resistant to DDT that about three-quarters of them could withstand severe exposure for two weeks. When they were removed from any exposure, resistance declined and was completely lost in 22 generations. Similarly, another strain lost its resistance to Lindane in 20 generations. What bearing have such results on the utilization of genetic resistance to disease in domestic animals?

Selected References

(For papers prior to 1960, see Chapter 22 in the first edition of this book.)

ADAMS, E. W., and C. G. RICKARD. 1963. The antistreptococcic activity of bovine teat canal keratin. *Amer. J. Vet. Res.* **24**:122–135. (Report of research that should lead to indicators of resistance to mastitis.)

BLOSSER, T. H. 1979. Economic losses from and the national research program on mastitis in the United States. *J. Dairy Sci.* **62**:119–27. (They exceed a billion dollars annually, and are also serious elsewhere).

BRILES, W. E., W. H. Mc GIBBON, and H. A. STONE. 1976. Effects of *B* alleles from Regional Cornell Randombred stock on mortality from Marek's disease. *Poult. Sci.* **55**:2011–12. (See text, p. 561.)

BRILES, W. E., H. A. STONE, and R. K. COLE. 1977. Marek's disease: Effects of *B* histocompatibility alloalleles in resistant and susceptible chicken lines. *Science* **195**:193–95. (See text, p. 561.)

COLE, R. K. 1967. Studies on genetic resistance to Marek's disease. *Avian Disease* **12**:9–28. (See text, pp. 559–60.)

COLE, R. K., and F. B. HUTT. 1973. Selection and heterosis in Cornell White Leghorns: A review with special consideration of interstrain hybrids. *An. Breed. Abstr.* 41:103–18. (Review of an experiment that lasted over 30 years.)

CRITTENDEN, L. B., W. E. BRILES, and H. A. STONE. 1970. Susceptibility to an avian leucosis-sarcoma virus: Close association with an erythrocyte antigen. *Science* **169**:1324–25. (Viruses of sub-group B, and the antigen R_1.)

DRINNAN, R. E., and L. A. ENGLAND. 1965. Further progress in rehabilitating oyster stocks. *Fisheries Board Canada, Gen. Series Circular* No. 48, St. Andrews, New Brunswick. (Use of genetically resistant oysters.)

EDWARDS, R. A., J. W. B. KING, and I. M. YOUSEF. 1973. A note on the genetic variation in the fatty acid composition of cow milk. *Anim. Prod.* 16:307–10. (In 25 pairs of Ayrshire twins.)

EUZÉBY, J., J. BUSSIÉRAS, P. MORRAILLON, and R. BOCCARD. 1961. Étude de la réceptivité comparée de races ovines a l'infestation par les nématodes de la famille des Trichostrongylides. *Soc. Sci. Vet et de Méd.* comparée de Lyon, No. 2:121–28. (Source of the data in Table 23-2.)

GROOTENHUIS, G. 1976. Mastitis onderzoek bij tien dochtergroepen. *Tijdschr. v. Diergeneeskunde* **101**:1375–77. (Detailed report of his procedures and findings.)

HUTT, F. B. 1958. *Genetic Resistance to Disease in Domestic Animals.* Ithaca, N.Y.: Cornell Univ. Press. London: Constable & Co., Ltd. (With details about examples barely mentioned in this chapter, and others; also references.)

JARVIK, L. F., A. FALEK, F. J. KALLMANN, and I. LORGE. 1960. Survival trends in a senescent twin population. *Amer. J. Human Genet.* **12**:170–79. (Use of the twin-study method to determine the influence of heredity on the duration of life.)

LONGENECKER, B. M., F. PAZDERKA, J. S. GAVORA, J. L. SPENCER, and R. F. RUTH. 1976. Lymphoma induced by herpesvirus: Resistance associated with a major histocompatibility gene. *Immunogenetics* **3**:401–7. (Source of data in Table 23-4.)

RYNIEWICZ, Z. 1972. Investigations upon environmental and genetic factors affecting the susceptibility of cows to mastitis. (Translated from Polish title.) *Inst. Genetyki i Hodowli Zwierzat PAN* Biuletyn 26. (With summary and headings of tables in English.)

SKOLASINSKI, W. Z., J. TYSZKA, and K. M. CHARON. 1977. Differences in the incidence of mastitis caused by various microorganisms in daughters of tested bulls. (Translated title.) *Prace i Materialy Zootech.* No. 14:131–40. (In Polish. Cited from *An. Breed. Abstr.* 46[4871], 1978.)

STAVÍKOVÁ, M., I. and L. LOJDA. 1975. Genetic control of the lipid composition of the lactosebum of the bovine teat canal. *Veterinární Medicína* **20**:679–87. (Title translated from Czech. Cited from *An. Breed. Abstr.* 45[697], 1977.)

INDEX OF NAMES

INDEX OF SUBJECTS